Advances in Mechanical Testing of Engineering Materials

Advances in Mechanical Testing of Engineering Materials

Editors

Stanisław Radkowski
Szymon Gontarz

MDPI • Basel • Beijing • Wuhan • Barcelona • Belgrade • Manchester • Tokyo • Cluj • Tianjin

Editors

Stanisław Radkowski
Warsaw University of Technology
Warsaw, Poland

Szymon Gontarz
Warsaw University of Technology
Warsaw, Poland

Editorial Office
MDPI
St. Alban-Anlage 66
4052 Basel, Switzerland

This is a reprint of articles from the Special Issue published online in the open access journal *Materials* (ISSN 1996-1944) (available at: https://www.mdpi.com/journal/materials/special_issues/test_engineering_materials).

For citation purposes, cite each article independently as indicated on the article page online and as indicated below:

LastName, A.A.; LastName, B.B.; LastName, C.C. Article Title. *Journal Name* **Year**, *Volume Number*, Page Range.

ISBN 978-3-0365-8258-0 (Hbk)
ISBN 978-3-0365-8259-7 (PDF)

Contents

materials

MDPI

Article

Innovative Controllable Torsional Damper Based on Vacuum Packed Particles

Dominik Rodak * and Robert Zalewski

Institute of Machines Design Fundamentals, Warsaw University of Technology, 02-254 Warsaw, Poland; Robert.Zalewski@pw.edu.pl

* Correspondence: Dominik.Rodak@pw.edu.pl; Tel.: +48-69-093-1301

Received: 13 July 2020; Accepted: 27 September 2020; Published: 30 September 2020

Abstract: In this paper a new concept of a controllable granular damper is presented. The introduced prototype works based on so-called vacuum packed particles (VPPs). Such structures are made of granular materials located in a soft and hermetic encapsulation. As a result of generating a partial vacuum inside the system, the structure starts to behave like a nonclassical solid body. The global physical (mechanical) features of VPPs depend on the level of internal underpressure. The introduced prototype of a controllable torsional damper exhibits various dissipative properties as a function of internal underpressure. The design details of the investigated device are presented. Basic laboratory tests results are discussed. To describe the hysteretic behavior of the device, the Bouc–Wen rheological model has been modified and adopted. Nonlinear functions of underpressure have been introduced to the initial model formulation. The developed Bouc–Wen model has been applied to capture the real response of the VPP torsional damper prototype.

Keywords: vacuum packed particles; torsional vibration damping; smart materials

1. Introduction

Damping of torsional vibrations is a highly important and challenging problem in many fields of industry. Protection from the negative impact of vibrations is one of the basic ways to extend machine life. Rotors, shafts and other rotational parts operate under a certain range of excitation frequencies, close to the natural frequency of a structure. These may cause a dangerous intensification of amplitudes, because of the resonance phenomenon. Damping of random excitation, noise reduction and preventing people from hazardous vibrations should be continuous motivations for development of innovative damping methods [1,2]. It is confirmed in [3] that application of intelligent dampers is recommended everywhere where such devices could be properly controlled. A parameter variation of the device is a crucial point in the control process of active and adaptive tuned mass dampers [4]. Various methods of control techniques are possible. The first group of methods involves a controlled change of the stiffness parameter of the device. Another one allows the user to change the dynamic response of the system by changing its inertial parameters. Such methods based on controlling the dynamic features of the structure can be commonly encountered in the world literature [5,6]. A different way to attenuate vibrations of the system is to dissipate its energy by changing its damping properties in real-time [7].

Popular intelligent materials which are often applied to adaptive torsional dampers are magnetorheological fluids (MRF) and magnetorheological elastomers (MRE) [8]. The shape of MREs is more specified than that of MRFs [9]. Moreover, MREs do not require hermetic spaces or limited containers [10]. To change their physical parameters such as shear modulus an external magnetic field is required [11]. Similar to other smart materials, MREs are applied to various types of structures. It is possible to find MRFs and MREs as a base of linear dampers, torsional dampers, or as a core in sandwich beams [12–14].

Recently, a new group of controllable structures called vacuum packed particles (VPPs) has attracted the interest of a wider society of researchers. Up until now, these structures were considered as linear dampers or multi-axial shock absorbers [15].

In this paper the authors propose an original and novel approach in the application of VPPs: the controllable torsional damper. The important feature of the VPP torsional damper prototype is a hysteretic behavior. The investigated device provides new opportunities in a semi-active damping of torsional vibrations strategies. The paper presents preliminary experimental results which lead to determination of basic mechanical properties of the discussed damper in the case of rotary movement.

In Section 1, basic information about active vibration damping is provided. In Section 2, the vacuum packed particles and their features are described. In the next section, the operating principles of the proposed device are presented and design details are depicted. Moreover, the research plan and details of experiments are provided. Quasi-static experiments for various underpressure values were carried out to collect original empirical data. Experimental results and estimated physical parameters are shown at the end of Section 3. In Section 4, the modified pressure dependent Bouc–Wen hysteresis model is introduced. In Section 5, obtained results are discussed. Experimental data are the basis for the model identification process. In the final stage of research, the real response of the prototype is verified with the numerical simulation results. Parameter characteristics are depicted in the Section 5. Conclusions are provided in the final section of the paper.

2. Vacuum Packed Particles

Vacuum packed particles (VPPs) are innovative structures pretending to be a group of materials, known as smart structures. VPPs are based on a loose material enclosed in a soft and hermetic envelope. When the pressure inside is close to atmospheric pressure, grains behave similarly to a dense and viscous fluid. When the partial vacuum is generated inside the structure, particles begin to come into contact with each other and with the flexible membrane separating them from the surroundings. This phenomenon strongly influences the macroscopic features of VPPs. In fact, the system is transformed into the semisolid state [16,17]. The value of underpressure (defined as the difference between atmospheric and internal pressures) and a type of granular material are parameters strongly affecting the stiffness and damping properties of the structure [18]. Vacuum packed particles exhibit several similarities to MRFs or MREs and can be modeled with similar rheological models [19,20].

An additional advantage of structures based on VPPs is that they can be customized to various external shapes. When the zero value underpressure (atmospheric pressure) is generated inside the system, VPPs can be treated as a kind of nonclassical plasticine, which easily allows the user to form any external shapes with it [17,21].

Thanks to the so-called jamming mechanism caused by the air being pumped out from the structure, the temporary shape of the VPPs can be permanently frozen [17]. Nowadays VPPs encounter many interesting practical applications.

For example, in [22] they were used as vacuum mattresses enabling injured patients to be safely transported to hospital. Another medical application of considered special granular structures can be found in controllable flexible endoscopes [23]. In the considered application, VPPs were used to design a special guide with variable rigidity, which facilitates endoscopic examinations that are not troublesome for the patient.

Vacuum packed particles are also applied in numerous controllable damping solutions. As an alternative to MRFs, VPPs are especially interesting as a basis of semi-active damping devices. Applying VPPs as a core in sandwich beams is an encouraging way to control their dynamics. In the paper [24], four different granular materials and classical rheological models were examined. Investigating the sandwich beam free vibrations, the authors showed the efficiency and accuracy of discussed models and characterized that the damping behavior depended on the grain type. In [25] or [26], free vibrations of a beam with a controllable core were examined. Presented results revealed the influence of underpressure on damping performance and stiffness. Experimental and numerical data confirmed that the logarithmic damping decrement depends on the underpressure value and increases with it. Both papers claim that VPPs are an interesting alternative for other, more sophisticated and expensive smart materials. In [25], the authors proved that the proper control of the granular core provides a significant reduction of amplitudes in the case of excited vibrations. The paper shows that the dynamic structure of a sandwich beam with a controllable granular core changes due to various values of underpressure.

It is also possible to find an application of VPPs as a linear damper of symmetrical and nonsymmetrical characteristics. According to the solution discussed in [21], the linear VPP vibration attenuator has a cylindrical shape. The single granular core is placed inside a steel spring that provides stiffness to the system. The controllable granular damper prototype was subjected to the sine excitation rule with various frequencies and values of partial vacuum. It was shown that the underpressure parameter is a suitable factor enabling the user to change the dissipative properties of the device. The biggest disadvantage of the presented VPP damper seems to be its nonsymmetrical damping characteristic. This fact means that the applicability of the discussed solution may be limited. To capture the recorded nonsymmetrical hysteresis loops, authors have proposed the modified Bouc–Wen model [27]. A similar topic of the Bouc–Wen model adaptation for cylindrical granular cores was presented in [28]. Using numerical methods for parameter identification the model was calibrated and verified with the real response. It can be concluded that typical mathematical models developed for magnetorheological (MR) fluids or MR elastomers are capable of capturing the nonlinear behavior of VPPs [28].

3. Experiments

3.1. VPP Torsional Damper Prototype

Experiments were conducted on the specially prepared VPP torsional damper prototype designed mainly using fused deposition modeling (FDM) technology. The scheme of the damper is depicted in Figure 1. It consists of the outer driver (I), the inner driver (II), two outer rings (III), two inner rings (IV), an outer flange (V), two sealing membranes (VI), two O-rings (VII) and a working chamber with granular material (VIII). The main part of the damper is an air-tight operating chamber built by the outer driver (I), inner driver (II) and sealing membranes (VI). It is filled with loose granular material (VIII). The air is pumped out of the chamber through the holes in the inner driver (II), causing suction of the membranes and change in the granular core macrostructure. The value of underpressure is controlled by a vacuum pump. The air is removed from the system through a hollow drive shaft mounted in the inner driver (II). It is possible to use various methods to drive the damper. In the discussed prototype, the rotary motion from the drive shaft is transmitted through a clamping sleeve fixed in the inner driver (II), through the deformable working chamber (VIII), to the outer driver (I), which is connected to the flange (V). The flange (V) enables the connection of the damper to the output shaft.

The granular material applied in experiments was a semi-finished product for an injection molding technology. It was made of ABS (acrylonitrile butadiene styrene) grains. Such a material was inexpensive and commonly available. The shape of a single grain was cylindrical and dimensions were 3 mm for length and 2.5 mm for diameter. The working chamber volume was 540 cm^3 with a filling ratio of 74.4% (426 g of grains). The filling ratio was taken empirically and was a constant parameter in all conducted tests.

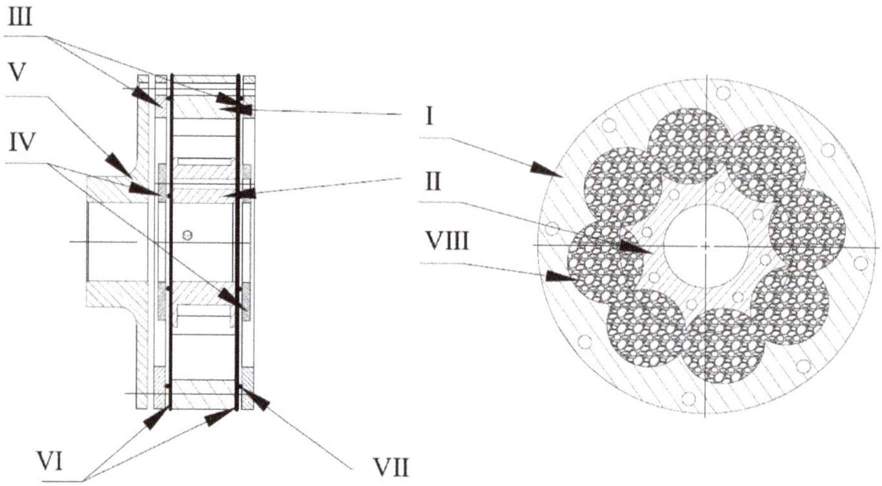

Figure 1. The scheme of the torsional controllable vacuum packed particle (VPP) damper prototype. (I) Outer driver, (II) inner driver, (III) outer rings, (IV) inner rings, (V) outer flange, (VI) sealing membranes, (VII) O-rings and (VIII) grains in the working chamber.

The detailed CAD model of the discussed torsional VPP damper is depicted in Figure 2.

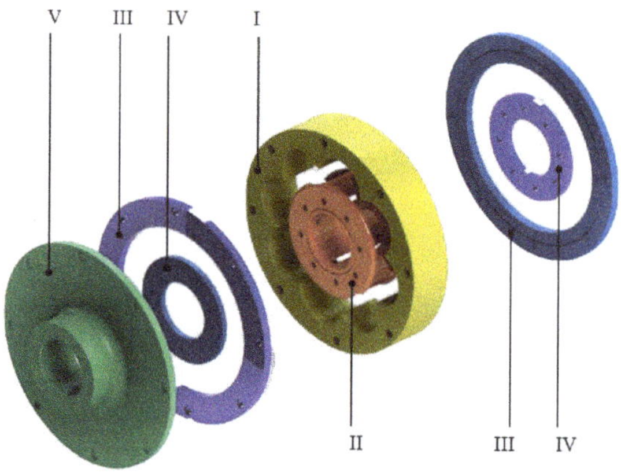

Figure 2. The CAD model of the investigated damper. (I) Outer driver, (II) inner driver, (III) outer rings, (IV) inner rings and (V) outer flange.

3.2. Research Plan

The main objective of the laboratory tests was to determine the basic dissipative properties of the innovative torsional vibration attenuator as a function of the internal underpressure. This paper presents the results of the quasi-static torsion tests. All experiments were carried out with the same range of applied torsion angles ($\Phi = \pm 5°$) (Figure 3) to ensure nondestructive deformations of the external wrap and to analyze mainly the underpressure influence on dissipative characteristics of the damper. Exceeding the threshold value of the angular loading would lead to elastic (in the first stage) and plastic deformations of the external encapsulation. The plastic strains of the envelope are underpressure independent and are unwanted phenomena in VPP devices. The range of underpressure values was the same for all tests, from 0.00 to 0.05 MPa with a step of 0.005 MPa. A higher underpressure intensifies the jamming mechanism, which is connected to strong adherence of grains to the envelope.

In such a case the loading caused micro perforations of the external wrap and it was difficult to control the underpressure parameter. Three loading cycles were applied to achieve a stabilized response of the device. It is worth mentioning that the loading cycles applied in the special granular structure experimental research excitation program (Figure 3) did not introduce inertial forces into the overall system, except at points where the velocity changed direction. This allowed for an accurate measurement of the torque response. The loading velocity was set as 0.5 deg/s. The strain rate was 0.0014 1/s. The recorded results were in the form of torque-angle of rotation characteristics (M = f(Φ)). Based on experimental data, it is possible to present hysteresis loops for various underpressure values. Such characteristics reflect controllable dissipative properties of the investigated VPP device.

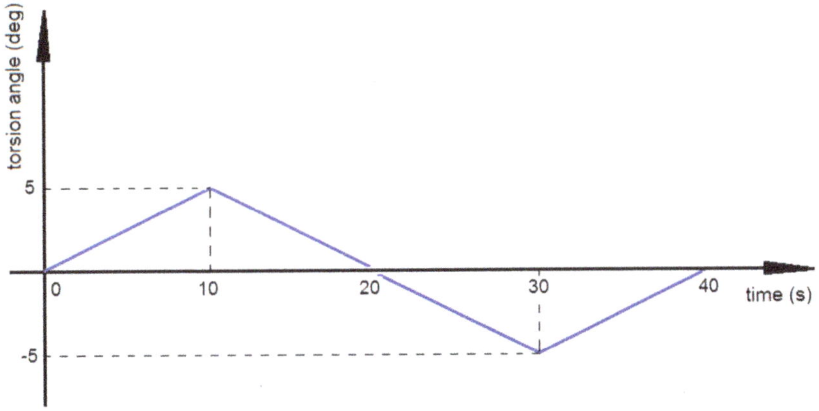

Figure 3. Single loading cycle.

3.3. Test Stand

Laboratory tests were carried out on the specially designed experimental stand. The test stand and its details are depicted in Figure 4. To provide an appropriate stiffness, the test stand consisted of aluminum alloy profiles. The torsional damper prototype was fixed in the bottom of the mounting set and connected with the encoder and the torque transducer in a way that allowed for rotation. The load was applied manually to the VPP damper by a lever (6) which rotated in the axis of the damper. The applied angles of rotation were limited by two pin-type limiters mounted on the plate (8) to keep the range of excitations constant. Signals from the torque transducer (4) and the encoder (2) were collected by a measuring card. Parts (1) and (5) were connection shafts which were coupled with sensors thanks to coupling sleeves (3). The whole measuring set was protected by the thrust washer (7) and fastened by mounting plates (8) and (9). The signal from the torque transducer was additionally amplified. All data was recorded on the PC thanks to the data acquisition system. The partial vacuum was generated by the external vacuum pump. The underpressure value was controlled by the potentiometer built into the pump. The torque was a resultant value of the assumed rotation range.

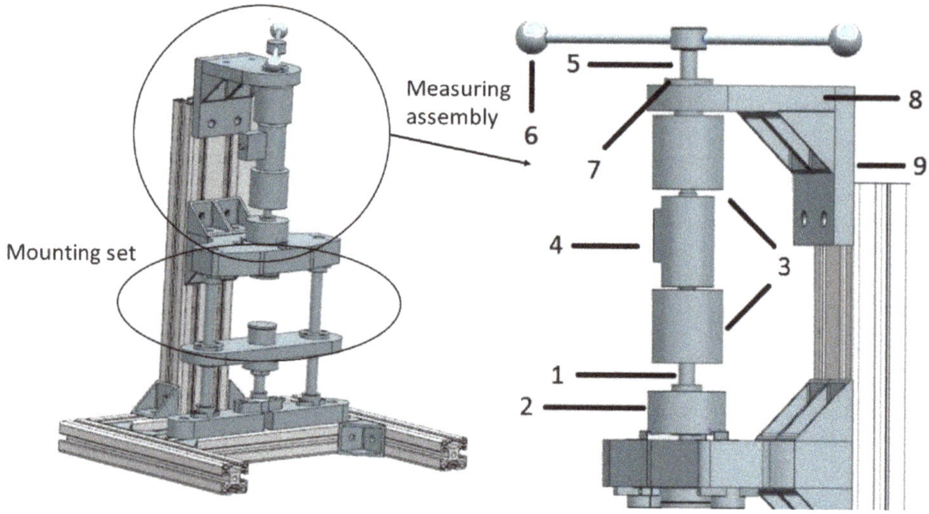

Figure 4. The test stand scheme. (1) Shaft, (2) encoder, (3) coupling sleeves, (4) torque transducer, (5) shaft, (6) lever, (7) thrust washer and (8,9) mounting plates.

Figure 5 depicts the complete test stand. It consisted of the PC (1), the vacuum pump with the controlling valve (2), the data acquisition card (3), the signal amplifier (4) and the investigated VPP damper (5). The load lever (6), coupling sleeves (7), torque transducer (8) and encoder (9) are also shown in Figure 5.

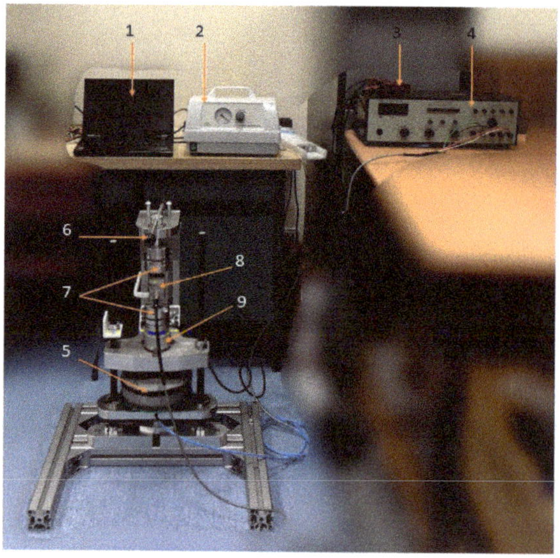

Figure 5. The test stand illustration. (1) PC, (2) vacuum pump, (3) data acquisition card, (4) signal amplifier, (5) VPP damper, (6) load lever, (7) coupling sleeves, (8) torque transducer and (9) encoder.

3.4. Results

A single measurement cycle included mounting the damper prototype in the test bench holder, setting the appropriate partial vacuum value, loading the damper and recording the device response.

Three separate measurement series were conducted to eliminate possible fatal errors. All data were collected without any additional filters.

Typical experimental dissipative characteristics (torque vs. torsion angle) for representative values of the partial vacuum are depicted in Figures 6 and 7. The influence of the underpressure value on the character of recorded characteristics is evident.

The first observation based on analysis of the data presented in Figures 6 and 7 is the dependence of the dissipative characteristics on the partial vacuum value generated inside the damper prototype. Torque–torsion angle characteristics are located higher for experiments conducted with greater underpressure values (Figure 6). Moreover, increased values of investigated parameters (equivalent stiffness and amount of dissipated energy) for higher partial vacuums have been observed. For the 0.00 MPa probe (atmospheric pressure), recorded values are noisy. In the case of no underpressure, membranes do not have stable contact with grains and they can easily reorganize. Small values of underpressure, like 0.01 MPa, provide more predictable behavior. Recorded hysteresis loops have nonclassical shapes because of the possible reorganization of grains and small clearances in the mounting point. This could be a reason for extension of the shape of the dissipative characteristics around the point 0.0.

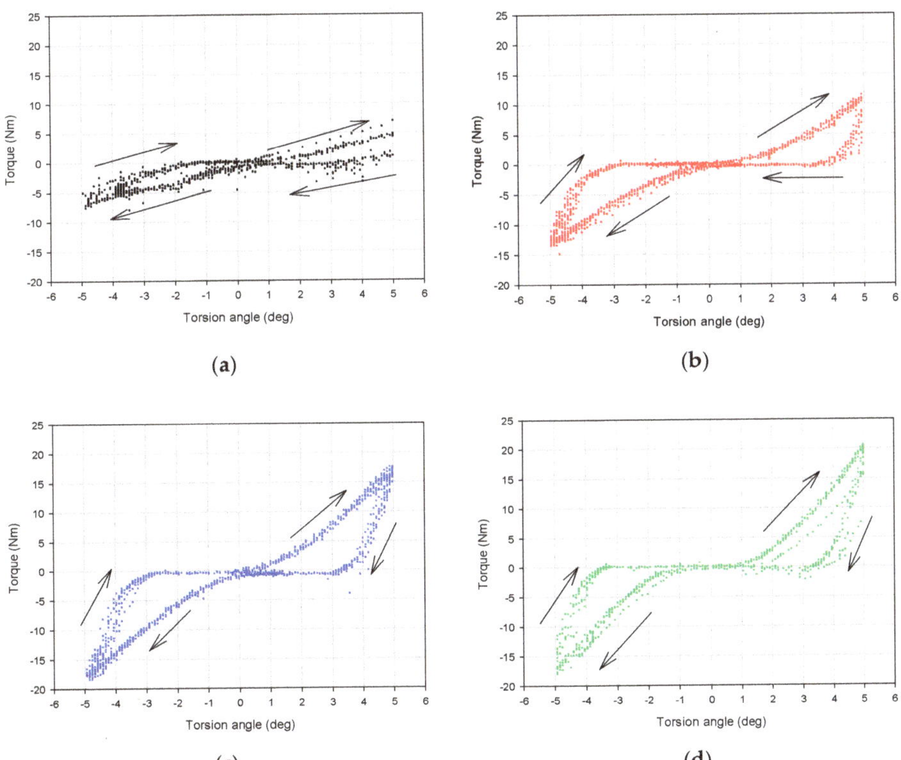

Figure 6. Damping characteristics for various underpressure values: (**a**) 0.00 MPa, (**b**) 0.01 MPa, (**c**) 0.02 MPa and (**d**) 0.05 MPa.

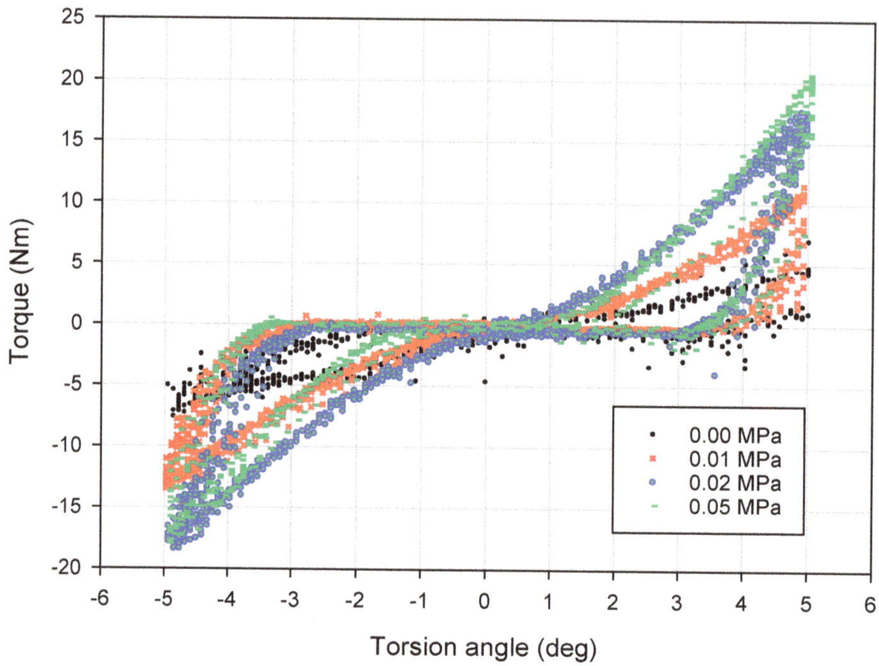

Figure 7. The influence of underpressure on damping properties of the VPP damper.

Loading-slope inclination increases following an exponential trend. The difference between 0.01 and 0.02 MPa probes is more evident than between 0.02 and 0.05 MPa. The inclination of the quasi-linear part of the 0.02 MPa test result is nearly the same as for the 0.05 MPa probe. Shapes of 0.02 and 0.05 MPa damping characteristics are similar to each other. The most substantial parameter revealing the impact of a partial vacuum is the recorded maximum torque value. This parameter reaches the highest value for the 0.05 MPa probe. The amount of dissipated energy is comparable for 0.02 and 0.05 MPa probes. This fact may lead to the conclusion that from 0.02 MPa, dissipative and stiffness properties do not change significantly.

In this paper the authors particularly focused on the impact of underpressure on three parameters: the VPP damper equivalent torsional stiffness (k_{eq_a}), maximum torque parameter (M_{max}) and amount of energy dissipated in a single loading cycle (E_d). Determining the equivalent stiffness value is a complex problem. The equivalent stiffness can be derived from potential energy and depends on damper movement direction. The detailed methodology of deriving the equivalent stiffness in systems with hysteresis can be found in [29]. In the initial stage of the loading and unloading process, $M = f(\Phi)$ characteristics follow a quasi-linear trend. The equivalent torsional stiffness parameter was defined as an average of k_{eq_l} and k_{eq_u} parameters. The process of determining the value of the parameters k_{eq_a}, k_{eq_l} and k_{eq_u} for different underpressures is illustrated in Figure 8.

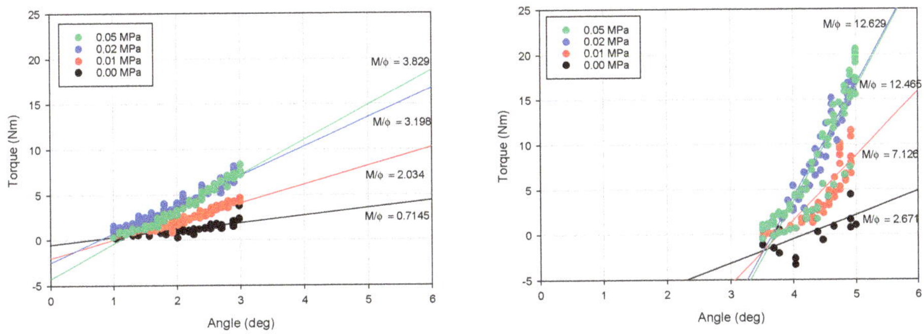

Figure 8. Determining the value of equivalent torsional stiffness parameters for various underpressures.

Experimentally-captured damping characteristics M = f(Φ) were recorded separately for the selected values of underpressure. Fragments of the characteristics from the range Φ < 1, 3 > (deg) and Φ < 5, 3.5 > (deg) were approximated with linear functions. The slopes of approximation lines were assumed as unknown k_{eq_l} and k_{eq_u} parameters defined as $k_{eq\_i} = dM/d\Phi$. The parameter k_{eq_l} value was estimated for a loading process (range Φ < 1, 3 > (deg)), similarly k_{eq_u} was defined using the unloading process (range Φ < 5, 3, 5 > (deg)). The region Φ < 0, 1 > (deg) was intentionally neglected to avoid possible disturbances related to grain reorganization in the changeable loading direction zone or technological clearances in the damper mounting point. The maximum torque parameter M_{max} was defined as a torque value measured for the Φ = 5 (deg) torsion angle.

The last parameter taken into consideration at this stage of research was related to the amount of energy dissipated in a single stabilized loading cycle. To define the E_d parameter value, the simplest numerical integration procedure based on the midpoint rule was used to calculate a numerical approximation for the area located inside experimentally captured dissipative characteristics.

The values of previously mentioned parameters were determined separately for each experimental probe conducted with a selected underpressure value. The obtained results are presented in Table 1.

Table 1. Selected mechanical features of VPP controllable torsional damper for representative underpressure values.

Underpressure (MPa)	Equivalent Loading Stiffness k_{eq_l} (Nm/deg)	Equivalent Unloading Stiffness k_{eq_u} (Nm/deg)	Equivalent Stiffness k_{eq_a} (Nm/deg)	Maximum Torque M_{max} (Nm)	Dissipated Energy (Nm/deg)
0.000	0.72	2.67	1.69	7.00	18.13
0.005	1.59	5.92	3.75	10.14	32.28
0.01	2.03	7.13	4.58	11.50	41.55
0.015	2.69	9.98	6.34	14.96	44.38
0.02	3.20	11.47	7.33	17.60	45.86
0.025	3.27	11.78	7.53	17.56	50.99
0.03	3.55	12.25	7.90	19.18	50.02
0.035	3.69	12.74	8.21	19.64	52.85
0.04	3.79	12.72	8.26	19.98	52.64
0.045	3.93	13.06	8.50	20.58	52.41
0.05	3.83	12.63	8.23	20.60	54.06

Detailed analysis of collected experimental data revealed that the variation of the M_{max}, k_{eq_a} and E_d parameters as a function of underpressure was nonlinear. Generally, the exponential upward trend for the VPP damper mechanical properties, during generation of higher partial vacuum values, has been discovered (Figures 9–11).

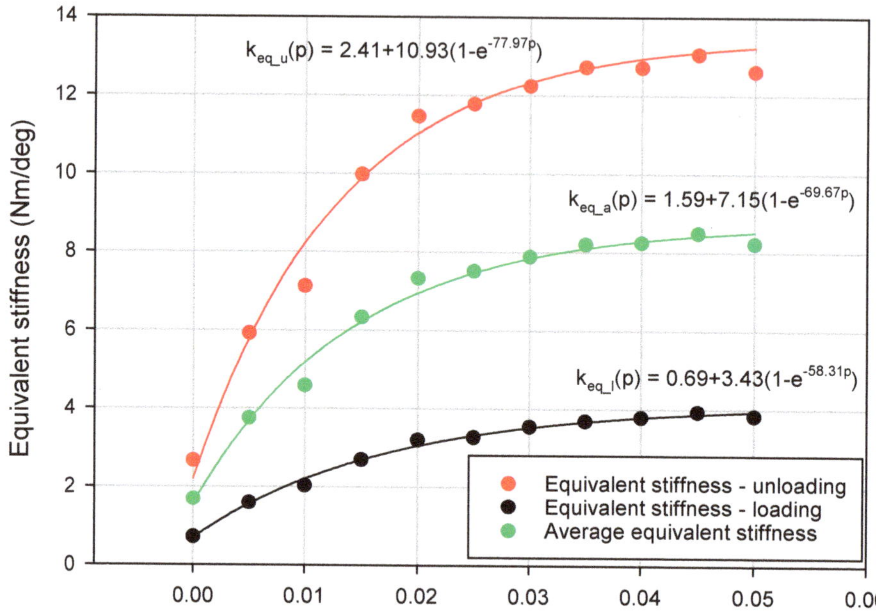

Figure 9. Exponential rise to maximum approximation for the k_{eq_a} parameter defined for various underpressures.

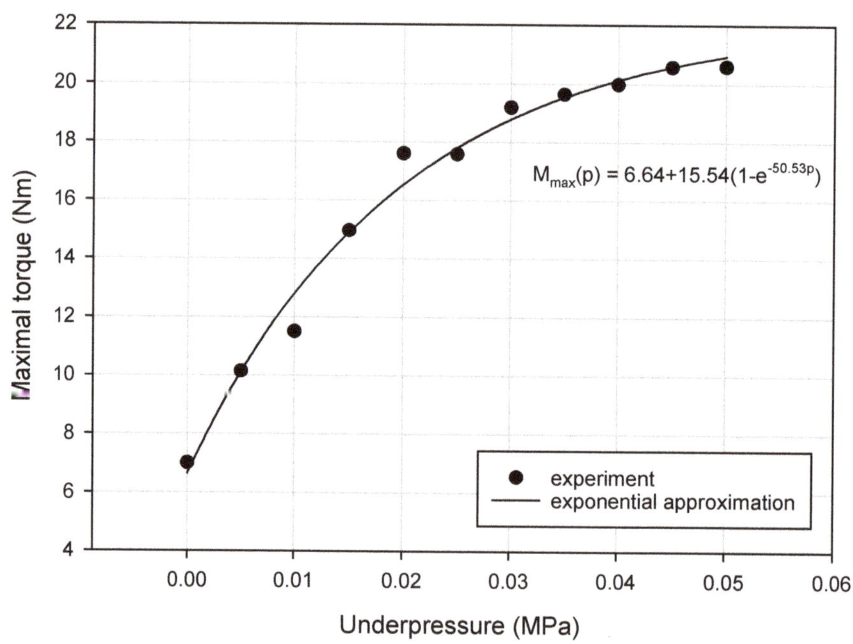

Figure 10. Exponential rise to maximum approximation for the M_{max} parameter defined for various underpressures.

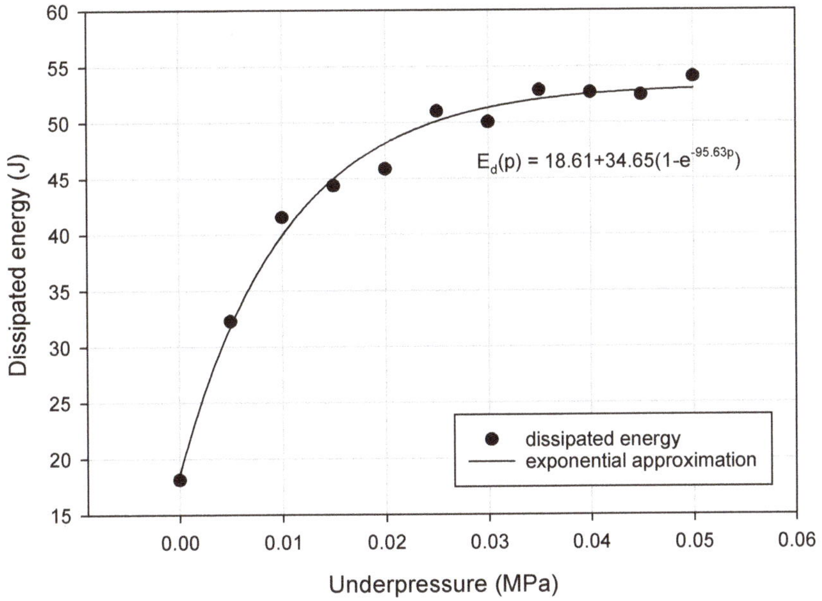

Figure 11. Exponential rise to maximum approximation of the E_d parameter defined for various underpressures.

The main objective of the experiments was to confirm the possibility of changing the energy dissipating properties of the VPP torsional damper prototype by controlling the underpressure parameter. Experimental data were used as the basis for the forthcoming modeling process, which was a classical inverse problem consisting of fitting the numerical model response into the real experimental results.

4. The Bouc–Wen Model

By analyzing the data depicted in Figure 6 it can be observed that the response of the VPP damper is nonlinear. In addition, the hysteretic type of behavior is confirmed. Recorded dissipative characteristics are generally symmetrical. Only in the case of atmospheric pressure (zero underpressure value, i.e., $p = 0$) was nonsymmetrical behavior observed (Figure 6a). Such a situation is predictable and has already been reported in the literature [30]. To assume VPPs as a continuous and homogeneous nonclassical solid body, a minimal underpressure threshold value must be applied. For example, in [31] it was shown that at least 0.005 MPa underpressure must be generated inside the granular core to provide a jamming mechanism. In [32] the authors examined a bended VPP beam and set the threshold partial vacuum value, enabling a solidification effect of the granular structure at 0.02 MPa. The modeling process of a fully loose granular material placed in a hermetic plastomer encapsulation is complex. It demands applying finite and discrete element methods together. Elastic membranes without tension caused by underpressure do not support grains in a determined way. Moreover, membranes do not exhibit explicit contact with the grains. In such a case the VPPs behave like a plastic mass that deforms locally rather than macroscopically. As a main part of the torsional damper the VPPs are expected to attenuate large amounts of energy. Consequently, the lowest ranges of underpressures are less important from an operational point of view.

As previously mentioned, the global behavior of VPPs and their controlling process in some aspects can be compared to magnetorheological fluids [33,34]. Expecting that in the near future VPPs could replace popular MR fluids in many engineering applications, especially in semi-active vibration damping, special attention was paid to rheological models of MRFs. Typically, the Bingham model is applied to model MRFs [35]. It is a comparatively simple model, making it easy to carry out numerical simulations. However, its biggest disadvantage is the impossibility for it to represent complex shapes of experimentally recorded hysteresis loops. Although there are many other, more complicated models, including the Gamota–Filisko [36], Jiles–Atherton [37,38] or Preisach hysteresis models [39], in the last decades the Bouc–Wen (B–W) model has become a common approach in hysteresis behavior modeling [40].

This model was initially proposed by Robert Bouc [27] and extended by Yi-Kwei Wen [41]. The complication level of the model is widely accepted in both computer simulation and control processes. One disadvantage of the B–W model is a relatively large number of parameters and the sensitivity of model responses to changes in their value.

The application of the B–W model is not only restricted to magnetorheological or electrorheological fluids. It is widely applied to capture the response of various nonlinear hysteretic systems. It has been modified and used to describe many engineering problems like base-isolation materials, multi-dimensional continuous systems, various viscous dampers, metals, masonry, timber and even transformers. It was also successfully applied in the structural control of nonlinear torsional dampers [42].

The mechanical scheme of the B–W model is depicted in Figure 12 and the restoring force is expressed as Equations (1) and (2) [28].

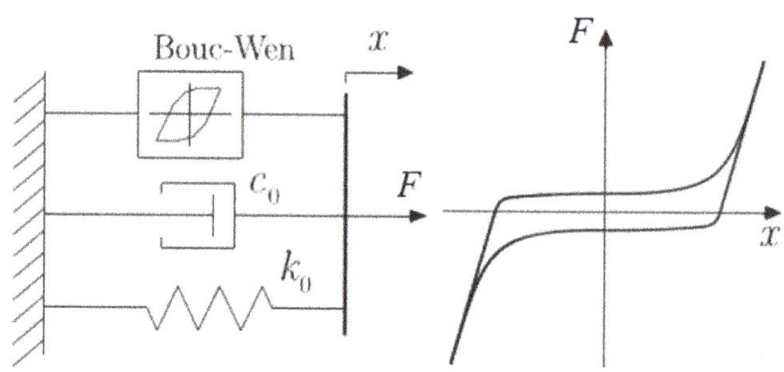

Figure 12. The scheme of the Bouc–Wen model.

Variable z is given by the following formula:

$$F(t) = \frac{d^2x}{dt^2} + 2\xi\omega_n\frac{dx}{dt} + \alpha{\omega_n}^2x + (1-\alpha)\omega_n z \tag{1}$$

$$\frac{dz}{dt} = -\gamma z\left|\frac{dx}{dt}\right||z|^{n-1} - \beta\frac{dx}{dt}|x|^n + A\frac{dx}{dt} \tag{2}$$

where

- *F(t)* = force function;
- ξ = linear viscous damping ratio;
- ω_n = pseudo-natural frequency of the system;
- n = degree of polynomial; and
- α; γ; β; A = parameters of hysteresis loop shape.

The model can be easily adapted to the rotational motion. Changing linear values to their rotational equivalents allows adoption of the B–W model to capture extraordinary features of the VPP torsional damper. Despite the compound form of the B–W model it can be represented in an analytical way. Equation (2) can be given by

$$\frac{dz}{dt} = A - ((\beta + \gamma \cdot \mathrm{sgn}(\frac{dx}{dt})))|z|^n \tag{3}$$

The parameter n is responsible for hysteresis loop radius of curvature [18]. Considering n = 2, the solution of Equation (3) has the following form:

$$z = \frac{\sqrt{A}}{\sqrt{\beta + \gamma}} \tanh(\sqrt{A(\beta + \gamma)} \cdot (x + C)) \tag{4}$$

The constant C can be derived from boundary conditions.

Although the analytical solution with parameter n = 2 is not suitable to describe hysteresis loops obtained from experiments carried out on the VPP damper prototype, the authors decided to apply the numerical B–W model calibration process to fit the experimental data correctly.

5. Result Discussion

The B–W model in the version given by Equations (1) and (2) consists of seven coefficients. Generally, determination of the model parameters is based on laboratory test results and consists of various identification techniques. The most popular procedures are derived from heuristic optimization methods such as simulated annealing [43], particle swarm [44], firefly algorithms [45] or evolutionary algorithms [46]. Also, least-squares methods are frequently implemented, where the most popular are the Marquard–Levenberg or Gauss–Newton techniques [47].

In this preliminary approach to the modeling process the authors did not use a sophisticated numerical identification procedure. A simple Monte Carlo method was implemented to capture the model's unknown parameter values. The applied technique consisted of a typical pattern: definition of the domain of possible inputs (the range of variability of parameter values of the Bouc–Wen model was taken from [31]), generation of random inputs over the uniform domain, carrying out simulation tests and comparing the model response with a real experimental data, and selecting the best solution. To provide a more accurate identification procedure, alternative sophisticated, nondeterministic optimization techniques are necessary. Genetic algorithms, simulated annealing or the swarm particles method can calibrate the discussed model with a better accuracy.

The numerical simulations of the VPP behavior were carried out for the B–W parameter values shown in Table 2. Table 2 also includes data reflecting the model accuracy, such as the coefficient of determination R^2 and error percentage defined as a relative error.

Table 2. The Bouc–Wen model parameters.

p (MPa)	ω_n (1/s)	ξ (Nm·s/°)	α (Nm/°)	γ (deg^{-2})	A (-)	β (deg^{-2})	n (-)	R^2 (-)	Error (%)
0.000	0.5000	2	13	24.01	3.00	−44.00	1.48	0.857	19.2
0.005	0.0503	106	2877	12.88	36.19	−21.25	1.48	0.902	18.1
0.01	0.0100	250	4259	11.00	75.83	−16.00	1.48	0.931	14.7
0.015	0.0065	292	5646	10.98	77.54	−13.92	1.48	0.950	13.5
0.02	0.0080	400	7511	9.00	90.50	−14.00	1.48	0.962	13.3
0.025	0.0066	603	7401	9.05	101.40	−13.64	1.48	0.970	16.1
0.03	0.0065	720	9235	9.04	112.17	−14.78	1.48	0.976	15.6
0.035	0.0062	887	9797	10.43	118.76	−15.07	1.48	0.979	13.5
0.04	0.0064	992	8660	10.82	122.51	−15.07	1.48	0.981	13.4
0.045	0.0063	1280	9646	10.33	134.74	−13.62	1.48	0.983	12.9
0.05	0.0050	1400	9630	10.8	144.70	−15.00	1.48	0.984	12.2

The numerical behavior of the investigated VPP damper simulated by the Bouc–Wen model correctly reflects the real data. The lowest precision of the B–W model is observed for the 0.00 MPa underpressure (Figure 13a), which was expected and discussed in Section 4. The large and uncontrolled scattering of experimental data in this case is caused by unpredictable local deformations of loose grains not subjected to the partial vacuum. For the higher underpressure values, much better agreement and precision of the numerical response was noticed. The average value of the coefficient of determination is greater than 0.95. This leads to the conclusion that the proposed model can be applied to describe the VPP torsional damper behavior. Relatively high values of percentage errors are caused by the unsophisticated numerical method used for the model identification.

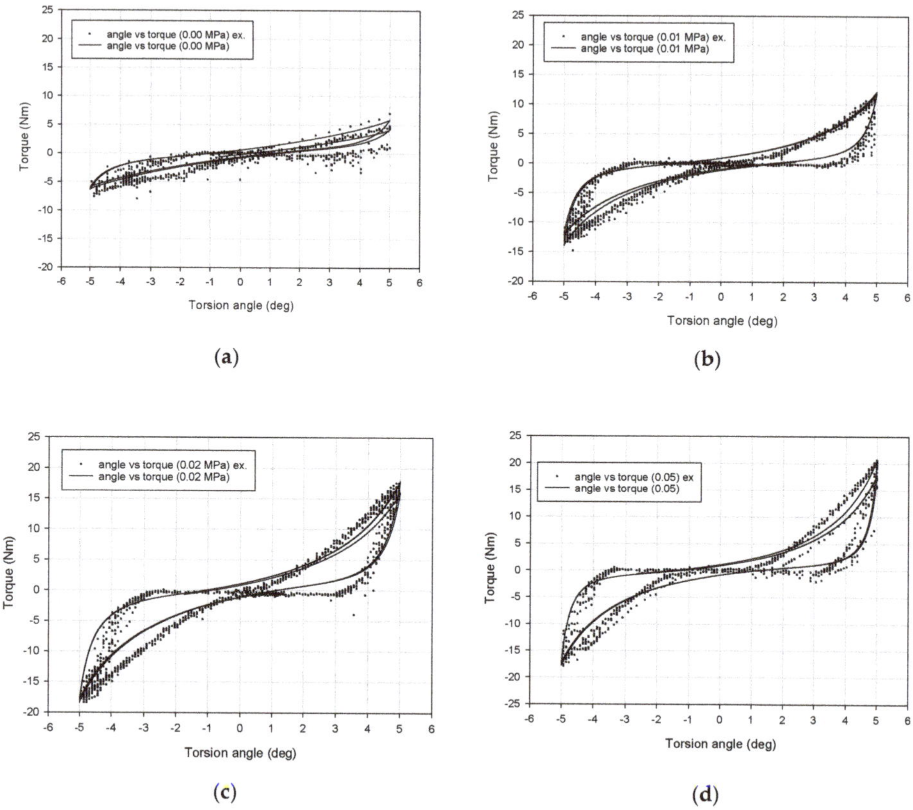

Figure 13. Verification of the numerical results and real VPP damper characteristics for representative probes for (**a**) 0.00 MPa, (**b**) 0.01 MPa, (**c**) 0.02 MPa and (**d**) 0.05 MPa values of underpressure.

For the accurate modeling of the VPP controllable torsional damper using the B–W model it is necessary to consider the partial vacuum effects.

In the following part of the paper the proposition of the modified Bouc–Wen model for the granular damper is presented. The modification process consists of replacing the seven model parameters with the empirically defined underpressure functions. Taking into account the partial vacuum parameter, the initial B–W model for the investigated device can be redefined into B–W-p model defined as

$$F(t) = \frac{d^2x}{dt^2} + 2\xi(p)\omega_n(p)\frac{dx}{dt} + \alpha(p)\omega_n(p)^2 x + (1-\alpha(p))\omega_n(p)z \tag{5}$$

$$\frac{dz}{dt} = -\gamma(p)z\left|\frac{dx}{dt}\right||z|^{n(p)-1} - \beta(p)\cdot\frac{dx}{dt}\cdot|x|^{n(p)} + A(p)\cdot\frac{dx}{dt} \quad (6)$$

Figure 14 depicts the verification process of numerical simulation results with real experimental data for representative values of underpressure.

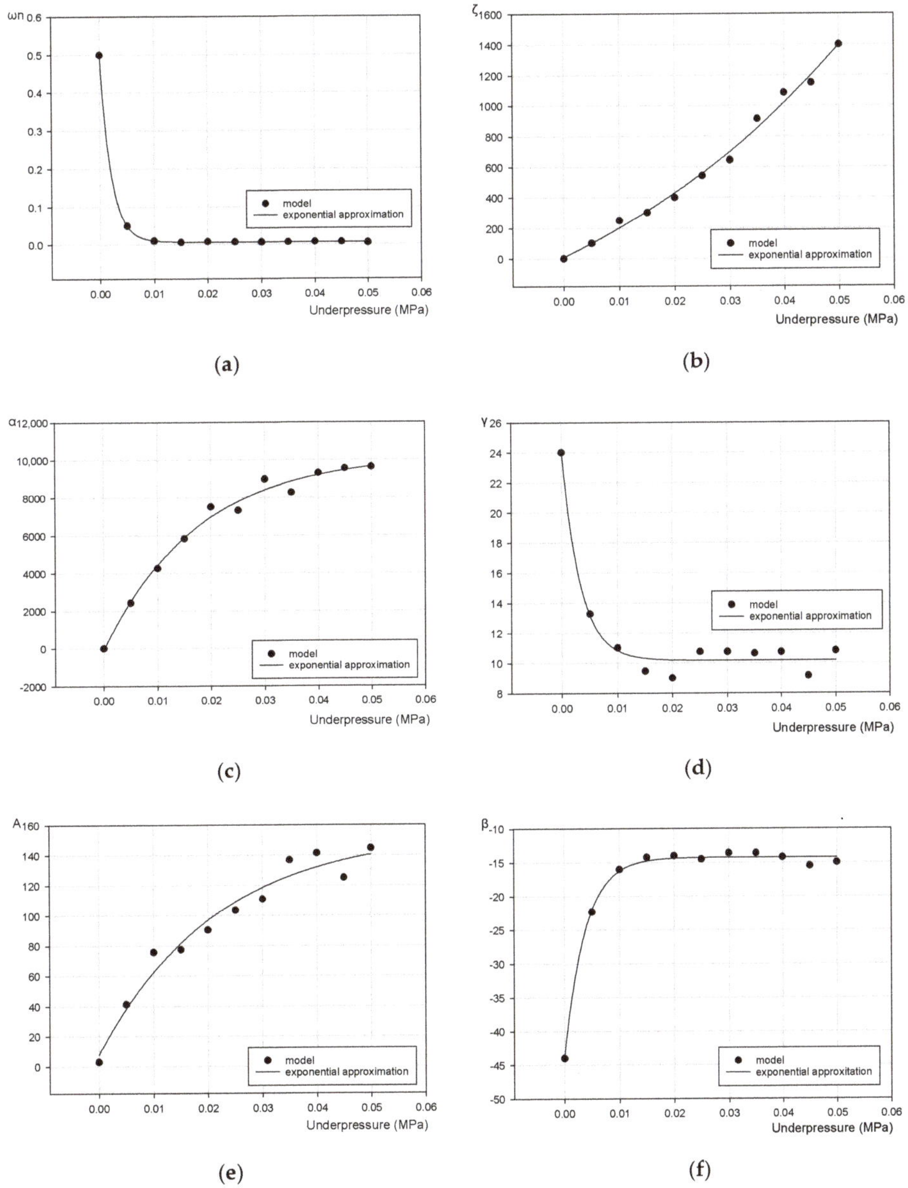

Figure 14. *Cont.*

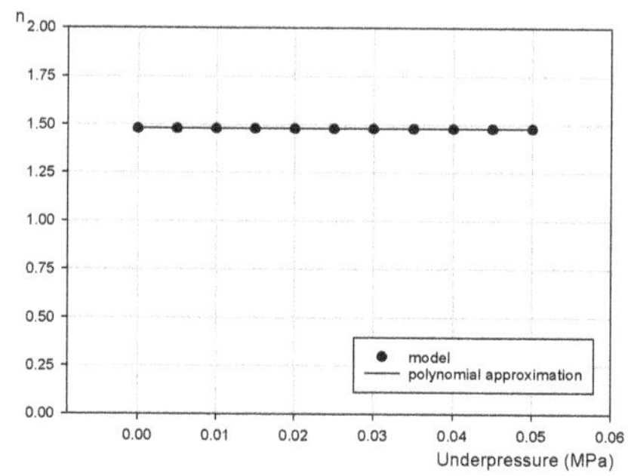

(g)

Figure 14. Variation of the Bouc–Wen model parameters in a function of partial vacuum. (**a**) ω_n(p), (**b**) ξ(p), (**c**) α(p), (**d**) γ(p), (**e**) A(p), (**f**) β(p) and (**g**) n(p).

Figure 14a–g presents graphical representations of the numerically determined parameters of the Bouc–Wen model for various partial vacuums.

General underpressure functions have been formulated on the basis of data presented in Figure 14. In the proposed approach, three various types of functions have been applied to capture the experimentally recorded behavior of the VPP torsional damper: an exponential rise to maximum three-parameter function $y = y_0 + \alpha(1-e^{(-\beta x)})$, an exponential decay three-parameter function $y = y_0 + \alpha e^{(-\beta x)}$ and an exponential growth three-parameter function $y = y_0 + \alpha e^{(\beta x)}$. It has additionally been observed that the n parameter is underpressure independent (n = 1.48). Finally, the proposed functions depending on the partial vacuum describing the Bouc–Wen rheological model material constants are

$$\omega_n(p) = \omega_{n0} + \alpha_{\omega n} \cdot e^{(-\beta_{\omega n} \cdot p)} \tag{7}$$

$$\xi(p) = \xi_0 + \alpha_\xi \cdot e^{(\beta_\xi \cdot p)} \tag{8}$$

$$\alpha(p) = \alpha_0 + \alpha_\alpha (1 - e^{(-\beta_\alpha \cdot p)}) \tag{9}$$

$$\gamma(p) = \gamma_0 + \alpha_\gamma \cdot e^{(-\beta_\gamma \cdot p)} \tag{10}$$

$$A(p) = A_0 + \alpha_A (1 - e^{(-\beta_A \cdot p)}) \tag{11}$$

$$\beta(p) = \beta_0 + \alpha_\beta (1 - e^{(-\beta_\beta \cdot p)}) \tag{12}$$

$$n(p) = 1.48 = const \tag{13}$$

Introducing functions of the partial vacuum (p) into the initial B–W model causes evident complications. They are connected to the necessity of dealing with a number of additional unknown coefficients that must be defined when basing on laboratory test results. The minimum number of experiments that have to be carried out to find the empirical form of Equations (7)–(13) is three experimental series conducted for different underpressure values.

The modified B–W-p model in its final form includes 19 unknown parameters. This multiparameter B–W-p formula is the cost of the reliable modeling of the discussed controllable VPP torsional damper. It should be emphasized that such a cost seems to be reasonable and acceptable when VPPs are considered as an innovative and competitive alternative to a much more advanced group of smart structures.

6. Conclusions

In this paper an innovative prototype of controllable VPP torsional damper has been discussed. A novel idea of using jammed state granular materials has been presented. The biggest advantage of the presented device over the traditional group of torsional dampers is the possibility of controlling its dissipative characteristics in real-time. The VPP reaction time issue is one of the future fields of research. It depends on a wide group of parameters. Examples like automotive airbags indicate the possibility of quick responses of inflated and deflated devices. The next benefit resulting from the introduced approach is an economic aspect. The grains that are the main part of the damper can be taken from the wide group of recycled materials (tires or plastics). The manufacturing costs of VPP dampers can therefore be significantly reduced (by up to 100 times) in comparison to classical energy attenuators. The next advantage is a simple controlling mechanism. The mechanical properties of the VPP damper, like stiffness and amount of dissipated energy, can be rapidly changed by generating a partial vacuum inside the system. Even the simplest vacuum pump is an efficient device to generate the internal underpressure and to increase the damping properties of the investigated device. The granular jamming phenomenon is fully reversible and allows for rapidly turning the solidification mechanism on and off.

The experimental results obtained in the discussed work confirmed the previous results, published in the authors' previous works, concerning simple cylindrical samples made of VPPs. Internal underpressure is a simple and convenient parameter that changes the global mechanical properties of the considered granular structures. As in the case of the uniaxial results, the discussed experimental data was modeled using the modified Bouc–Wen model. This paper confirms that generating higher underpressure value intensifies the grain locking mechanism resulting in increasing the dissipative properties of the VPP torsional damper prototype. The experimentally obtained dissipative characteristics of the VPP torsional damper are highly nonlinear. To capture extraordinary features of the device, a multiparameter rheological model has to be implemented. In this paper the authors chose the Bouc–Wen model of hysteresis. To extend the model enabling taking into account the controlling parameter (partial vacuum), the empirically determined underpressure functions have been introduced to the basic model equations. The efficiency of the modified B–W-p model has been experimentally verified and confirmed.

Further laboratory tests of granular materials subjected to the underpressure are intended. They will focus on investigating the influence of grain type and their microfeatures, like stiffness or roughness, on the macroscopic response of the VPP torsional damper. Prospective trends in the research of granular controllable dampers should also include the implementation of their numerical models into a finite element method code. Moreover, in order to improve the model prediction, more sophisticated model calibration methodologies have to be implemented.

Author Contributions: Conceptualization, D.R. and R.Z.; methodology, D.R. and R.Z.; investigation, D.R. and R.Z.; resources, D.R. and R.Z.; data curation, D.R. and R.Z.; writing—original draft preparation, D.R. and R.Z.; writing—review and editing, D.R. and R.Z. All authors have read and agreed to the published version of the manuscript.

Funding: This research received no external funding.

Conflicts of Interest: The authors declare no conflict of interest.

References

1. Dyniewicz, B.; Pręgowska, A.; Czesław Bajer, I. Adaptive control of a rotating system. *Mech. Syst. Signal Process.* **2014**, *43*, 90–102. [CrossRef]
2. Pisarski, D.; Szmidt, T.; Bajer, C.I.; Dyniewicz, B.; Bajkowski, J.M. Vibration control of double beam system with multiple smart damping members. *Shock Vib.* **2016**, 2438902. [CrossRef]
3. Spencer, B.F., Jr. *Reliability of Randomly Excited Hysteretic Structures*; Springer Science & Business Media: Berlin/Heidelberg, Germany, 2012; Volume 21.
4. Shi, W.; Wang, L.; Lu, Z. Study on self-adjustable tuned mass damper with variable mass. *Struct. Control Health Monit.* **2018**, *25*, e2114. [CrossRef]
5. Gao, P.; Xiang, C.; Liu, H.; Walker, P.; Zhang, N. Design of the frequency tuning scheme for a semi-active vibration absorber. *Mech. Mach. Theory* **2019**, *140*, 641–653. [CrossRef]

6. Lin, G.L.; Lin, C.C.; Chen, B.C.; Soong, T.T. Vibration control performance of tuned mass dampers with resettable variable stiffness. *Eng. Struct.* **2015**, *83*, 187–197. [CrossRef]
7. Wasilewski, M.; Pisarski, D.; Konowrocki, R.; Bajer, C.I. A new efficient adaptive control of torsional vibrations induced by switched nonlinear disturbances. *Int. J. Appl. Math. Comput. Sci.* **2019**, *29*, 285–303. [CrossRef]
8. Szmidt, T.; Pisarski, D.; Konowrocki, R.; Awietjan, S.; Boczkowska, A. Adaptive Damping of a Double-Beam Structure Based on Magnetorheological Elastomer. *Shock Vib.* **2019**, *2019*, 8526179. [CrossRef]
9. Szmidt, T. Shear deformation damping of a double-beam structure. *J. Sound Vib.* **2016**, *370*, 163–175. [CrossRef]
10. Nayak, B.; Dwivedy, S.K.; Murthy, K.S.K.R. Dynamic analysis of magnetorheological elastomer-based sandwich beam with conductive skins under various boundary conditions. *J. Sound Vib.* **2011**, *330*, 1837–1859. [CrossRef]
11. Boczkowska, A.; Awietjan, S.F.; Pietrzko, S.; Kurzydłowski, K.J. Mechanical properties of magneto-rheological elastomers under shear deformation. *Compos. Part B Eng.* **2012**, *43*, 636–640. [CrossRef]
12. Wen, J.M.; Shen, D.Z.; Wang, R.M.; Li, J.P.; Ma, J.J. A two-fixed-end beam piezoelectric inertial actuator using electromagnet controlled magnetorheological fluid (MRF) for friction regulation. *Smart Mater. Struct.* **2020**, *29*, 065011. [CrossRef]
13. Khanouki, M.A.; Sedaghati, R.; Hemmatian, M. Multidisciplinary Design Optimization of a Novel Sandwich Beam-Based Adaptive Tuned Vibration Absorber Featuring Magnetorheological Elastomer. *Materials* **2020**, *13*, 2261. [CrossRef]
14. Bartkowski, P.; Zalewski, R. A concept of smart multiaxial impact damper made of vacuum packed particles. In *MATEC Web of Conferences*; EDP Sciences: Les Ulis, Frances, 2018; Volume 157, p. 05001.
15. Jiang, A.; Aste, T.; Dasgupta, P.; Althoefer, K.; Nanayakkara, T. Granular Jamming with Hydraulic Control. In Proceedings of the International Design Engineering Technical Conferences and Computers and Information in Engineering Conference, Portland, OR, USA, 4–7 August 2013.
16. Mozeika, A.; Steltz, E.; Jaeger, M.H. The first steps of a robot based on jamming skin enabled locomotion. In Proceedings of the IEEE/RSJ International Conference on Intelligent Robots and Systems, St. Louis, MO, USA, 11–15 October 2009; pp. 408–409.
17. Brown, E.; Rodenberg, N.; Amend, J.; Mozeika, A.; Steltz, E.; Zakin, M.R.; Lipson, H.; Jaeger, H.M. Universal robotic gripper based on the jamming of granular material. *Proc. Natl. Acad. Sci. USA* **2010**, *107*, 18809–18814. [CrossRef]
18. Zalewski, R.; Szmidt, T. Application of special granular structures for semi-active damping of lateral beam vibrations. *Eng. Struct.* **2014**, *65*, 13–20. [CrossRef]
19. Skalski, P.; Zalewski, R. Viscoplastic properties of an MR fluid in a damper. *J. Theor. Appl. Mech.* **2014**, *52*, 1061–1070. [CrossRef]
20. Wang, J.; Meng, G. Magnetorheological fluid devices: Principles, characteristics and applications in mechanical engineering. *Proc. Inst. Mech. Eng. L Mater. J. Des. Appl.* **2001**, *215*, 165–174. [CrossRef]
21. Zalewski, R.; Chodkiewicz, P. Semi-active linear vacuum packed particles damper. *J. Theor. Appl. Mech.* **2016**, *54*, 311–316. [CrossRef]
22. Luscombe, M.; Williams, J. Comparison of a long spinal board and vacuum mattress for spinal immobilization. *Emerg. Med. J.* **2003**, *20*, 476–478. [CrossRef]
23. Loeve, A.J.; van de Ven, O.S.; Vogel, J.G.; Breedveld, P.; Dankelman, J. Vacuum packed particles as flexible endoscope guides with controllable rigidity. *Granul. Matter* **2010**, *12*, 543–554. [CrossRef]
24. Szmidt, T.; Zalewski, R. Inertially excited beam vibrations damped by vacuum packed particles. *Smart Mater. Struct.* **2014**, *23*, 105026. [CrossRef]
25. Bajkowski, J.M.; Bajer, C.I.; Dyniewicz, B.; Pisarski, D. Vibration control of adjacent beams with pneumatic granular coupler: An experimental study. *Mech. Res. Commun.* **2016**, *78*, 51–56. [CrossRef]
26. Bajkowski, J.; Dyniewicz, B.; Bajer, C. Corrigendum to "damping properties of a beam with vacuum-packed granular damper". *J. Sound Vib.* **2015**, *341*, 74–85. [CrossRef]
27. Bouc, R. Forced vibrations of mechanical systems with hysteresis. In Proceedings of the Fourth Conference on Nonlinear Oscillations, Prague, Czech Republic, 5–9 September 1967.
28. Bartkowski, P.; Zalewski, R.; Chodkiewicz, P. Parameter identification of Bouc-Wen model for vacuum packed particles based on genetic algorithm. *Ach. Civil. Mech. Eng.* **2019**, *19*, 322–333. [CrossRef]

29. Weber, F.; Maślanka, M. Precise Stiffness and Damping Emulation with MR Dampers and its Application to Semi-active Tuned Mass Dampers of Wolgograd Bridge. *Smart Mater. Struct.* **2014**, *23*, 015019. [CrossRef]
30. Zalewski, R.; Chodkiewicz, P.; Shillor, M. Vibrations of a mass spring system using a granular-material damper. *Appl. Math. Model.* **2016**, *40*, 8033–8047. [CrossRef]
31. Zalewski, R. Analysis of Mechanical Properties of Vacuum Packed Particles. Ph.D. Thesis, Warsaw University of Technology, Warsaw, Poland, 2005. (In Polish).
32. Bartkowski, P. Empirical Research and Modeling of Energy Dissipators Working on the Basis of Vacuum Packed Particles. Ph.D. Thesis, Warsaw University of Technology, Warsaw, Poland, 2019. (In Polish)
33. Spencer, B., Jr.; Dyke, S.; Sain, M.; Carlson, J. Phenomenological model for magnetorheological dampers. *Eng. J. Mech.* **1997**, *123*, 230–238. [CrossRef]
34. Yang, G.; Spencer, B.F., Jr.; Jung, H.-J.; Carlson, J.D. Dynamic modeling of large-scale magnetorheological damper systems for civil engineering applications. *Eng. J. Mech.* **2004**, *130*, 1107–1114. [CrossRef]
35. Holnicki-Szulc, J.; Graczykowski, C.; Mikułowski, G.; Mróz, A.; Pawłowski, P.; Wiszowaty, R. Adaptive Impact Absorption—The Concept and Potential Applications. *Int. J. Prot. Struct.* **2015**, *6*, 357–377. [CrossRef]
36. Gamota, D.; Filisko, F. Dynamic mechanical studies of electrorheological materials: Moderate frequencies. *Rheol. J.* **1991**, *35*, 399–425. [CrossRef]
37. Lederer, D.; Igarashi, H.; Kost, A.; Honma, T. On the parameter identification and application of the Jiles-Atherton hysteresis model for numerical modeling of measured characteristics. *IEEE Trans. Magnet.* **1999**, *35*, 1211–1214. [CrossRef]
38. Asif Zaman, M.; Hansen, C.; Neustock, P.T.; Padhy, L.; Hesselink, P.L. Adjoint Method for Estimating Jiles-Atherton Hysteresis Model Parameters. *J. Appl. Phys.* **2016**, *120*, 093903. [CrossRef]
39. Bertotti, G. Dynamic generalization of the scalar Preisach model of hysteresis. *IEEE Trans. Magnet.* **1992**, *28*, 2599–2601. [CrossRef]
40. Ismail, M.; Ikhouane, F.; Rodellar, J. The hysteresis Bouc-Wen model, a survey. *Arch. Comput. Methods Eng.* **2009**, *16*, 161–188. [CrossRef]
41. Wen, Y.-K. Method for random vibration of hysteretic systems. *Eng. J. Mech. Div.* **1976**, *102*, 249–263.
42. Dong, H.; Han, Q.; Du, X. Application of an extended Bouc-Wen model for hysteretic behavior of the RC structure with SCEBs. *Struct. Eng. Mech.* **2019**, *71*, 683–697.
43. Kirkpatrick, S.; Gelatt, D., Jr.; Vecchi, P.M. Optimization by simulated annealing. *Science* **1983**, *220*, 671–680. [CrossRef]
44. Ye, M.; Wang, X. Parameter estimation of the Bouc-Wen hysteresis model using particle swarm optimization. *Smart Mater. Struct.* **2007**, *16*, 2341. [CrossRef]
45. Asif Zaman, M.; Sikder, U. Bouc-Wen hysteresis model identification using modified firefly algorithm, Magnet. *J. Magnet. Mater.* **2015**, *395*, 229–233. [CrossRef]
46. Pyrz, M.; Krzywoblocki, M. Crashworthiness optimization of thin-walled tubes using macro element method and evolutionary algorithm. *Thin-Walled Struct.* **2017**, *112*, 12–19. [CrossRef]
47. Nocedal, J.; Wright, S. *Numerical Optimization*; Springer: New York, NY, USA, 1999; ISBN 0-387-98793-2.

materials

MDPI

Article

A Novel Method for Changing the Dynamics of Slender Elements Using Sponge Particles Structures

Mateusz Żurawski *, Bogumił Chiliński and Robert Zalewski

Institute of Machine Design Fundamentals, Warsaw University of Technology, 02-524 Warsaw, Poland; Bogumil.Chilinski@pw.edu.pl (B.C.); Robert.Zalewski@pw.edu.pl (R.Z.)

* Correspondence: mateusz.zurawski@pw.edu.pl; Tel.: +48-535-278-333

Received: 11 September 2020; Accepted: 27 October 2020; Published: 30 October 2020

Abstract: The paper concerns problems related to controlling the dynamic properties of beam-like elements. The parameters of the investigated system can be changed by external factors, resulting in partial changes in the system mass redistribution. It is assumed that it is possible to control the system dynamics by shaping the object frequency structure. The paper introduces the mathematical model of the investigated cantilever beam filled with a Sponge Particle Structure. The continuous model has been simplified to a discrete multi-degree of freedom system. The influence of the system parameters on its behavior is discussed in details. The possible applications of the presented concept are proposed. The spectral vibration analyses were carried out. Theoretical considerations enabled the use of the preliminary semi-active method for controlling the vibration frequencies through a mass redistribution. Experimental studies were carried out to verify the proposed mathematical model.

Keywords: mass redistribution; sponge particles structure; semi-passive damping; experimental research

1. Introduction

Vibrations are important phenomenon which cause many undesired effects. There are a lot of problems related to operating various types of devices in a resonant range [1]. Harmful and dangerous vibrations are generally attenuated using passive [2,3] and active methods [4,5]. Nowadays, semi-active vibrations attenuation techniques are also getting popular [6]. The most sophisticated, complex and, in consequence, expensive technique is an active method [7]. Unfortunately, it requires an additional power supply, sophisticated measuring system and very often does not fulfill costs and environment limitations. The obvious alternative for a previously mentioned method is a passive strategy. Various approaches to passive vibration attenuation exist. The most popular is frictional, viscous and impact damping [8]. The main disadvantages of the viscous (viscoelastic) and frictional damping are nonlinear characteristics or strong degradation effects [9]. Concluding, the most interesting and universal seem to be impact dampers. In this group of devices, Particle Impact Dampers (PIDs) play the most important role [10–12]. Problems related to the damping of vibrations are popular in publications in the field of civil and mechanical engineering. Therefore, in this paper the authors will limit themselves to a brief description of vibration attenuation methods using granular materials [13]. A granular damping technique shows some similarities to classical impact damping. When the impact damping involves the movement of a single additional mass in the vibrated system, the particles damping refers to many auxiliary masses with relatively small sizes, moving in a specially-shaped cavity. The principle of the Particle Impact Dampers is the dissipation of energy of the vibrating element through inelastic collisions occurring between individual grains and between grains, and the container in which they are located [11]. Another phenomenon existing in described energy attenuators is friction occurring as an "intergranular" force. The main advantages of granular dampers are low sensitivity to temperature changes [14], long life, reliability in a wide range of frequencies, reduction in the global mass of the attenuated system (the grains often have lower mass than the mass of the initial structure they replace), and low sensitivity for operation in a harsh environment [15,16]. The main disadvantage of PIDs is the passive means of vibration damping [17]. The adjustment of the damping properties is acquired by the selection of a proper size, material and filling ratio of particles. Detailed analyses of PIDs behavior is mainly based on empirical tests [18]. The results of experiments are the base for numerical simulations carried out mainly using the Discrete Element Method (DEM) or Finite Element Method (FEM) [19]. Analytical approaches seem to be less popular. Although the grain interaction mechanics are quite simple, such an approach demands a high computing power. Conducting numerical tests for thousands of grains is a challenge even for modern and effective computers. Over the last three decades, many papers have revealed the results of research carried out in the field of granular energy attenuators. Interesting research results were discussed in [20], where the system consisting of 10,000 particles was taken into consideration. Loose grains were moving in the cavity accordingly to various external loading forces. The damping efficiency of such a system was investigated in detail. In [21], a mathematical model was developed that allows one to predict the behavior of granular damper. The applied mathematical formulation describes the grains dynamics and takes into account the viscous properties of the grain materials. Additionally, a friction phenomenon occurring in the contact zones was involved. An interesting application of the granular damper for the vibrations attenuation of a cantilever beam is presented in [22]. The authors revealed preliminary laboratory tests results. Based on experimental data, it can be stated that capturing the real behavior of the highly nonlinear granular system is challenging. The verification of simulation results carried out, based on the proposed mathematical model with real empirical data, seems to be promising. A PID design methodology has been discussed in [23,24]. The authors proposition limits and many typical problems encountered in such devices. It is underlined that granular dampers have strongly nonlinear damping characteristics. Interesting standards for designing particles dampers are presented. The guidelines simplifying the selection of the proper granular damper for the vibrated systems are presented and verified experimentally.

Not only are loose granular materials applied in damping strategies. An interesting proposition of the semi-active damping of vibrations strategy, including so called Vacuum Packed Particles, are presented in previous works of the authors. In [25,26] it is confirmed that the underpressure value can be an efficient parameter of controlling the dissipative properties of a cantilever beam subjected to various types of excitations.

In this paper, the original method of slender elements (shafts, beams or rods) natural frequency shaping with an application of the local mass redistribution is presented. The global concept is to change the vibrating object mass distribution and to affect its dynamics. Such a solution seems to be possible thanks to the application of innovative granular structures. The idea of the proposed method is to divide the beam into several hermetic sections. Each segment is filled with a structure that enables its density to be changed. In this paper, the authors assumed that Sponge Particles Structures (SPS) can be a filling material. Such a structure consists of loose grains embedded in soft sponge layers that exhibit various mechanical properties according to changes in the external pressure. The proposed shaping of the dynamic features strategy is novel and promising. In this paper, the authors assume that the external pressure provided to the structure by the system of balloons can be an efficient parameter to reorganize the grains inside each section, causing the changes in the mass redistribution of the investigated element.

The concept introduced in the paper involves a rectangular shell-like cantilever beam. Inside the object, two fixed partitions are mounted. The system additionally consists of a Sponge Particles Structure and six balloons. The discussed idea is depicted in Figure 1.

Inflating balloons (3) results in compressing the SPS (2) placed between them. Changing the internal pressure in the balloons enables the granular mass embedded in a sponge matrix to be compressed and redistributed. This phenomenon is a convenient way to change the dynamical properties of the investigated system. For the purpose of mass redistribution, not only SPS can be involved. Additionally, other unconventional structures, such as magnetorheological fluids or elastomers can be useful [27,28]. The volume of the internal sections can also be changed mechanically (by screws), hydraulically, or by the external magnetic field. The paper presents a novel adaptive–passive (semi-active) method for damping of vibrations. The main objective is to confirm the effectiveness of the mass redistribution method for a controlling process of the beam structures dynamics. The proposed methodology can be applied in various practical engineering applications. It is worth mentioning that such an adaptive damping strategy can be potentially used in structures, such as in the wings of aircrafts, slender skyscrapers, long bridge spans, or supports of wind turbines.

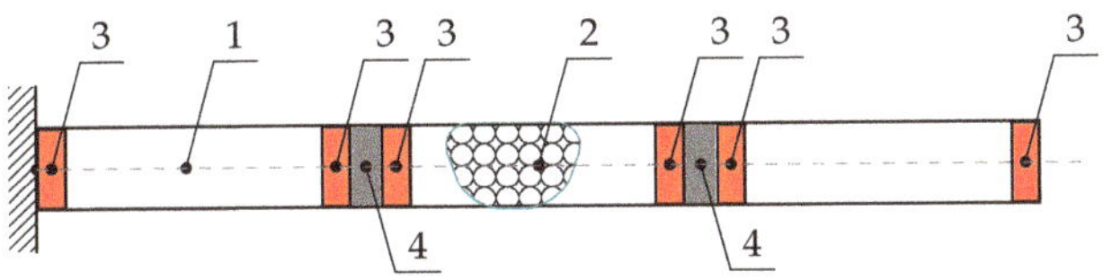

Figure 1. The idea of the investigated system (1—shell type beam structure; 2—Sponge Particles Structure (SPS); 3—balloons; 4—fixed diaphragms).

2. Preliminary Modeling and Analysis of Mass Redistribution on the System Dynamics

Based on the mathematical analysis of the continuous systems, it is possible to show that the models with infinite degrees of freedom can be reduced to the finite number of nodes. Such a simplification allows for easy and quick dynamics analyses. However, an inappropriate discretization of the problem will lead to incorrect results. Therefore, in the preliminary stage of the presented research, the impact of Degree of Freedom (DOF) number on eigenvalues and eigenmodes was numerically tested. In order to perform eigenfrequencies validation, the beam was divided into arbitrary given number of segments. For each segment, equivalent mass and stiffness were obtained. Based on determined data, inertia and stiffness matrices were computed. It enables one to find eigenfrequencies of the given system and to investigate the impact of node numbers on the computational results. Based on the aforementioned methodology, the natural vibration frequency of the fixed beam reduced to 3, 5, 7, 9 and 10 nodes was calculated.

Governing equations were rearranged to the eigenvalue problem form, in order to determine natural frequencies of the system. The solution was predicted in the form of Formula (1).

$$X(t) = A \cdot \sin(\omega t + \varphi) \tag{1}$$

After necessary substitutions and rearrangements, the following form of the eigenvalue equation can be obtained:

$$(M^{-1} \cdot K)A = \omega^2 \cdot A \tag{2}$$

The roots of the Equation (2) yield the natural system frequencies. The results of computations are shown in Table 1.

Table 1. Natural frequencies of beam for various discretization variants—uniform mass distribution.

No		3 Nodes	5 Nodes	7 Nodes	9 Nodes	10 Nodes
f_1	Hz	8.1	8.5	8.9	9.2	9.23
f_2	Hz	53.1	54.1	56.4	57.8	58.4
f_3	Hz	142.7	153.4	159.1	162.6	163.9

Analyzing the results, it can be concluded that the vibration frequencies of a system with 3 and 10 degrees of freedom differ by less than 15 %. The authors assumed that such values in the context of further calculations are acceptable at the initial stage of investigations. To ensure the possibility of using a simplified beam model with 3-DOF, additional mathematical analyses were carried out. The frequency of natural vibrations for the system was determined and compared in which 60 % of the entire system mass was moved to the end of the beam. The results are shown in Table 2.

Table 2. A List of results of beam's natural frequency for various discretization variants—the mass redistributed to the end of the beam.

No		3 Nodes	5 Nodes	7 Nodes	9 Nodes	10 Nodes
f_1	Hz	11	11.2	11.3	12.7	13.2
f_2	Hz	52.1	55	57.3	64.5	68.2
f_3	Hz	149.6	160.5	166.2	168.6	169.8

Analyzing the data presented in Table 2 the authors assumed that for the preliminary numerical tests simple 3-DOF beam model is sufficient. The difference between responses of 3-DOF and 10-DOF models differs by less than 15%.

Every single linear dynamic system can be described by a second order differential equations. In the case of a multi DOF system it is possible to perform a modal decomposition and transform the problem to a simpler form obtaining the system of uncoupled equations. Then every mode can be treated as a single degree of freedom. A governing equation is given by (3).

$$\ddot{x} + \omega_0^2 \cdot x = A \cdot sin(\Omega \cdot t) \tag{3}$$

where:

$\ddot{x}$—acceleration;
ω_0—natural frequency;
x—displacement;
A—excitation amplitude;
Ω—excitation frequency.

Dynamics of the system with varying parameters is important from the practical point of view. In the initial stage of investigations, the following problem of step changing parameters was analyzed. The transient process is described by Heaviside's step function, which provides appropriate differential properties (for the adopted definition of the distributional derivative). It represents a sudden change of the system parameters, what is the system's worst case scenario. Although it is a simple model of such a phenomenon, it allows for an efficient analysis of the case under investigation (from the analytical and numerical point of view). The formula describes a mass change which is presented in (4).

$$m(t) = m_0 \cdot (1 + H(t - t_0)) \tag{4}$$

where:

$m(t)$—varying mass;
m_0—initial mass;
$H(t - t_0)$—Heaviside function;
t_0—redistribution process starting time.

For that case, the governing equation is as follows:

$$\ddot{x} + \omega_0^2 \cdot (1 + H(t - t_0))^{-1} \cdot x = A \cdot (1 + H(t - t_0))^{-1} \cdot \sin(\Omega \cdot t) \tag{5}$$

Equation (5) is an ordinary differential equation with discontinuous parameters. It is a distributional problem which can be solved using generalized functions. For the problem under consideration, the following particular solution was predicted:

$$x(t) = (A_1 \cos(\Omega \cdot t) + B_1 \sin(\Omega \cdot t)) \cdot H(t_0 - t) + \\ + (A_2 \cos(\Omega \cdot t) + B_2 \sin(\Omega \cdot t)) \cdot H(t - t_0) \quad (6)$$

Based on the predicted solution (6), the parameters A_1, B_1,A_2 and B_2 were found.

$$A_1 = \frac{a_0}{\Omega^2 - \omega_0^2}$$

$$B_1 = 0$$

$$A_2 = \frac{a_0}{\Omega^2 - 0.5 \cdot \omega_0^2} \cdot \tan^{-1}\left(\frac{\sin(t_0)}{\cos(t_0)}\right) \quad (7)$$

$$B_2 = \frac{a_0}{\Omega^2 - 0.5 \cdot \omega_0^2} \cdot \cot^{-1}\left(\frac{\sin(t_0)}{\cos(t_0)}\right)$$

It should be noted that the damping parameter does not appear in the general mathematical solution (Equation (5)) of the proposed phenomenon. The term "damping" is often used in many papers and handbooks to describe the process of the reduction in vibration amplitudes. The idea of the proposed method is exactly the same as a working principle of vibration absorbers. These devices are often called Tuned Mass Dampers. That is why the authors decided to use the term "damping" to call the phenomenon of resonant frequency shifting.

3. A Proposition of Modeling and Shaping of Beams Dynamics

In general, case beam-like elements should be treated as continuous systems. Unfortunately, such an approach is quite complicated from the computational point of view. There are closed solutions for continuous shaft dynamics, but only for simple cases. For this purpose, it is necessary to find an equivalent inertia and stiffness matrix to start with a preliminary analysis of the continuous system. In the case under consideration, fundamental laws of linear-elasticity—e.g., Castigliano theorem or Maxwell-Mohr method—were used. Using Equations (8) and (9) it is possible to find flexibility coefficients:

$$\delta_{ij} = \int_0^l \frac{M'_{bi} M'_{bj}}{EJ} dl \quad (8)$$

where:

δ_{ij}—flexibility matrix;
M'_{bi}—unit bending moment;
M_{bj}—bending moment;
E—Young modulus;
J—moment of inertia.

$$K = \delta^{-1} \quad (9)$$

where:

K—stiffness matrix.

The Inertia matrix M can be derived with the application of mass conservation law or energy conservation of entire system. In the general case, it is possible that the matrix M will be diagonal (10).

$$M = \text{diag}(m_1, m_2, ..., m_n) \quad (10)$$

where:

M—inertia matrix;

n—number of nodes;
m_i—nodal masses.

Natural frequencies can be found as eigenvalues of the matrix $M^{-1}K$. Four various inertia matrices were taken into account in the paper. The fixed beam with a circular cross-section was adopted as a model. The masses were concentrated at three different points as shown in Figure 2.

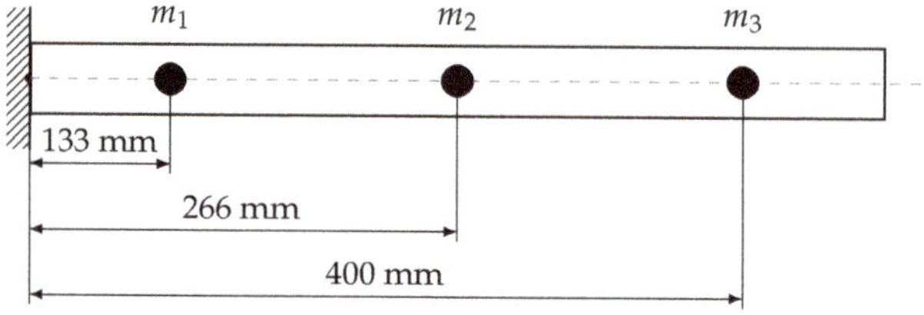

Figure 2. Mathematical model of the cantilever beam.

For further calculations, the following mass distribution values for individual nodes were adopted:

Basic mass distribution: $M_{base} = diag(2,2,2)$ kg;
First mass distribution: $M_1 = diag(4,1,1)$ kg;
Second mass distribution: $M_2 = diag(1,4,1)$ kg;
Third mass distribution: $M_3 = diag(1,1,4)$ kg;
Youngs modulus: $E = 2.1 \times 10^{11}$ Pa;
Moment of inertia: $I = 4.43 \times 10^{-6}\,\text{m}^4$;
Inner diameter of the beam: $d = 100$ mm;
Outer diameter of the beam: $D = 116$ mm.

To determine the influencing factors, the bending moment must be determined in advance. In the case under consideration (Figure 2), it can be described using the Heaviside function:

$$M_b = (x - L) \cdot H(x - L) \tag{11}$$

The governing Equations (12)–(14) were determined based on obtained forms of the inertia and the stiffness matrices. The Lagrangian mechanics were utilized for this reason. The carried out computations revealed the influence of the mass changes. It has to be noticed that there is the impulse force, which comes from a sudden change in the mass during the redistribution process.

$$m_1\ddot{x}_1 + \dot{m}_1\dot{x}_1 + c_1\dot{x}_1 + k_{11}x_1 + \frac{k_{12}x_2}{2} + \frac{k_{13}x_3}{2} + \frac{k_{21}x_2}{2} + \frac{k_{31}x_3}{2} = f(t) \tag{12}$$

$$m_2\ddot{x}_2 + \dot{m}_2\dot{x}_2 + c_2\dot{x}_2 + \frac{k_{12}x_1}{2} + \frac{k_{21}x_1}{2} + k_{22}x_2 + \frac{k_{23}x_3}{2} + \frac{k_{32}x_3}{2} = 0 \tag{13}$$

$$m_3\ddot{x}_3 + \dot{m}_3\dot{x}_3 + c_3\dot{x}_3 + \frac{k_{13}x_1}{2} + \frac{k_{23}x_2}{2} + \frac{k_{31}x_1}{2} + \frac{k_{32}x_2}{2} + k_{33}x_3 = 0 \tag{14}$$

Frequencies of natural vibrations in the investigated cantilever beam, including various mass redistribution cases, are presented in Figure 3.

Analysis of the results shows that the mass redistribution affects the dynamics of the entire system. However, it turns out that about 30% change in the inertia matrix shifts some frequencies by about 20%. In addition, it can be observed that the redistribution of mass to a particular location causes a significant change in the frequency of vibrations in other nodes.

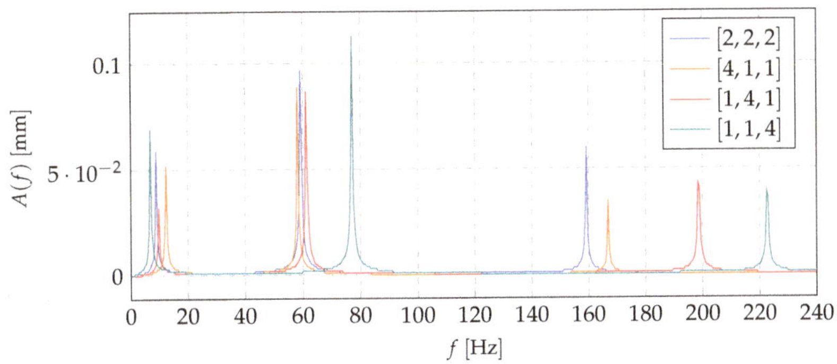

Figure 3. Resonance frequency for all cases of mass distribution.

4. Simulations

Obtained results confirmed the fact that the change in mass distribution will shift the resonant range. In the simulations, the harmonic force was introduced into the system. The investigated cantilever beam was subjected to various resonance loading frequencies. This allows us to observe the form of vibrations and to present a way for the mass distribution to influence the dynamics of the structure during operation. In Figure 3a, Frequency Response Functions (FRFs) for all cases under consideration are presented.

Based on the mathematical model, and taking into account the excitation with the first natural frequency, the first forms of vibration were obtained. Shifting the mass to the proper node (choice of the node depending on the system dynamics and its initial mass distribution) decreases the amplitude of the dynamic response. It is possible to use that concept to switch the system states depending on the loading force pulsation. This approach ensures that the excitation forces will not cause a resonance or that it would appear in the short period of time. Figures 4 and 5 present simulations for beam vibrations for all four redistribution cases, based on the resonance frequency of the $[1, 4, 1]$ case of the first and second eigenvalue (9.05 and 59.26 Hz). Several integration procedures were applied, in order to carry out numerical simulations. The best results were obtained with Runge–Kutta fourth order algorithm.

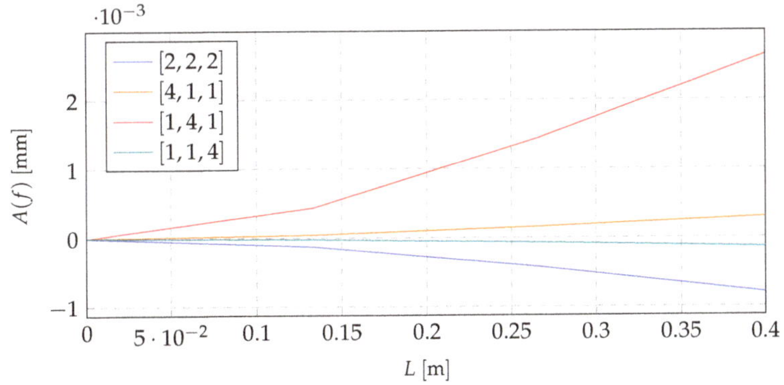

Figure 4. Resonance vibrations for 9.05 Hz and various mass redistribution cases.

Similar data for the third natural frequency are depicted in Figure 6.

Analyzing the data depicted in Figures 4–6, it can be stated that there is a possibility to avoid the resonant range by shifting the mass. It means that the mass redistribution process can be the basis for the efficient method of controlling the dynamics of the vibrating system. In the proposed method the mass transfer is carried out in a static way. Furthermore, simulations are based on the beam vibration analysis during the mass redistribution process. It means that the moving of the mass took a certain time. This causes the need for the investigation of the system behavior, depending on how the process of mass redistribution is carried out. Figure 7 presents time plot of masses changes at nodes during the system operation.

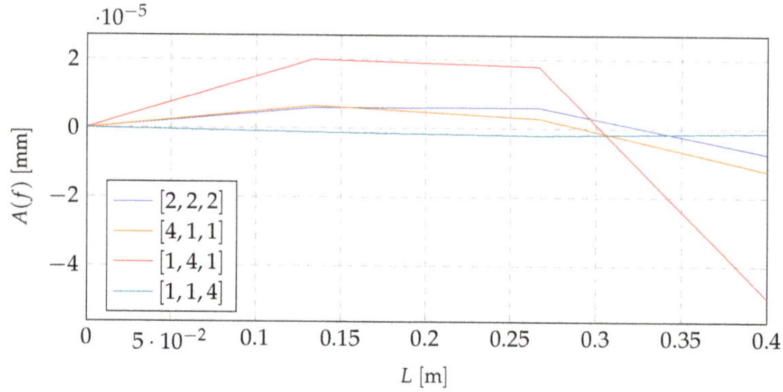

Figure 5. Resonance vibrations for 59.26 Hz and various mass redistribution cases.

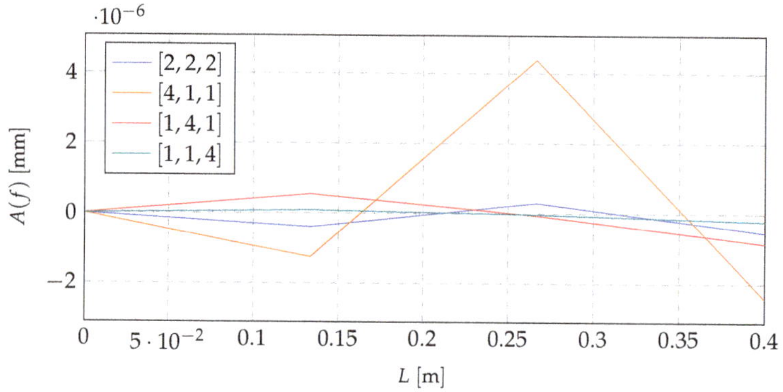

Figure 6. Resonance vibrations for 159.22 Hz and various mass redistribution cases.

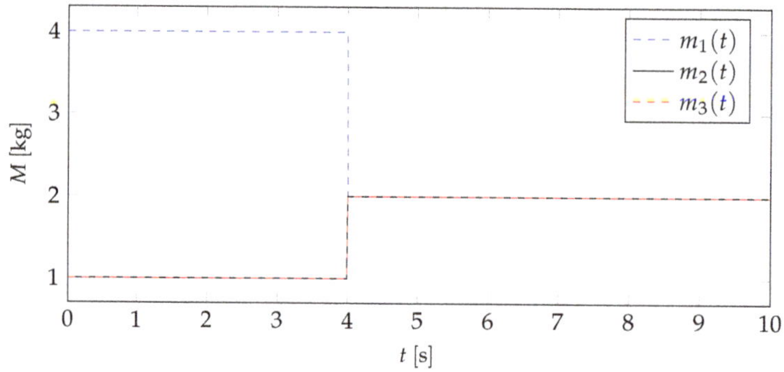

Figure 7. Functions of the masses changes for every node.

Typical results of vibration displacements of the beam end for a selected mass redistribution case ($M_1 \rightarrow M_{base}$) are depicted in Figure 8.

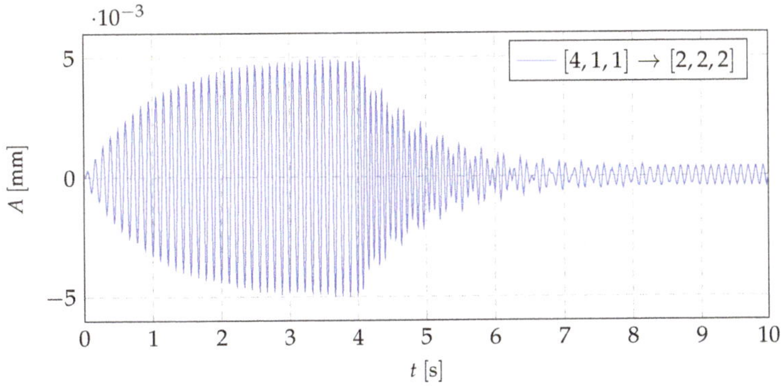

Figure 8. Deflection of the end of the beam during the process of mass redistribution in the neighborhood of the resonant vibration.

The spectral analysis of vibrations displacements of the investigated cantilever beam during the dynamic mass redistribution process is depicted in Figure 9. Three types of vibrations can be distinguished: 1—vibrations of the beam with the mass being shifted to a specific point (node 1) (from 0 s to 4 s of simulation); 2—Beam vibrations during mass transfer (from 4 s to 8 s of simulation); 3—Beam vibrations with a homogeneous mass distribution (from 8 s to 10 s of simulation). Analyzing the data depicted in Figure 8a significant change in the displacements of the beam end can be observed. It could mean that the vibration frequencies during this process reach unexpected values. An amplitude spectrum of the beam vibrations (Figure 9) shows that the mass redistribution phenomenon causes "blurring" of natural frequencies. It means that there is an intermediate process causing a smooth transition from the initial (in the resonance) to the final (beyond resonance) dynamic state. This is crucial because there is no risk of falling into other, unpredictable resonant ranges.

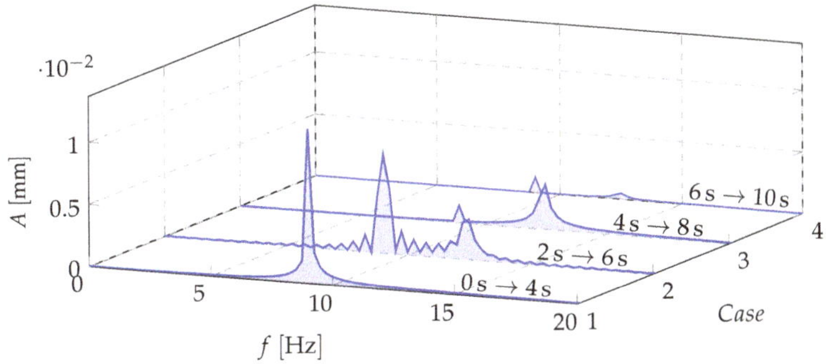

Figure 9. Short time spectrum of deflection vibrations—STFT.

5. Test Stand

In order to validate the mathematical model, the special research stand was designed. The main concept assumed a light and thin beam filled by special Sponge Particle Structure. In such a case SPS divides the beam into three separate sections by fixed elastic plates, balloons, and movable plates. Each section has a structure consisting of sponge sheets and uniformly distributed particles. The sponge parts are connected by glue, and the heavy steel grains are placed between them (Figure 10). Pumping the balloon allows one to move the sliding plates. As a result, sponges with particles in each section can change their volume and position which can be described as a mass redistribution. It was assumed that the masses of the SPS sections are much larger than the entire beam. It allows us to discretize the continuous system to three nodes. It corresponds to the model assumptions and simplifications adopted at the preliminary stage.

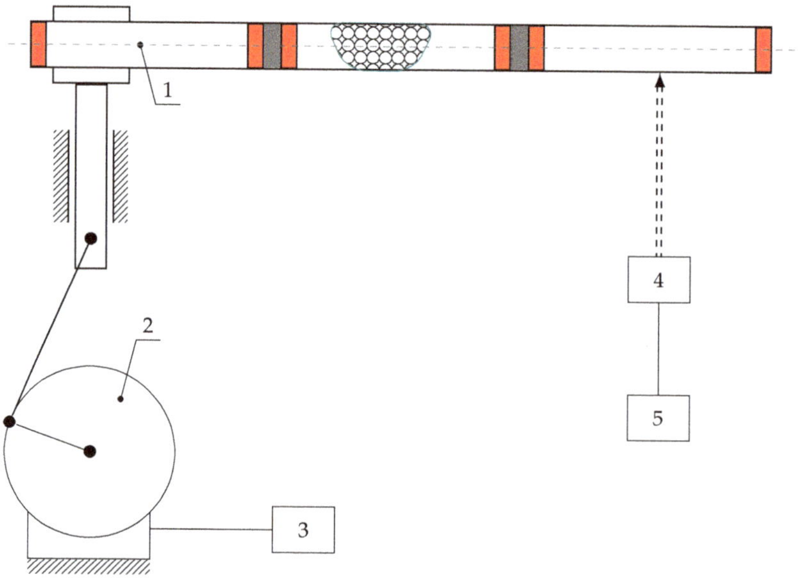

Figure 10. Scheme of the test stand (1—sponge particles beam; 2—motor; 3— inverter; 4—laser sensor; 5—data acquisition system).

The designed test stand can be used for various types of the excitation. Vibration measurements with different distribution of masses along the beam were carried out. Test stand parameters are presented in Table 3 where V_{max} means maximum volume of the balloon necessary for the redistribution of the whole grains mass to one node, and P_{max} is the pressure of the balloon when the volume reaches its maximum value V_{max}.

Table 3. Test stand parameters.

Component	Parameter	Value
Cantilever beam	Size	Length: 400 mm Inner diameter: 100 mm Outer diameter: 116 mm
	Material	Steel
Sponge	Density	25 kg/m^3
	Material	Poliuretan
Grains	Size	Diameter: 4 mm
	Material	Steel
	Density	7900 kg/m^3
Balloon	Size	$V_{max} = 8.35 \times 10^{-3}$ m^3
	Pressure	$P_{max} \approx 100$ Pa
	Material	Rubber

Figures 11–13 show the amplitude spectra for the acceleration vibrations for various cases of the mass redistribution. Based on that, it is possible to compare the natural frequency of the beam for the initial case (uniform mass distribution) and for the mass being redistributed.

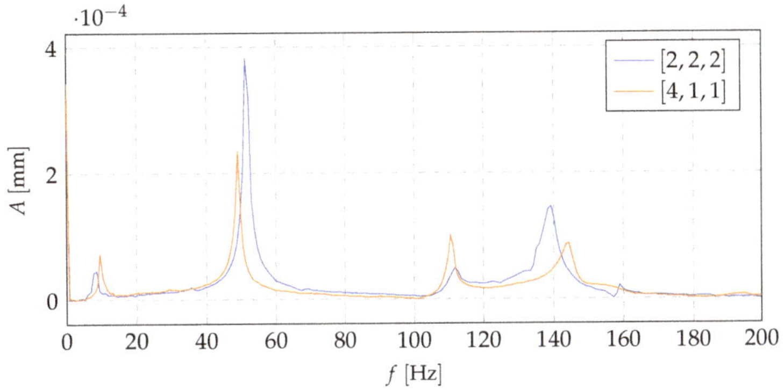

Figure 11. The amplitude spectrum of vibration accelerations (comparison of basic case and first mass distribution case.

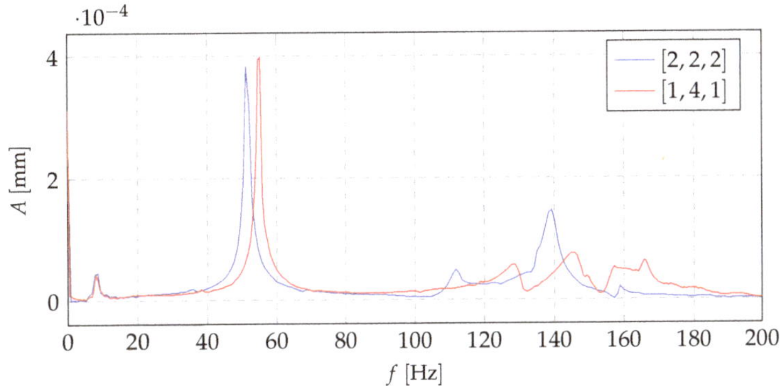

Figure 12. The amplitude spectrum of vibration accelerations (comparison of basic case and the second mass distribution case.

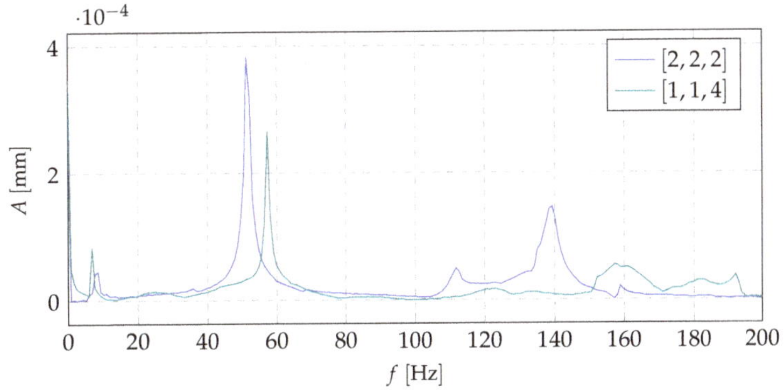

Figure 13. The amplitude spectrum of vibration accelerations (comparison of basic case and the third mass distribution case).

Analyzing the data presented in Figures 11–13, it can be concluded that the proposed shaping method of the frequency structure causes the observable change in a dynamic behavior of the whole system. It can be stated that, for the each mass distribution case, there is a corresponding eigenfrequency which will change during the redistribution process. Increasing the mass density at a selected node allows one to change the natural vibration frequency up to 25% in comparison to the remaining configurations (distributions of the mass). Such changes in a frequency structure cause various frequency responses under the harmonic excitation. It means that, even if the system reaches its natural frequency, it is possible, thanks to the SPS, to change the mass distribution and, in consequence, limit the value of vibration amplitude. Moreover, theoretical and experimental analyses indicate that particular changes affect only selected eigenfrequencies. It is very important from the practical point of view because it allows us to control only a narrow band of the FRF. The remaining part of the spectrum will not be changed. It provides predictable dynamics of the system and makes the control process easier.

6. Summary

Slender elements, such as shafts, rods and beams, often work in a resonant range, in many engineering applications. Therefore, it is crucial to take into account this dangerous phenomenon in the design process. In this paper the authors proposed a novel method for adaptive–passive (semi-active) damping of vibration techniques, thanks to introducing an innovative structure based on granular materials. The paper also presents a discretized mathematical model, allowing for a quick determination of the natural frequency of a cantilever beam. This was applied to determine the value of frequency changes when 30% of the total mass of the structure was redistributed. Two theoretical methods of the mass relocation were taken into account: a step function and continuous change in the beam interior density. Both cases showed that there is a rational approach to the possible resonant range shifting. In addition, simulations of the dynamic mass redistribution allowed us to reveal a smooth transition from the initial to the expected frequency structure. Although the step change in vibrations is observed, the variations in a frequency structure do not appear and the system dynamic behavior is stable. Experimental tests carried out on the special test stand confirmed the correctness of numerical calculations. For the mass redistribution, not only SPS can be involved. Additionally, other unconventional structures, such as magnetorheological fluids or elastomers, can be useful [27,28]. The volume of the internal sections can also be changed mechanically (by screws), hydraulically or by the external magnetic field. Novel methods will be developed for a rapid mass change with the application of smart materials, such as Vacuum Packed Particles or magnetorheological fluids. Such an approach would allow for "in real time" changes between different values of natural frequencies and vibration amplitudes.

Author Contributions: Conceptualization, M.Ż., B.C. and R.Z.; methodology, M.Ż. and B.C.; software, B.C.; validation, M.Ż. and B.C.; formal analysis, M.Ż. and B.C.; investigation M.Ż. and B.C.; data curation M.Ż. and B.C.; writing—original draft preparation, M.Ż., B.C. and R.Z.; writing—review and editing, B.C. and R.Z.; visualization, M.Ż. and B.C.; supervision, R.Z. All authors have read and agreed to the published version of the manuscript.

Funding: This research received no external funding. The APC was funded by Warsaw University of Technology.

Conflicts of Interest: The authors declare no conflict of interest.

References

1. Osiński, J. *Damping of Vibrations*; A.A.Balkema: Rotterdam, The Netherlands, 1998.
2. Parulekar, Y.M. Passive response control systems for seismic response reduction: A state-of-the-art review. *Int. J. Struct. Stab. Dyn.* **2009**, *9*. [CrossRef]
3. Diez-Jimenez, E.; Rizzo, R. Review of Passive Electromagnetic Devices for Vibration Damping and Isolation. *Shock Vib.* **2019**, *11*. [CrossRef]
4. Bankar, V.K.; Aradhye, A. A Review on Active, Semi-active and Passive Vibration Dampin. *Int. J. Curr. Eng. Technol.* **2016**, *6*. [CrossRef]
5. Chuaqui, T.R.; Roque, C.M.; Ribeiro, P. Active vibration control of piezoelectric smart beams with radial basis function generated finite difference collocation method. *J. Intell. Mater. Syst. Struct.* **2018**, *29*, 2728–2743. [CrossRef]
6. Allien, J.V.; Kumar, H.; Desai, V. Semi-active vibration control of SiC-reinforced Al6082 metal matrix composite sandwich beam with magnetorheological fluid core. *Proc. Inst. Mech. Eng. Part L J. Mater. Des. Appl.* **2019**, *234*, 408–424. [CrossRef]

7. Mojtaba, B.; Gromada, M.S.F. Optimal configuration of piezoelectric sensors and actuators for active vibration control of a plate using a genetic algorithm. *Acta Mech.* **2015**, *226*, 3451–3462. [CrossRef]
8. Kumar, K.; Krishna, Y.B.P. Damping in beams using viscoelastic layers. *Proc. Inst. Mech. Eng. Part L J. Mater. Des. Appl.* **2015**, *229*, 117–125. [CrossRef]
9. Ashkan, H.; Quenneville, P. Large-scale testing of low damage rocking Cross Laminated Timber (CLT) wall panels with friction dampers. *Eng. Struct.* **2020**, *206*. [CrossRef]
10. Cherif, S.; Trigui, M. Design parameters optimization of a particles impact damper. *Int. J. Interact. Des. Manuf.* **2018**, *12*, 1283–1297. [CrossRef]
11. Lu, Z.; Wang, Z., Masri, S.F.; Lu, X. Particle impact dampers: Past, present, and future. *Struct. Control Health Monit.* **2018**, *25*, e2058. [CrossRef]
12. Gagnon, L.; Morandini, M. A review of particle damping modeling and testing. *J. Sound Vib.* **2019**, *459*. [CrossRef]
13. Du, Y.; Guan, Z. Research on a New Particle Impact Damper. In Proceedings of the 4th Annual International Workshop on Materials Science and Engineering (IWMSE2018), Xi'an, China, 18–20 May 2018; Volume 381, pp. 1283–1297. [CrossRef]
14. Panossian, H. Structural damping enhancement via non-obstructive particle damping technique. *J. Vib. Acoust.* **1992**, *114*, 101–105. [CrossRef]
15. Ehrgott, R.H. Panossian, G.D. Modelling techniques for evaluating the effectiveness of particle damping in turbomachinery. *Pratt Whitney Rocket.* **2009**. [CrossRef]
16. Simonian, S. Particle beam damper. *Proc. SPIE* **1995**, *2445*, 149–160. [CrossRef]
17. Ashley, S. A new racket shakes up tennis. *Mech. Eng.* **1995**, *117*, 80–81.
18. Mohamed, G.; Karkoub, M. Experimental Investigation of Linear Particle Chain Impact Dampers in Free-Vibration Suppression. *J. Struct. Eng.* **2017**, *143*, 407–417. [CrossRef]
19. Toshihiko, K.; Iwata, Y. A combined approach for modeling particle behavior in granular impact damper using discrete element method and cellular automata. *Int. J. Mech. Mater. Des.* **2017**, *13*, 407–417. [CrossRef]
20. Panossian, H. Non-obstructive Particle Damping Experience and Capabilities. *Proc. SPIE Int. Soc. Opt. Eng.* **2002**, *2*, 936–941. [CrossRef]
21. Olson, S.E. An analytical particle damping model. *J. Sound Vib.* **2003**, *264*, 1155–1166. [CrossRef]
22. Fowler, B.L.; Flint, E.M.; Olson, S.E. Effectiveness and Predictability of Particle Damping. *Smart Struct. Mater.* **2000**. [CrossRef]
23. Fowler, B.L.; Flint, E.M.; Olson, S.E. Design Methodology for Particle Damping. *SPIE Conf. Smart Struct. Mater.* **2001**. [CrossRef]
24. Mao, K.; Wang, M.Y. DEM simulation of particle damping. *Powder Technol.* **2004**, *142*, 154– 165. [CrossRef]
25. Szmidt, T.; Zalewski, R. Inertially excited beam vibrations damped by Vacuum Packed Particles. *Smart Mater. Struct.* **2014**, *23*. [CrossRef]
26. Szmidt, T.; Zalewski, R. Application of Special Granular Structures for semi-active damping of lateral beam vibrations. *Eng. Struct.* **2014**, *65*, 13–20. [CrossRef]
27. Ahamed, R.; Choi, S.B.; Ferdaus, M.M. A state of art on magneto-rheological materials and their potential applications. *J. Intell. Mater. Syst. Struct.* **2018**, *29*, 2051–2095. [CrossRef]
28. Mojtaba, A.; Kouchakzadeh, M.A. Aeroelastic characteristics of magneto-rheological fluid sandwich beams in supersonic airflow. *Compos. Struct.* **2016**, *143*, 93–102. [CrossRef]

Publisher's Note: MDPI stays neutral with regard to jurisdictional claims in published maps and institutional affiliations.

materials

MDPI

Article

Deformation of Bioinspired MXene-Based Polymer Composites with Brick and Mortar Structures: A Computational Analysis

Shreyas Srivatsa [1,*], Paweł Paćko [2], Leon Mishnaevsky, Jr. [3], Tadeusz Uhl [1,*] and Krzysztof Grabowski [1]

1 Academic Center for Materials and Nanotechnology, AGH University of Science and Technology, 30-059 Krakow, Poland; kgrabow@agh.edu.pl
2 Department of Robotics and Mechatronics, AGH University of Science and Technology, 30-059 Krakow, Poland; pawel.packo@agh.edu.pl
3 Department of Wind Energy, Technical University of Denmark, 4000 Roskilde, Denmark; lemi@dtu.dk
* Correspondence: sshreyas@agh.edu.pl (S.S.); tuhl@agh.edu.pl (T.U.); Tel.: +48-730-610-199 (S.S.); +48-602-398-564 (T.U.)

Received: 21 October 2020; Accepted: 13 November 2020; Published: 17 November 2020

Abstract: In this work, the deformation behavior of MXene-based polymer composites with bioinspired brick and mortar structures is analyzed. MXene/Polymer nanocomposites are modeled at microscale for bioinspired configurations of nacre-mimetic brick-and-mortar assembly structure. MXenes (brick) with polymer matrix (mortar) are modeled using classical analytical methods and numerical methods based on finite elements (FE). The analytical methods provide less accurate estimation of elastic properties compared to the numerical one. MXene nanocomposite models analyzed with the FE method provide estimates of elastic constants in the same order of magnitude as literature-reported experimental results. Bioinspired design of MXene nanocomposites results in an effective increase of Young's modulus of the nanocomposite by 25.1% and strength (maximum stress capacity within elastic limits) enhanced by 42.3%. The brick and mortar structure of the nanocomposites leads to an interlocking mechanism between MXene fillers in the polymer matrix, resulting in effective load transfer, good strength, and damage resistance. This is demonstrated in this paper by numerical analysis of MXene nanocomposites subjected to quasi-static loads.

Keywords: MXenes; biomimicry; micromechanical models; finite element method; brick-and-mortar structures; computational analysis; effective interface model

1. Introduction

The discovery of nanomaterials in the last few decades has led to numerous applications of these nanomaterials in the fields of battery technology [1], sensors [2,3], wireless communication [4], and shock absorption [5]. Various nanomaterials like carbon nanotubes (CNTs), graphene, molybdenum di-sulfide (MoS_2), and boron nitride (BN) were used as fillers with polymer matrices to form nanocomposites with new desired functionalities. Graphene was the first two-dimensional (2D) nanomaterial discovered in 2004 [6]. Along with graphene, other 2D materials (MoS_2 and BN) were also used for various nanocomposites [7]. In 2011, a new 2D material was reported, namely, MXenes ($Ti_3C_2T_x$, where surface termination T_x can be –O, –OH or –F) [8]. Almost immediately, MXenes attracted a great deal of interest in various fields of applications due to their unique physical properties such as good conductivity [8], film-forming ability and good elasticity [9]. Moreover, it has been reported that MXenes are environmental friendly materials (low toxicity [10] and biodegradable [11]), thus showing their potential in biosensing applications.

MXenes are inorganic compounds of metal carbides or nitrides. The surface termination and highly electro-positive edges of MXene materials result in hydrophilic behavior [12]. The presence of metal atoms in MXene results in good conductivity [8]. The hydrophilic behavior exhibited by MXenes has an advantage over other nanomaterials such as unfunctionalized CNTs, which tend to form agglomerates resulting in unpredictable behavior (non-uniform distribution of CNTs within the polymer composite leads to uncertain elastic and electrical properties resulting in unpredictable stress-strain or strain-resistance behavior) [8,13]. Moreover, hydrophilicity allows MXenes to be uniformly dispersed within a polymer matrix, therefore enabling a highly repeatable fabrication procedure. On the other hand, mechanical properties of monolayer MXenes are reported to be better than commercially available reduced-graphene oxides (r-GO), which are used extensively for applications with graphene [9]. These advantages of MXenes over other conventional nanomaterials provide unique and promising opportunities to progress in the current state of the art in nanocomposites.

Similar to other nanocomposites, MXene monolayers (delaminated form or flakes) are often used as fillers with polymer matrix materials resulting in MXene Nanocomposites (MXNC) [13,14]. Though pure MXene films have good mechanical properties and conductivity [13,15], these are not chemically stable for a long duration due to oxidation [16]. Unfortunately, this affects the conductivity and mechanical properties of MXenes. The same literature work also indicates the use of polymer material as a matrix for MXene nanomaterials to reduce the oxidation rate. Varying the MXene-to-polymer weight fraction and fabrication processes [13,14,17] results in a wide range of electrical and mechanical properties, creating space for tailoring an MXNC for specific application. This complex design process requires numerous and costly experimental procedures to find the proper combination of process and material parameters for obtaining the desired functionality of composites. The recently demonstrated brick-and-mortar (or layer-by-layer) assembly process of MXenes with controlled polymer intercalation [18] provides a solution for the controllable tailor-made fabrication process. Thus, this specific structure of MXene forms the main scope of this work, where analytical and numerical methods will be used in order to model such structures.

Pre-design of MXNC using analytical and numerical models plays a major role in overcoming the challenge posed by extensive physical testing of nanocomposites with different nanocomposite constituents. Models not only help in estimating the effective (overall) nanocomposite physical properties but also aid in predicting the behavior (response) of the nanocomposite to various loading types and scenarios (e.g., stress-strain response, strain-resistance response etc.). Developing models for such nanocomposites involves complexity in geometry and material properties from the nano to macro scale due to the size- and scale-effect phenomenon observed in composites, calling for multiscale modeling approaches. Several multiscale modeling strategies, as well as numerical techniques dedicated for nano-, micro-, meso- and macro-scale, and coupling procedures exist. Among the coupling procedures, the hierarchical approach [19–22] of scaling microscale models to macroscale models and the concurrent approach of modeling [23] without scaling are popular. Despite a wide variety of existing numerical tools for predicting the mechanical properties of materials in multiscale, their application in predicting the behavior of MXenes is very limited. Among the approaches developed so far, work on nanoscale modeling and estimating mechanical properties of MXenes with molecular dynamic techniques can be mentioned [24,25]. Recently, some work on microscale modeling of MXene/Polymer nanocomposite has been reported that uses the finite element method to analyze such micromechanical models [26,27]. Therefore, noting that there is a very limited number of works on modeling of MXene nanomaterials, we develop microscale mechanical models for bioinspired nacre-mimetic assembly MXNC.

In this paper, we consider the bioinspired brick-and-mortar structures of MXene/Polymer nanocomposites. Biomaterials in nature have inspired engineers for ages to develop composite materials for various applications [28]. Among these bioinspired materials and related pre-design, there have been efforts to mimic the bio-composites and develop man-made composites with enhanced strength, toughness, elasticity, and damage resistance capabilities [29,30]. The biocomposites considered for pre-design in this paper is nacre (known as mother of pearls). Nacre has more strength and toughness than its main constituent material of calcium carbonate. Investigations of nacre over the years have led to micro and meso scale designs biomimicking the brick and mortar structure [31]. The studies into tensile deformation of nacre at micro and macroscales [32] have led to nacre-mimetic pre-design of nanocomposites [33,34]. The brick-and-mortar assembly of the nacre provides topological and structural assembly advantages (interlocking mechanism) for the biocomposite to have greater strength and toughness than its main constituent [35]. We utilize these design benefits and develop pre-design configurations for MXene/Polymer nanocomposites. The paper provides a novel pre-design approach of micromechanical modeling and analysis study incorporating the bioinspired nacre-mimetic assembly of MXene/Polymer nanocomposites.

A literature review of the modeling of nanomaterial-based composites and particularly MXene nanocomposites emphasizes the need for a multiscale modeling approach to nanomaterial-based composites. The MXene nanocomposites developed over the last few years have indicated a stacking assembly of MXenes in a polymer matrix. Modeling methods that provide consistent elastic property estimation need to be explored. Pre-design and modeling of MXene nanocomposites for bioinspired nacre-mimetic assembly has not been explored in the literature yet, and so it is expected to provide better mechanical properties with effective load transfer between MXene fillers and polymer matrix through such work. Moreover, microscale model could potentially be included in further studies involving multi-scale approaches (such as nano and macro scales).

The structure of this paper is organized and discussed henceforth. First, the geometric and material properties of MXenes (titanium carbide, $Ti_3C_2T_x$) are summarized based on a literature review outcome along with the experimental characterization tests carried out by the authors (provided in Section 3.1). Then, approaches for modeling a single MXene layer (flake) and MXene/polymer nanocomposite based on nacre-mimetic structure are discussed. MXNC films formed via brick-and-mortar (layer-by-layer) [33] assembly (same models can be applied to vacuum-assisted filtration [13,14,36] process formed MXNCs but the stacking and layer-by-layer formation is not controlled) are the focus of these model developments. The consistent stack formation of MXene with polymer material provides the basis for the deterministic model configurations considered later. Subsequently, analytical and numerical methods with potential applicability in terms of modeling MXNC—like effective interface model (EIM) [37], classical laminate plate theory (CLPT) [16,38], and finite element analysis (FEA)—are discussed and implemented. Finally, a comparison is made between the reported experimental results and results obtained from all the MXNC models employed in this paper. This leads to discussions on the use of the brick-and-mortar assembly during fabrication of MXNC and their influence on the effective nanocomposite behavior. The paper concludes with a discussion of the results.

For the purpose of modeling and property estimation, weight fractions of MXenes and polymer will be used throughout this paper. Weight fraction can be converted to volume fraction based on the density of the filler and matrix, and the volumes depend on the models developed in this paper. The units used in the paper are mm, N, t/mm^3 unless otherwise stated.

2. Modeling and Methods

2.1. Mechanical Properties of MXenes and Polymers

2.1.1. MXenes

Geometrical properties of MXene ($Ti_3C_2T_x$) monolayers have been studied extensively. The results [9,11] of morphology and characterization of MXenes ($Ti_3C_2T_x$), developed in the last few years obtained using a synthesis process of in-situ hydrogen fluoride (HF) formation with the minimally intensive layer delamination (MILD) method, indicate an average lateral dimension of MXene ($Ti_3C_2T_x$) monolayers to be in the range of a few microns (1–12 μm, without sonification) and the thickness to be in the range of few nanometers (1–10 nm) [39]. Lateral dimensions of 2 microns and a thickness of 2 nm is used for all the MXene monolayer models in this paper.

The elastic properties of the MXene monolayer are determined through experiments like nanoindentation using Atomic Force Microscopy [9] and various computational processes in literature on Molecular Dynamics (MD) [16,24] and Density Functional Theory (DFT) [40,41] were used. The Young's modulus of MXenes ($Ti_3C_2T_x$) estimated using DFT, molecular dynamics, and experimental results are 312.5 GPa (with Poisson's ratio of 0.2265), 502 GPa and 330 ± 30 GPa, respectively. As the prediction of DFT studies are close to the experimental results, we consider the material properties from the DFT studies in this paper (provided in Table 1).

Table 1. MXene's physical properties [40].

Mechanical Properties	Symbol	Values	Units
Young's Modulus in planar direction	E	312.5	GPa
Poisson's ratio in the planar direction	μ	0.2265	-
In-plane shear modulus	G	141	GPa
Density	ρ	~3.2×10^{-9}	tmm^{-3}
Maximum allowable Tensile stress	σ	(17.3 ± 1.6)	GPa

2.1.2. Polymers

Two polymers are considered in the process of analysis in this paper, namely, epoxy-resin and polyvinyl alcohol (PVA). The material properties of PVA are: Young's modulus of 1 GPa, Poisson's ratio of 0.42, density of 1.19×10^{-9} tmm^{-3} and allowable maximum stress of 30 MPa. The material properties of epoxy-resin are: Young's modulus of 3.0741 GPa, Poisson's ratio of 0.29, density of 1.1×10^{-9} tmm^{-3} and allowable maximum stress of 49.9 MPa [13,33,39,42,43].

2.2. *Analytical Methods*

Once the properties of single-layer MXene model are defined, models for MXNC are built by distributing several flakes within the polymer. The topology is based on deterministic configurations derived from MXNC samples fabricated by brick-and-mortar (layer-by-layer) assembly (a similar method can be used for vacuum-assisted filtration fabricated samples as well). The polymer is modeled as a representative volume cube within which the MXene monolayers are orderly distributed by defining the weight fractions of each constituent in the composite used in the analytical methods. The analytical methods of EIM and CLPT—used in this study—consider the interface between a MXene monolayer and polymer. A generic MXNC model with a set of MXene flakes with polymeric matrix is shown in Figure 1. The model assumes the interface layer to have the same shape as the filler.

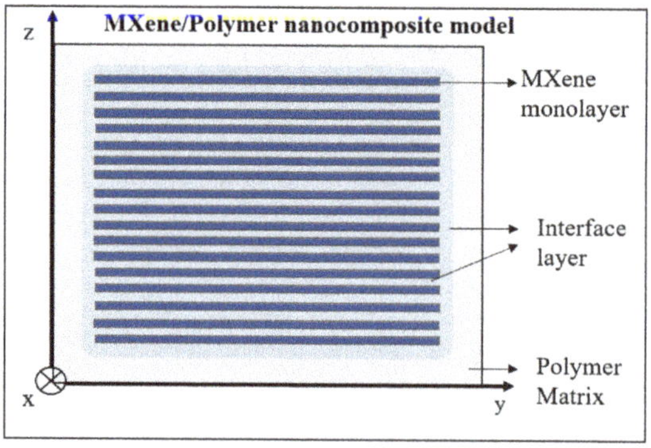

Figure 1. MXene/Polymer nanocomposite model representation at micro scale without MXene overlaps and coordinate axes of X, Y, and Z, which are orthogonal to each other.

There are two interfaces defined in this paper—namely, the filler-matrix-filler interface (in case of a stack formed with MXenes and polymer) and the filler-matrix (between MXene monolayer and polymer) interface. Both these interfaces are assumed to have the same elastic properties for the EIM and CLPT method. The analytical models help in studying the effective MXNC elastic properties and comparing them with the reported experimental results.

Effective interface model: The Effective Interface Model is a modified approach of the continuum mechanics-based Mori-Tanaka model [37]. The latter is based on analytical considerations of Eshelby's inclusion principle [44,45]. The model considers the filler, matrix, and interface material of the MXNC. In determining the mechanical response of the MXNC, MXene monolayers are assumed to be distributed in an infinite space of polymer matrix material. With this assumption, the MXene flake is considered as the inclusion in the current model [46]. Based on the dimensions of MXenes considered in Section 2.1.1, the MXene flake is approximated according to Eshelby's inclusion as a penny shaped inclusion (L = B> > h) for the purpose of forming Eshelby's tensor, which is essentially a tensor based on geometric properties of the inclusion or filler (in this case, the MXene flake). Here the filler-matrix-filler interface and filler-matrix interface are assumed to have the same properties as previously discussed. The drawback of the method is that a single filler material with interface is assumed in an infinite matrix space around it, thus, the stacking and MXene assembly effect due to the fabrication process cannot be captured from this model. The equations for EIM formulation are provided in Appendix A.

Classical laminate plate theory: Although a single MXene layer behaves like a membrane, the resulting bioinspired nacre-mimetic MXNCs display mechanical properties that may be effectively modeled via continuum-based approaches for plates. This is valid for a stacked sequence of MXene flakes embedded in a polymeric matrix, making such a composite setup very similar to a multilayered composite (shown in Figure 1). The MXNC is modeled using the CLPT to estimate the effective elastic properties of the MXNC and incorporate the interface layers (and provide estimations of their elastic properties). Here, the Kirchhoff's Plate theory—assuming the normal material line being infinitely rigid along its own length, normal material line of the plate remaining a straight line after deformation, and normal material line being normal to the deformed plane of the plate—is considered for modeling MXNCs [38]. The method considers MXene flakes as thin plates (with small displacement or rotations and small strains) with polymer material between them. The assumptions for displacements and strains leading to the equations are provided in Appendix B. The in-plane stiffness matrix of the formulation provided in Appendix B is considered for this paper as we only consider tensile loading of the MXNC model.

2.3. Numerical Methods

The limitations of analytical modeling techniques related to shapes, thicknesses, distribution, interactions between the inclusions and other aspects lead to the application and developments of numerical methods for predicting mechanical responses of MXenes and bioinspired MXNCs. A common choice is the widely used and versatile finite element method. Among many tools and techniques available in finite elements, the so-called multi-point constraints (MPCs) can be effectively used for modeling MXenes with polymer material. In the proposed numerical model, MXene flakes and the polymer matrix are meshed independently, while the two sets of meshes do not need to be congruent (i.e., no common nodes are required). Next, the nodes of the MXene flakes are tied to the nearest nodes of the polymer matrix via MPCs. This allows the field variables (displacements, temperatures, currents etc.) at nodes of MXene flakes to be linked with the field variable at the nodes of the polymer. For setting up the MPC equations, the interface is assumed to be perfectly bonded between the filler (MXene) and matrix (polymer). Figure 2a–c illustrates the configurations used for modeling MXNC. These topological distributions are based on the bioinspired nacre-mimetic brick-and-mortar assembly. Configuration 1 consists of a simple MXene/Polymer nanocomposite model, while configuration 2 and 3 utilize nacre-mimetic nanocomposite assembly. The MXene/Polymer nanocomposites, modeled as RVEs [47,48] with a cube configuration at the micro scale, can later be used for hierarchical material framework for multi-scale analysis. A representative volume cube of 3.3 μm as the side length is considered.

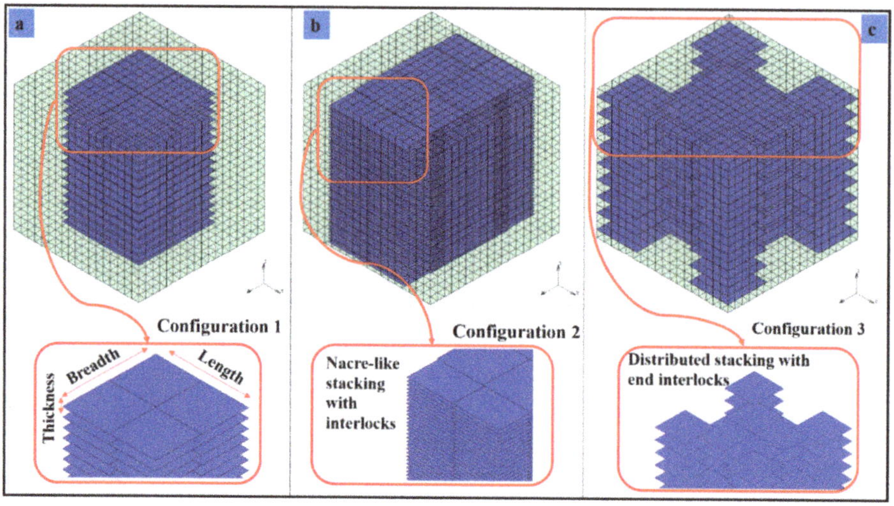

Figure 2. RVE configurations of MXene/Polymer nanocomposite. (**a**) Configuration 1, (**b**) Configuration 2, (**c**) Configuration 3.

Element selection: A single MXene layer in the RVE is modeled using thick plate elements. The current models were implemented in MSC Marc software; therefore, the shape function details are provided in [49]. The material properties used for the model are given in Section 2.1. Unlike the CLPT analytical model, which has a thin plate assumption, the thick plate element in the numerical model uses a modified version of the Mindlin-Reissner plate model [50] whose original version is also sometimes referred to in literature as the First-Order Shear Deformation Theory (FSDT) [51]. The FSDT theory assumes the normal material line, initially normal to the mid-plane of the plate, remains straight and unstretched after deformation, but not necessarily normal to the mid-plane of the plate. This leads to the finite rotations of the cross-section of the plate to be considered. The present modification [50] of the thick plate element theory consists of the formulation of parabolic distribution of transverse shear strains and satisfies the zero transverse shear stress requirements on the plate surfaces. The set of assumptions about the strain of these thick plate elements is provided in Appendix C for further reference. The polymeric matrix is represented by eight-node three-dimensional brick elements with a trilinear interpolation. These elements have three global displacements as degrees of freedom at each node along with eight-point Gaussian integration. Details on the element's shape function formulation are given in Appendix C.

Boundary conditions: Numerical models of MXNCs are subjected to test conditions similar to experimental works in literature [13] in order to estimate their mechanical properties via a virtual tensile test [52]. Subjecting the microscale models in the form of RVE to a 1-D virtual tensile test, the scaling-up of microscale models to macroscale to analyze the elastic properties may be omitted and the results from the RVEs can be directly analyzed (following Saint-Venant's principle). This helps in the study of micromechanical model response subjected to quasi-static uniaxial displacement boundary condition at one end and fixed boundary condition at the opposite end.

Furthermore, for the estimation of elastic properties, particularly, the Young's modulus of the MXNC from the RVEs, periodic boundary conditions along with the Hill-Mandel condition [48,53,54] are applied. This was implemented in MSC Marc through in-house Fortran subroutines and quantitative results are provided in subsequent sections.

3. Results and Discussions

3.1. Experimental Characterization of MXene Samples

The MXene ($Ti_3C_2T_x$; supplied by Materials Research Centre, Kiev, Ukraine) sample morphology and structure were investigated using transmission electron microscopy (TEM) (manufacturer details: Tecnai TF 20 X-TWIN, FEI Company (subsidiary of Thermo Fisher Scientific), Hillsboro, OR, USA) and scanning electron microscopy (SEM) (manufacturer details: Versa 3D scanning electronic microscope with field electron gun and ion, FEI Company (subsidiary of Thermo Fisher Scientific), Hillsboro, OR, USA). The lateral dimensions and single layer thicknesses of MXene ($Ti_3C_2T_x$) were measured. Figure 3a,b show the stack of pure MXenes flakes observed using high resolution TEM (HR-TEM, FEI Company (subsidiary of Thermo Fisher Scientific), Hillsboro, OR, USA). The thicknesses of these MXene samples were approximately 1–2 nm, confirming other literature studies (a thickness of 1–10 nanometers and lateral dimensions in the order of 1–10 microns). Figure 4 provides energy dispersion X-ray (EDX, FEI Company (subsidiary of Thermo Fisher Scientific), Hillsboro, OR, USA) spectroscopic data of MXene samples confirming the presence of titanium carbide with surface termination ($Ti_3C_2T_x$). Combining these results with the other geometrical features of MXenes (see e.g., [13]) the yielded data to be used for MXNC models is developed in Section 2.

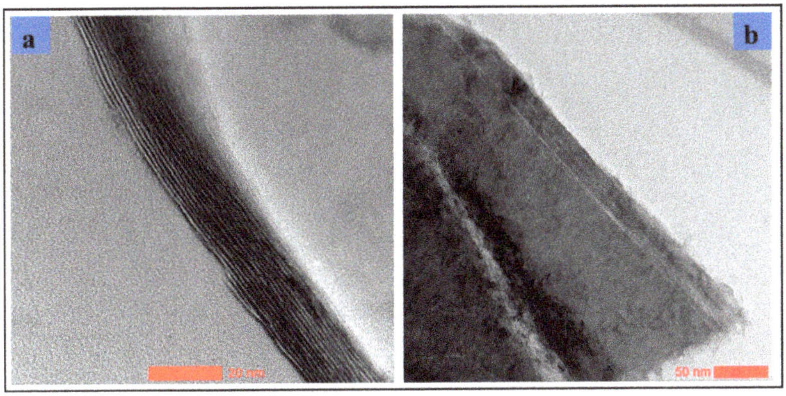

Figure 3. (**a**,**b**) HR-TEM images of stacks of pure MXene samples.

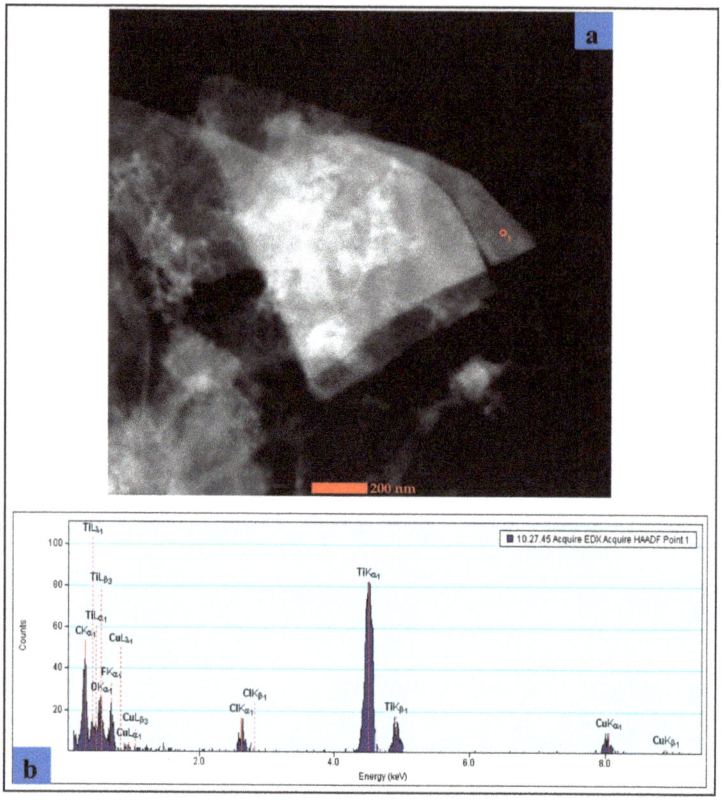

Figure 4. (**a**) TEM image of MXene flakes (**b**) EDX analysis of the elemental composition.

3.2. *Estimations of Effective Young's Modulus of MXNC*

The two analytical models are implemented for the various weight fractions of MXenes in polymer. The model parameters are shown in Table 2. Various deterministic configuration implemented for numerical models are marked (Table 2) for the same weight fractions. MXene/Epoxy-resin nanocomposite simulation results and the quantitative comparison of these results with reported experimental ones are shown in Table 3, while the results for the MXene/PVA nanocomposite are shown in Table 4.

Table 2. Implementation of the three configuration models using the numerical method (Configurations and the corresponding weight fraction of filler is denoted by "X" mark.).

Volume Fraction of Filler vol.%	Weight Fraction of Filler wt.%	Number of MXene Layers	Configuration 1	Configuration 2	Configuration 3
0.37844	1.093	17			X
1	2.96	45	X		
1.8031	5.0709	81			X
2	5.67	91		X	
5.6988	14.9518	256			X
10	24.42	449	X		
20	42.12	899		X	X

Table 3. (**a**) MXene/Epoxy-resin nanocomposite analytical results; (**b**) MXene/Epoxy-resin nanocomposite numerical results compared with reported experimental results; (**c**) MXene/Epoxy-resin nanocomposite numerical results with maximum stress and strain for each configuration implemented.

(a)

MXene/Epoxy	Weight Fraction (wt.%)	E Experimental [14] (GPa)	EIM (GPa)	CLPT (GPa)
Interface Layer E (GPa)			1×10^{-3}	3.25×10^{3}
	1.093	3.62	3.998	0.1378
	2.96	4.1	5.7463	0.3517
	5.0709	4.37	7.74	0.5941
	5.67	Not Available	8.3274	0.6654
	14.9518	3.42	18.069	1.8419
	24.42	Not Available	29.578	3.2179
	39.52	Not Available	52.295	5.8916

(b)

MXene/Epoxy	Weight Fraction (wt.%)	E Numerical (GPa)	E Experimental (GPa) [14]	Error in Estimation (%)
Configuration 1	2.96	3.2554	4.1	20.60
Configuration 1	24.42	3.5006	Not Available	Not Available
Configuration 2	5.67	3.4028	Not Available	Not Available
Configuration 2	42.12	3.346	Not Available	Not Available
Configuration 3	**1.093**	**3.1769**	**3.62**	**12.24**
Configuration 3	**5.0709**	**3.4484**	**4.37**	**21.09**
Configuration 3	**14.9518**	**3.8474**	**3.42**	**12.50**
Configuration 3	42.12	3.9769	Not Available	Not Available

(c)

MXene/Epoxy	Weight Fraction (wt.%)	E Numerical (GPa)	Maximum Stress (MPa)	Maximum Strain
Configuration 1	2.96	3.2554	63.92	0.0207
Configuration 1	24.42	3.5006	65.02	0.0204
Configuration 2	5.67	3.4028	57.37	0.0189
Configuration 1	42.12	3.346	81.89	0.0258
Configuration 3	1.093	3.1769	63.99	0.0202
Configuration 3	5.0709	3.4484	65.97	0.0207
Configuration 3	14.9518	3.8474	66.56	0.0206
Configuration 3	42.12	3.9769	71.61	0.05716

Table 4. MXene/PVA nanocomposite results.

MXene/PVA	E Numerical (GPa)	E Experimental (GPa) [13]	EIM (GPa)	CLPT (GPa)
	[wt.% = 42.12]	[wt.% = 40]	[wt.% = 38.06]	[wt.% = 38.06]
Interface Layer EI (GPa)	Not considered		1×10^{-3}	3.25×10^{3}
Configuration 2	1.4414	3.7	43.777	5.8876
Configuration 3	42.12			

Analytical models of EIM and CPLT are implemented based on their respective formulation in MATLAB code form. These models are tuned for the weight fractions of MXene in polymer. The total volume of these models is defined using a representative volume cube for the purpose of providing volume constrains (thereby, weight constraints) on the models developed. The number of MXene layers depend on the weight (volume) fraction of MXene and is given in Table 2. The interface volume fraction is defined as a function of the MXene weight fraction. Based on these two weights (volume) fraction values, the matrix weight (volume) fraction is calculated. These weight (volume) fractions are normalized for the purpose of generalization. Individual subroutine codes (in MATLAB code) are written for each analytical model and the results of stiffness matrix estimation is used to finally obtain the effective Young's modulus values of the MXNC. The equations provided in Appendices A and B are implemented and the Young's modulus is derived from the in-plane stiffness matrix for CLPT. The resulting estimations of the effective Young's modulus from the EIM and CLPT for MXene/Epoxy-resin nanocomposite are an overestimation and underestimation, respectively, compared to the experimental results (literature-based) and numerical results (obtained in this paper). In the MXene/PVA nanocomposite case, both EIM and CLPT methods overestimate the effective Young's modulus compared to other results. These estimations from the analytical methods clearly indicate oversimplification of the models developed using these methods. They are less effective unless they can be modified and the underlying assumptions can be improved to capture the physics (close to experimental conditions) of the material being modeled.

The numerical model consists of RVE built with 8000 ($20 \times 20 \times 20$) polymeric matrix brick elements created in finite element method. The number of MXene layers are chosen based on the weight fraction of MXenes defined by the configurations. The configurations given in Figure 2 and Table 2 are under consideration here. The MXene monolayer modeled as plate elements have their nodal field variable linked to the nearest matrix brick element nodes using MPCs discussed in Section 2.3.

The results of numerical models are provided in Table 3b. For configuration 1, the increase of MXene weight fraction in epoxy-resin from 2.96 to 24.42% leads to an increase of the effective Young's modulus values by 7.5321%. For the configuration 2, a decrease in effective Young's modulus values by 1.66% is observed for the same change in weight fraction of MXene in epoxy-resin. Configuration 1 has a simple distribution of MXenes while configuration 2 has an edge overlap (interlock) nacre-mimetic brick-and-mortar assembly of MXenes. Configuration 3 models the nacre-mimetic brick-and-mortar assembly process with end interlock regions. Both configuration 2 and 3 can have controlled polymer interaction and this has been demonstrated by experiments in literature [33]. For configuration 3, the increase in weight fraction of MXene in epoxy-resin from 1.093 to 5.0709%, 5.0709 to 14.9518% and 14.9518 to 42.12% results in an increase in effective Young's modulus of MXene/Epoxy-resin nanocomposites from 8.546 to 11.05705 and to 3.3659%, respectively. For an increase of MXene weight fraction from 1.093 to 14.9518% and 1.093 to 42.12%, the effective Young's modulus of configuration 3 increases by 21.1105 and 25.1817%, respectively. The strength (maximum stress capacity within elastic limits) of the RVEs are given in Table 3c. In configurations 1, 2, and 3, with the increase in MXene weight fraction in epoxy-resin from 2.96 to 24.42%, 5.67 to 42.12%, and 1.093 to 42.12%, results in the increase in strength by 1.7209, 42.374, and 11.9081%, respectively.

Along with the increase in MXene weight fraction, the assembly pattern of MXenes in the epoxy-resin results in improving the load bearing capacity and effective transfer of load from the polymer material to the MXene filler material. The bioinspired nacre-mimetic brick-and-mortar configurations used in configurations 2 and 3 indicate these enhanced elastic behaviors of the MXene nanocomposite. Figure 5a–c indicates that these enhanced elastic behaviors in which the polymer experiences less stress, and the applied load is distributed among the MXenes according to the assembly pattern. Interestingly, configuration 2 has the MXenes stacked along the direction of the quasi-static load and not sparsely distributed as in configuration 3, and this results in greater strength for the RVE for configuration 2 (Figure 6 and Table 3b,c) provides illustrations of stress fields of the three configurations. The sparsely distributed configuration 3 also has an increase in strength as well as an increase in the effective Young's modulus values. The interlocking mechanism between the MXene fillers due to the nacre-mimetic assembly allows for effective load transfer from polymer to MXenes and with increasing weight/volume fraction of MXenes, the strength and effective Young's modulus increases (configuration 3 shown in Figure 6 & Table 3b,c). The nacre-mimetic brick-and-mortar assembly process offers the advantage of tailoring the MXene assembly with polymer intercalation and this ensures an effective load transfer, as seen from the results in Table 3c. With controlled assembly of the MXenes with the polymer, along with an increase in the load-bearing capacity, the damage propagation path might also be controlled. The crack propagation in such nanocomposites are along the fillers surfaces and their interfaces with polymer matrix [55].

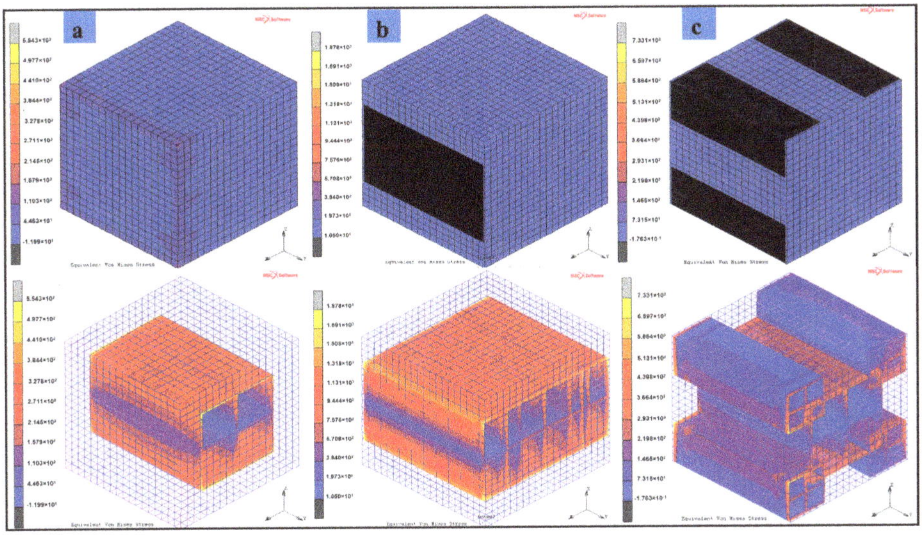

Figure 5. Stress field due to quasi-static tensile loading (X-direction) for various configurations (**a**) Configuration 1–24.42 wt.% of MXene in epoxy-resin, (**b**) Configuration 2–42.12 wt.% of MXene in epoxy, (**c**) Configuration 3–42.12 wt.% of MXene in epoxy.

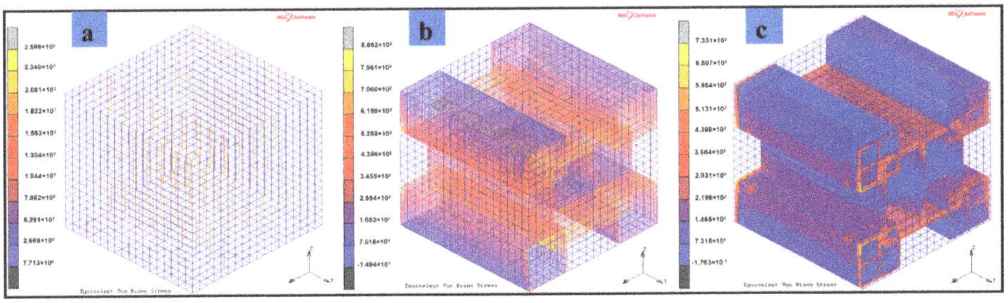

Figure 6. Quasi-static tensile loading (X-direction) applied to configuration 3 with different weight fractions of MXene in epoxy-resin. The resulting stress fields are shown in (**a**) 5.0709 wt.% of MXene in epoxy, (**b**) 14.9518 wt.% of MXene in epoxy, (**c**) 42.12 wt.% of MXene in epoxy.

3.3. Comparisons with Literature-Based Experimental Results

For MXene/Epoxy-resin nanocomposites shown in Table 3a,b, the EIM estimates the Young's modulus values of the nanocomposite to be of an order of magnitude higher than the reported experimental results (for the weight fraction of MXene filler above 10%). The EIM results has overestimations compared to the reported experimental results [9]. The results from CLPT underestimates the effective nanocomposite Young's modulus values by an order of magnitude lesser than reported experimental results (except for weight fraction close to 40 wt.%) as seen in Table 3a. The interface Young's modulus employed in the EIM and CLPT methods is documented in Table 3a.

The FEA results estimate the Young's modulus values to be of the same order of magnitude as the reported experimental results. The FEA-based results provide a more consistent estimation close to the reported experimental results for the MXene/Epoxy-resin nanocomposite. The numerical results of configuration 3 shown in bold in Table 3b are compared with results presented in [14]. Error in estimation from numerical models in comparison with reported experimental results is in the range of 12.4–21.09%. The quality of estimation of Young's modulus values in comparison with reported experimental results are shown in Figure 7. The numerical results show an increase in Young's modulus values of MXNC with an increase in weight fraction of MXene in the composite. The reported experimental result indicates an increase in the Young's modulus value, reaching a maximum value of 4.37 GPa and then a reduction in the Young's modulus with respect to weight fraction of MXene in the composite. The numerical model results provide a more consistent prediction compared to the classical analytical models.

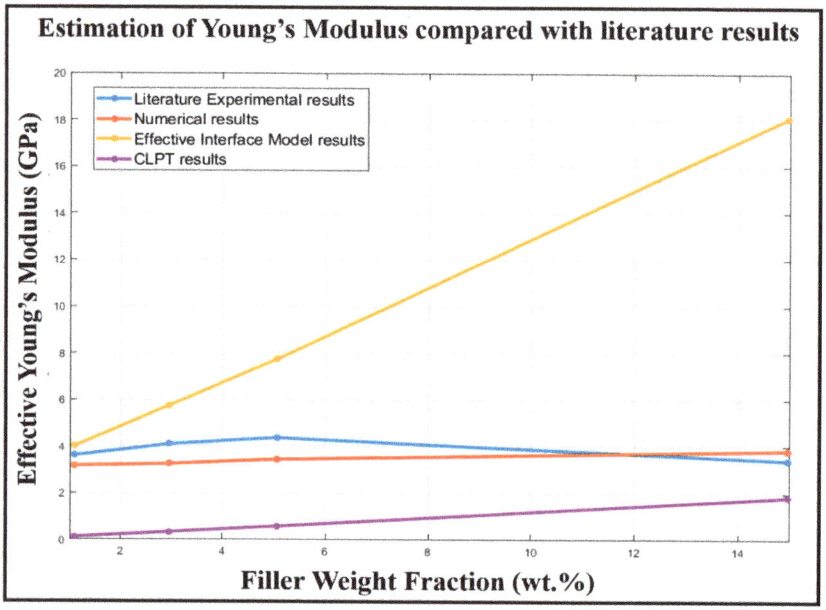

Figure 7. MXene/Epoxy-resin nanocomposite modeling results compare to reported experimental results.

The same interface conditions presented in Table 3 are employed for the MXene/PVA nanocomposite results shown in Table 4. EIM, CLPT, and modified CLPT reached the volume constraint condition (restricting the fillers from being modeled outside the RVE volume) imposed on the RVE models in the MATLAB code and hence the weight fraction values were at 38.06% for EIM and CLPT and 38.09% for the modified CLPT method. The MXene/PVA nanocomposite numerical results are of the same order as those obtained experimentally in literature [8] but the reported experimental result value is more than twice that of the numerical one.

The various analytical and numerical models developed and tested against the experimental results reported in this paper provide a wide range of possibilities for further model refinements. A designer based on the requirement can use the models during the process of pre-designing the MXNC. An estimation of nanocomposite properties from these models will allow the MXNC designer to select a model for composite properties simulation and compare it with the models developed in this paper. The advantages and drawbacks of these models concluded in this paper can support the design process.

3.4. Discussions on the Influence of the Interface between MXenes and Polymer

The assumption of a perfect bonding condition between MXene and polymers (in the numerical model) seems to be the cause of some variation in the estimation of the results in the case of epoxy-resin and PVA matrix-based nanocomposites. In both the cases, the surface termination of MXenes plays a crucial role in bonding with the polymer chain in the fabrication of MXNC and modeling it within the numerical model framework is an important step. The results also indicate that the numerically obtained Young's modulus values of MXene/PVA nanocomposite are half of the experimental results obtained in literature. The increased strength of the composite compared with the pure PVA (an approximate Young's modulus of 1 GPa) indicates a strong bond between the hydroxyl group of PVA polymer chain and the surface termination of MXene. The effect of the size of MXene monolayers on the composite and interface bonding also requires further investigation. The surface termination of MXene can lead to two types of bonding as per the chemical bonding study [56]. The surface termination of MXenes might feature either hydrogen or van-der-Waal type of chemical bonding with the polymers. MD simulations and experimental investigation of the effect of bonding can be adopted to the models presented in this paper. Several new MXenes [57] that are being discovered can be adopted to the models developed in this paper as well, particularly, the numerical models. Finally, the micromechanical models developed in this paper form a basis for a multi-scale study of MXene mechanical properties in the future. The models developed with numerical methods provide a consistent estimation of the Young's modulus of the nanocomposites compared to the analytical methods.

4. Conclusions

The paper focuses on developing models for bioinspired nacre-mimetic brick-and-mortar assembly of MXene ($Ti_3C_2T_x$) with polymer matrix. Nacre-mimetic composites have had a history of providing good strength, toughness, damage resistance, and effective load-bearing capacity. These advantages of nacre-mimetic assembly led to the development of the MXene/Polymer nanocomposite configurations discussed in this paper. An experimental characterization of pure MXene ($Ti_3C_2T_x$) samples were made to study the morphology and dimensionality of MXene samples and validate the observations contained in the literature. The material properties of MXenes ($Ti_3C_2T_x$) and polymers were determined from the literature. MXene micromechanical models are developed using a simple arrangement as shown in configuration 1 and with nacre-mimetic assembly (with interlocking mechanism at the edges of the MXenes) as shown in configuration 2 and 3. The models were analyzed by using two analytical methods, namely EIM and CLPT, with consideration of the interface regions between MXene and the polymer. Numerical models based on the FEM with the MPC method considered the perfect bonded condition between MXene and the polymer. The EIM overestimated the MXene/Epoxy-resin nanocomposite effective Young's modulus value by an order of magnitude compared to reported experimental results (up to 40% weight fraction of MXene). The CLPT method estimated the MXene/Epoxy-resin nanocomposite effective Young's modulus value to be of the same order of magnitude or an order lesser compared to the reported experimental results. The numerical results estimated the MXene/Epoxy-resin nanocomposite effective Young's modulus values to be of the same order of magnitude as that of the reported experimental results and more consistent than the analytical methods. The estimation errors for the numerical models of the MXene/Epoxy-resin nanocomposite compared to the reported experimental results were less than 21%. The results indicate that better models for bonding conditions between MXene/Epoxy-resin and MXene/PVA may lead to close quantitative estimations to the reported experimental results. For configurations 1, 2, and 3, the increase in MXene weight fraction in epoxy-resin from 2.9 to 24.4%, 5.6 to 42.1% and 1 to 42.1% results in an increase in the effective Young's modulus by 7.5, 1.6, and 25.1%, respectively.

The results provide an interesting insight that the nacre-mimetic brick-and-mortar assembly configuration of MXene with polymer intercalation can also play a role in controlling the elastic properties of the nanocomposite. This is also helpful in controlling the effective load transfer from the polymer material to the MXene filler material. With the three configurations, we observe that the nacre-mimetic configurations 2 and 3 with an interlocking mechanism assembly of MXenes with polymer have a different load bearing capacity and the strength depends on the assembly pattern. The controlled assembly of MXene with polymer intercalation can result in a better load-bearing capacity of the nanocomposite by effective transfer of the load to the MXene filler and the damage propagation path may also be controlled with further optimization of the brick-and-mortar assembly process.

Author Contributions: Conceptualization, S.S., P.P., L.M.J., T.U. and K.G.; methodology, S.S., P.P., L.M.J. and K.G.; software, S.S.; validation, S.S.; formal analysis, S.S., P.P., L.M.J., and K.G.; investigation, S.S., P.P., L.M.J., and K.G.; resources, T.U.; writing—original draft preparation, S.S. and K.G.; writing—review and editing, P.P., L.M.J., and K.G.; visualization, S.S.; supervision, T.U. and K.G.; project administration, T.U.; funding acquisition, T.U. All authors have read and agreed to the published version of the manuscript.

Funding: The authors at AGH UST received funding for this project from the European Union's Horizon 2020 research and innovation programme (DyVirt—Dynamic Virtualization: modeling performance of engineering structures) under the Marie Skłodowska-Curie grant agreement No. 764547.

Acknowledgments: The authors at AGH UST would like to acknowledge the Material Research Centre, Ukraine, for the MXene samples supply and Katarzyna Berent and Marta Gajewska at AGH UST for their support in the experimental characterization of MXenes presented in the paper.

Conflicts of Interest: The authors declare no conflict of interest.

Appendix A

Implementation of EIM: The implementation of EIM equations used in the paper are given here. Equation (A1) provides the overall nanocomposite stiffness matrix C. Here, the filler-matrix-filler interface and filler-matrix interface are assumed to have the same properties. The tensor T_{fi} is the dilute strain concentration tensor of the filler considering the interface in the matrix and T_f is the dilute strain concentration tensor of the filler alone and these are given in Equations (A2) and (A3). V_i, V_f and V_m are the volume fractions of the interface, filler, and matrix, respectively. This interface volume fraction has been made a function of the filler volume fraction considering the interface to be layers above and below each MXene monolayer, as shown in Figures 2 and 4 in the paper. The interface is modeled by using an equivalent continuum approach there by providing the equations given below:

$$C = C_m + \left[\left(V_f + V_i\right)\left(C_i - C_m\right) + V_f\left(C_f - C_i\right)T_f\right]\left[\left(V_m I + \left(V_f + V_i\right)T_{fi}\right)\right]^{-1} \tag{A1}$$

$$T_f = I - S_f\left[S_f + C_m\left(C_f - C_m\right)^{-1}\right]^{-1} \tag{A2}$$

$$T_{fi} = I - S_f\left\{\left(\frac{V_f}{V_i + V_f}\right)\left[S_f + C_m\left(C_f - C_m\right)^{-1}\right]^{-1} + \left(\frac{V_i}{V_i + V_f}\right)\left[S_f + C_m\left(C_i - C_m\right)^{-1}\right]^{-1}\right\} \tag{A3}$$

Appendix B

Implementation of CLPT: The displacement and strain fields of the Kirchhoff's plate theory are given by Equations (A4)–(A6) and Equations (A7)–(A10), respectively. Displacements at any point on the plate are $u_1(x_1,x_2,x_3)$, $u_2(x_1,x_2,x_3)$ and $u_3(x_1,x_2,x_3)$, and displacements of the normal material line are $\bar{u}_1(x_1,x_2,x_3)$, $\bar{u}_2(x_1,x_2,x_3)$ and $\bar{u}_3(x_1,x_2,x_3)$, where (x_1,x_2,x_3) are the positions along the orthogonal unit basis vector (b_1, b_2, b_3) forming the user-defined coordinate axis. The rotations of the cross-section of the plate (φ_1, φ_2) are assumed to be derivatives of the out-of-plane (direction b3) displacement of the mid-plane of the plate. Based on these assumptions, displacement, strain, and stress fields of the laminate can be computed as shown in [51,58]. The in-plane stiffness (A), bending stiffness (D), and coupled stiffness (B) matrices are calculated based on Equations (A11)–(A13). Because of the 2D plate structure assumption, the matrices are reduced to a 3×3 matrix. The overall stiffness matrix (Q) is represented by Equation (A14). The constitutive relationship based on in-plane forces and deformation as well as bending moments and curvatures of the laminated stack can be represented by Equation (A15). N and M are in-plane forces per unit length and bending moment per unit length, while ε_0 and κ are the mid-plane strains and curvature of the laminate stack. Permutation matrix (S) of Equation (A16) is introduced solely to change signs and reorder the curvatures in Equation (A15). Equation (A17) is the reduced stiffness matrix of each layer (can be of filler, matrix, or interface). The term with the notation $(.)_{ij}$ indicates the layer number considering up to n layers (k = 1,2, 3, … , n). The in-plane stiffness matrix (A) depends on the thickness of each layer and this helps in considering any shape of the inclusion as discussed in Section 3.2. The vectors and matrices are represented with bold letters as per the notation followed. The matrix S called the permutation matrix is introduced solely for the purpose of sign conventions.

$$u_1(x_1,x_2,x_3) = \overline{u_1}(x_1,x_2) - x_3\varphi_1;\ \varphi_1 = \left(\frac{\partial\overline{u_3}}{\partial x_1}\right) \tag{A4}$$

$$u_2(x_1,x_2,x_3) = \overline{u_2}(x_1,x_2) - x_3\varphi_2;\ \varphi_2 = \left(\frac{\partial\overline{u_3}}{\partial x_2}\right) \tag{A5}$$

$$u_3(x_1,x_2,x_3) = \overline{u_3}(x_1,x_2) \tag{A6}$$

$$\gamma_{13} = \gamma_{23} = 0 \tag{A7}$$

$$\epsilon_1 = \frac{\partial\overline{u_1}}{\partial x_1} - x_3{\varphi_1}^2; {\varphi_1}^2 = \left(\frac{\partial^2\overline{u_3}}{\partial x_1^2}\right) \tag{A8}$$

$$\epsilon_2 = \frac{\partial\overline{u_2}}{\partial x_2} - x_3{\varphi_2}^2; {\varphi_2}^2 = \left(\frac{\partial^2\overline{u_3}}{\partial x_2^2}\right) \tag{A9}$$

$$\gamma_{13} = \frac{\partial\overline{u_1}}{\partial x_2} + \frac{\partial\overline{u_2}}{\partial x_1} - 2x_3\left(\frac{\partial^2\overline{u_3}}{\partial x_1\partial x_2}\right) \tag{A10}$$

$$\mathbf{A} = \begin{bmatrix} A_{11} & A_{12} & A_{13} \\ A_{21} & A_{22} & A_{23} \\ A_{31} & A_{32} & A_{33} \end{bmatrix} \tag{A11}$$

$$\mathbf{B} = \begin{bmatrix} B_{11} & B_{12} & B_{13} \\ B_{21} & B_{22} & B_{23} \\ B_{31} & B_{32} & B_{33} \end{bmatrix} \tag{A12}$$

$$\mathbf{D} = \begin{bmatrix} D_{11} & D_{12} & D_{13} \\ A_{21} & A_{22} & D_{23} \\ D_{31} & D_{32} & D_{33} \end{bmatrix} \tag{A13}$$

$$\mathbf{Q} = \left[\begin{array}{cccccc} A_{11} & A_{12} & A_{13} & B_{11} & B_{12} & B_{13} \\ A_{21} & A_{22} & A_{23} & B_{21} & B_{22} & B_{23} \\ A_{31} & A_{32} & A_{33} & B_{31} & B_{32} & B_{33} \\ \hline B_{11} & B_{12} & B_{13} & D_{11} & D_{12} & D_{13} \\ B_{21} & B_{22} & B_{23} & A_{21} & A_{22} & D_{23} \\ B_{31} & B_{32} & B_{33} & D_{31} & D_{32} & D_{33} \end{array}\right] \tag{A14}$$

$$\left\{ \begin{array}{c} \mathbf{N} \\ \mathbf{SM} \end{array} \right\} = \begin{bmatrix} \mathbf{A} & \mathbf{B} \\ \mathbf{B} & \mathbf{D} \end{bmatrix} \left\{ \begin{array}{c} \boldsymbol{\varepsilon}_\mathbf{o} \\ \mathbf{S}\kappa \end{array} \right\} \tag{A15}$$

$$\mathbf{S} = \begin{bmatrix} 0 & 1 & 0 \\ -1 & 0 & 0 \\ 0 & 0 & -1 \end{bmatrix} \tag{A16}$$

$$\mathbf{C} = \begin{bmatrix} C_{11} & C_{12} & C_{16} \\ C_{21} & C_{22} & C_{26} \\ C_{61} & C_{62} & C_{66} \end{bmatrix} \tag{A17}$$

Appendix C

Element 75 of MSC Marc: Equations (A18)–(A22) (h is the single plate thickness) provides the strain expressions equations used to model the thick-plate elements of MXenes in numerical analysis. The displacement field are same as Equations (A4)–(A6).

$$\epsilon_1 = \frac{\partial \overline{u_1}}{\partial x_1} + x_3\widehat{\varphi_2};\ \widehat{\varphi_2} = \left(\frac{\partial \varphi_2}{\partial x_1}\right) \tag{A18}$$

$$\epsilon_2 = \frac{\partial \overline{u_2}}{\partial x_2} - x_3\widehat{\varphi_1};\ \widehat{\varphi_1} = \left(\frac{\partial \varphi_1}{\partial x_2}\right) \tag{A19}$$

$$\epsilon_{12} = \frac{1}{2}\left[\left(\frac{\partial \overline{u_1}}{\partial x_2} + \frac{\partial \overline{u_2}}{\partial x_1}\right) + x_3\left(\frac{\partial \varphi_2}{\partial x_2} - \frac{\partial \varphi_1}{\partial x_1}\right)\right] \tag{A20}$$

$$\epsilon_{23} = \left(\frac{\partial \overline{u_3}}{\partial x_2} - \varphi_1\right)\left(1 - \frac{4}{h^2}{x_3}^2\right) \tag{A21}$$

$$\epsilon_{23} = \left(\frac{\partial \overline{u_3}}{\partial x_1} - \varphi_2\right)\left(1 - \frac{4}{h^2}{x_3}^2\right) \tag{A22}$$

Element 7 of MSC Marc: The displacement assumption and mapping from x-y-z space or b1-b2-b3 space into a cube (Element 7) in the ξ, η, ζ space is given below. Equations (A23) and (A24) are the coordinate transformation equations used in MSC Marc and provided by the developer [46].

$$x = a_0 + a_1\xi + a_2\eta + a_3\zeta + a_4\xi\eta + a_5\zeta\xi + a_6\xi\zeta + a_7\zeta\xi\eta \tag{A23}$$

$$\psi = b_0 + b_1\xi + b_2\eta + b_3\zeta + b_4\xi\eta + b_5\zeta\xi + b_6\xi\zeta + b_7\zeta\xi\eta \tag{A24}$$

Function or coordinate expressed through nodal quantities using integration function are given below. Equation (A25) provides the elemental to global displacement transformation and Equations (A26)–(A33) are the shape functions used in the process of the transformation:

$$x = \sum_{i=1}^{8} x_i \rho_i \tag{A25}$$

$$\rho_1 = \frac{1}{8}(1-\xi)(1-\eta)(1-\zeta) \tag{A26}$$

$$\rho_2 = \frac{1}{8}(1+\xi)(1-\eta)(1-\zeta) \tag{A27}$$

$$\rho_3 = \frac{1}{8}(1+\xi)(1+\eta)(1-\zeta) \tag{A28}$$

$$\rho_4 = \frac{1}{8}(1-\xi)(1+\eta)(1-\zeta) \tag{A29}$$

$$\rho_5 = \frac{1}{8}(1-\xi)(1-\eta)(1+\zeta) \tag{A30}$$

$$\rho_6 = \frac{1}{8}(1+\xi)(1-\eta)(1+\zeta) \tag{A31}$$

$$\rho_7 = \frac{1}{8}(1+\xi)(1+\eta)(1+\zeta) \tag{A32}$$

$$\rho_8 = \frac{1}{8}(1-\xi)(1+\eta)(1+\zeta) \tag{A33}$$

References

1. Han, Y.; Ge, Y.; Chao, Y.; Wang, C.; Wallace, G.G. Recent progress in 2D materials for flexible supercapacitors. *J. Energy Chem.* **2018**, *27*, 57–72. [CrossRef]
2. Sadeghi, F.; Sarvi, A.; Sundararaj, U. PVDF/Carbonnanotubes/Nanoclay Composites for Piezoelectric Applications. *Int. Polym. Process.* **2014**, *29*, 81–87. [CrossRef]
3. Cullinan, M.A.; Culpepper, M.L. Carbon nanotubes as piezoresistive microelectromechanical sensors: Theory and experiment. *Phys. Rev. B - Condens. Matter Mater. Phys.* **2010**, *82*, 1–6. [CrossRef]
4. Sarycheva, A.; Polemi, A.; Liu, Y.; Dandekar, K.; Anasori, B.; Gogotsi, Y. 2D titanium carbide (MXene) for wireless communication. *Sci. Adv.* **2018**, *4*, 1–9. [CrossRef]
5. Chinke, S.L.; Sandhu, I.S.; Saroha, D.R.; Alegaonkar, P.S. Graphene-Like Nanoflakes for Shock Absorption Applications. *ACS Appl. Nano Mater.* **2018**, *1*, 6027–6037. [CrossRef]
6. Novoselov, K.S. Electric Field Effect in Atomically Thin Carbon Films. *Science* **2004**, *306*, 666–669. [CrossRef]
7. Liu, W.; Ullah, B.; Kuo, C.-C.; Cai, X. Two-Dimensional Nanomaterials-Based Polymer Composites: Fabrication and Energy Storage Applications. *Adv. Polym. Technol.* **2019**, *2019*, 1–15. [CrossRef]
8. Naguib, M.; Kurtoglu, M.; Presser, V.; Lu, J.; Niu, J.; Heon, M.; Hultman, L.; Gogotsi, Y.; Barsoum, M.W. Two-Dimensional Nanocrystals Produced by Exfoliation of Ti3AlC2. *Adv. Mater.* **2011**, *23*, 4248–4253. [CrossRef]
9. Lipatov, A.; Lu, H.; Alhabeb, M.; Anasori, B.; Gruverman, A.; Gogotsi, Y.; Sinitskii, A. Elastic properties of 2D Ti 3 C 2 T x MXene monolayers and bilayers. *Sci. Adv.* **2018**, *4*, eaat0491. [CrossRef]
10. Rasool, K.; Helal, M.; Ali, A.; Ren, C.E.; Gogotsi, Y.; Mahmoud, K.A. Antibacterial Activity of Ti3C2TxMXene. *ACS Nano* **2016**, *10*, 3674–3684. [CrossRef]
11. Rastin, H.; Zhang, B.; Mazinani, A.; Hassan, K.; Bi, J.; Tung, T.T.; Losic, D. 3D bioprinting of cell-laden electroconductive MXene nanocomposite bioinks. *Nanoscale* **2020**, *12*, 16069–16080. [CrossRef] [PubMed]
12. Natu, V.; Hart, J.L.; Sokol, M.; Chiang, H.; Taheri, M.L.; Barsoum, M.W. Edge Capping of 2D-MXene Sheets with Polyanionic Salts To Mitigate Oxidation in Aqueous Colloidal Suspensions. *Angew. Chem. Int. Ed.* **2019**, *58*, 12655–12660. [CrossRef] [PubMed]
13. Ling, Z.; Ren, C.E.; Zhao, M.-Q.; Yang, J.; Giammarco, J.M.; Qiu, J.; Barsoum, M.W.; Gogotsi, Y. Flexible and conductive MXene films and nanocomposites with high capacitance. *Proc. Natl. Acad. Sci. USA* **2014**, *111*, 16676–16681. [CrossRef]
14. Wang, L.; Chen, L.; Song, P.; Liang, C.; Lu, Y.; Qiu, H.; Zhang, Y.; Kong, J. Fabrication on the annealed Ti 3 C 2 T x MXene / Epoxy nanocomposites for electromagnetic interference shielding application. *Compos. Part B* **2019**, *171*, 111–118. [CrossRef]
15. Srivatsa, S.; Belthangadi, P.; Ekambaram, S.; Pai, M.; Sen, P.; Uhl, T.; Kumar, S.; Grabowski, K.; Nayak, M.M. Dynamic response study of Ti3C2-MXene films to shockwave and impact forces. *RSC Adv.* **2020**, *10*, 29147–29155. [CrossRef]
16. Habib, T.; Zhao, X.; Shah, S.A.; Chen, Y.; Sun, W.; An, H.; Lutkenhaus, J.L.; Radovic, M.; Green, M.J. Oxidation stability of Ti3C2Tx MXene nanosheets in solvents and composite films. *npj 2D Mater. Appl.* **2019**, *3*, 8. [CrossRef]
17. Weng, G.M.; Li, J.; Alhabeb, M.; Karpovich, C.; Wang, H.; Lipton, J.; Maleski, K.; Kong, J.; Shaulsky, E.; Elimelech, M.; et al. Layer-by-Layer Assembly of Cross-Functional Semi-transparent MXene-Carbon Nanotubes Composite Films for Next-Generation Electromagnetic Interference Shielding. *Adv. Funct. Mater.* **2018**, *28*. [CrossRef]
18. Lipton, J.; Weng, G.M.; Röhr, J.A.; Wang, H.; Taylor, A.D. Layer-by-Layer Assembly of Two-Dimensional Materials: Meticulous Control on the Nanoscale. *Matter* **2020**, *2*, 1148–1165. [CrossRef]
19. Runesson, K.; Larsson, F. *Computational Homogenization and Multiscale Modeling*; Chalmers University of Technology: Gothenburg, Sweden, 2011.
20. Pan, Y.; Iorga, L.; Pelegri, A.A. Numerical generation of a random chopped fiber composite RVE and its elastic properties. *Compos. Sci. Technol.* **2008**, *68*, 2792–2798. [CrossRef]
21. Wang, C.; Lyu, D. Multiscale cohesive zone modeling and simulation of high-speed impact, penetration, and fragmentation. *J. Micromechanics Mol. Phys.* **2018**, *03*, 1850003. [CrossRef]
22. Kochmann, D.M.; Hopkins, J.B.; Valdevit, L. Multiscale modeling and optimization of the mechanics of hierarchical metamaterials. *MRS Bull.* **2019**, *44*, 773–781. [CrossRef]

23. Ren, X.; Seidel, G.D. Concurrent Multiscale Modeling of Coupling between Continuum Damage and Piezoresistivity in CNT-Polymer Nanocomposites. In Proceedings of the 56th AIAA/ASCE/AHS/ASC Structures, Structural Dynamics, and Materials Conference, Kissimmee, FL, USA, 5–9 January 2015; American Institute of Aeronautics and Astronautics: Reston, Virginia, 2015.
24. Borysiuk, V.N.; Mochalin, V.N.; Gogotsi, Y. Molecular dynamic study of the mechanical properties of two-dimensional titanium carbides Tin+1Cn(MXenes). *Nanotechnology* **2015**, *26*, 1–10. [CrossRef] [PubMed]
25. Borysiuk, V.N.; Mochalin, V.N.; Gogotsi, Y. Bending rigidity of two-dimensional titanium carbide (MXene) nanoribbons: A molecular dynamics study. *Comput. Mater. Sci.* **2018**, *143*, 418–424. [CrossRef]
26. Monastyreckis, G.; Mishnaevsky, L., Jr.; Hatter, C.B.; Aniskevich, A.; Gogotsi, Y.; Zeleniakiene, D. Micromechanical modeling of MXene-polymer composites. *Carbon N. Y.* **2020**, *162*, 402–409. [CrossRef]
27. Srivatsa, S.; Kumar, S.; Grabowski, K.; Jain, P.; Nayak, M.M.; Uhl, T.; Sen, P. *Numerical and Experimental Investigations of Pure MXene (Ti3C2Tx) Film and MXene Nanocomposites for Structural Health Monitoring (Conference Presentation)*; SPIE: Bellingham, WA, USA, 2020; p. 99.
28. Mishnaevsky, L.; Tsapatsis, M. Hierarchical materials: Background and perspectives. *MRS Bull.* **2016**, *41*, 661–664. [CrossRef]
29. Luz, G.M.; Mano, J.F. Biomimetic Design of Materials and Biomaterials Inspired by the Structure of Nacre. *Philos. Trans. Math. Phys. Eng. Sci.* **2009**, *367*, 1587–1605. [CrossRef]
30. Mishnaevsky, L., Jr. Nanostructured interfaces for enhancing mechanical properties of composites: Computational micromechanical studies. *Compos. Part B Eng.* **2015**, *68*, 75–84. [CrossRef]
31. Smith, B.L.; Schäffer, T.E.; Vlani, M.; Thompson, J.B.; Frederick, N.A.; Klndt, J.; Belcher, A.; Stucky, G.D.; Morse, D.E.; Hansma, P.K. Molecular mechanistic origin of the toughness of natural adhesives, fibres and composites. *Nature* **1999**, *399*, 761–763. [CrossRef]
32. Qi, H.J.; Bruet, B.J.F.; Palmer, J.S.; Ortiz, C.; Boyce, M.C. Micromechanics and Macromechanics of the Tensile Deformation of Nacre. In *Mechanics of Biological Tissue*; Holzapfel, G.A., Ogden, R.W., Eds.; Springer: Berlin/Heidelberg, Germany, 2006; pp. 189–203, ISBN 978-3-540-31184-3.
33. Lipton, J.; Weng, G.-M.; Alhabeb, M.; Maleski, K.; Antonio, F.; Kong, J.; Gogotsi, Y.; Taylor, A.D. Mechanically strong and electrically conductive multilayer MXene nanocomposites. *Nanoscale* **2019**, *11*, 20295–20300. [CrossRef]
34. Shi, X.; Wang, H.; Xie, X.; Xue, Q.; Zhang, J.; Kang, S.; Wang, C.; Liang, J.; Chen, Y. Bioinspired Ultrasensitive and Stretchable MXene-Based Strain Sensor via Nacre-Mimetic Microscale "brick-and-Mortar" Architecture. *ACS Nano* **2019**, *13*, 649–659. [CrossRef]
35. Katti, K.S.; Katti, D.R.; Pradhan, S.M.; Bhosle, A. Platelet interlocks are the key to toughness and strength in nacre. *J. Mater. Res.* **2005**, *20*, 1097–1100. [CrossRef]
36. Shahzad, F.; Alhabeb, M.; Hatter, C.B.; Anasori, B.; Man Hong, S.; Koo, C.M.; Gogotsi, Y. Electromagnetic interference shielding with 2D transition metal carbides (MXenes). *Science* **2016**, *353*, 1137–1140. [CrossRef] [PubMed]
37. Odegard, G.M.; Clancy, T.C.; Gates, T.S. Modeling of the Mechanical Properties of Nanoparticle/Polymer Composites. *Polymer* **2005**, *46*, 553–562. [CrossRef]
38. Love, A.E.H. On the small free vibrations and deformations of elastic shells. *Philos. Trans.* **1888**, *179*, 491–549.
39. Maleski, K.; Ren, C.E.; Zhao, M.Q.; Anasori, B.; Gogotsi, Y. Size-Dependent Physical and Electrochemical Properties of Two-Dimensional MXene Flakes. *ACS Appl. Mater. Interfaces* **2018**, *10*, 24491–24498. [CrossRef]
40. Fu, Z.H.; Zhang, Q.F.; Legut, D.; Si, C.; Germann, T.C.; Lookman, T.; Du, S.Y.; Francisco, J.S.; Zhang, R.F. Stabilization and strengthening effects of functional groups in two-dimensional titanium carbide. *Phys. Rev. B* **2016**, *104103*, 1–10. [CrossRef]
41. Kurtoglu, M.; Naguib, M.; Gogotsi, Y.; Barsoum, M.W. First principles study of two-dimensional early transition metal carbides. *MRS Commun.* **2012**, *2*, 133–137. [CrossRef]
42. Lipatov, A.; Alhabeb, M.; Lukatskaya, M.R.; Boson, A.; Gogotsi, Y.; Sinitskii, A. Effect of Synthesis on Quality, Electronic Properties and Environmental Stability of Individual Monolayer Ti 3 C 2 MXene Flakes. *Adv. Electron. Mater.* **2016**, *2*, 1600255. [CrossRef]
43. Mark, J.E. Polymer Data Handbook, 2nd ed. *J. Am. Chem. Soc.* **2009**, *131*, 16330. [CrossRef]
44. Eshelby, J.D. The Determination of the Elastic Field of an Ellipsoidal Inclusion, and Related Problems. *Proc. R. Soc. A Math. Phys. Eng. Sci.* **1957**, *241*, 376–396. [CrossRef]

45. Duan, H.L.; Wang, J.; Karihaloo, B.L. Theory of Elasticity at the Nanoscale. *Adv. Appl. Mech.* **2009**, *42*, 1–68. [CrossRef]
46. Weinberger, C.; Cai, W.; Barnett, D. Stanford University ME340B Lecture Notes—Elasticity of Microscopic Structures. Available online: http://micro.stanford.edu/~{}caiwei/me340b/content/me340b-notes_v01.pdf (accessed on 16 November 2020).
47. Kouznetsova, V.; Brekelmans, W.A.M.; Baaijens, F.P.T. An approach to micro-macro modeling of heterogeneous materials. *Comput. Mech.* **2001**, *27*, 37–48. [CrossRef]
48. *Krzysztof Grabowski Design and Development of the Sensors for Structural Health Monitoring (SHM) Based on the Carbon Nanomaterials*; AGH University of Science and Technology: Krakow, Poland, 2017.
49. Software, M. *Marc 2013 Volume B: Element Library*, 2013th ed.; MSC Softwares: Santa Ana, CA, USA, 2013.
50. Tanov, R.; Tabiei, A. A Simple Correction to the First Order Shear Deformation Shell Finite Element Formulations. Available online: https://www.dynalook.com/conferences/international-conf-2000/session8-1.pdf (accessed on 16 November 2020).
51. Chopra, I.; Sirohi, J. *Smart Structures Theory*; Cambridge Aerospace Series; Cambridge University Press: Cambridge, UK, 2013; ISBN 9781139025164.
52. Mishnaevsky, L. *Computational Mesomechanics of Composites*; John Wiley & Sons, Ltd: Chichester, UK, 2008; ISBN 9780470513170.
53. Kochmann, D.M. *Computational Multiscale Modeling*; ETH Zurich: Zurich, Switzerland, 2018.
54. Li, X.; Liu, Q.; Zhang, J. A micro-macro homogenization approach for discrete particle assembly - Cosserat continuum modeling of granular materials. *Int. J. Solids Struct.* **2010**, *47*, 291–303. [CrossRef]
55. Dai, G.; Mishnaevsky, L. Graphene reinforced nanocomposites: 3D simulation of damage and fracture. *Comput. Mater. Sci.* **2014**, *95*, 684–692. [CrossRef]
56. Liu, Y.; Zhu, X.; Pan, L. Hybrid Architectures based on 2D MXenes and Low-Dimensional Inorganic Nanostructures: Methods, Synergies, and Energy-Related Applications. *Small* **2018**, *1803632*, 1–22. [CrossRef] [PubMed]
57. Lipatov, A.; Alhabeb, M.; Lu, H.; Zhao, S.; Loes, M.J.; Vorobeva, N.S.; Dall'Agnese, Y.; Gao, Y.; Gruverman, A.; Gogotsi, Y.; et al. Electrical and Elastic Properties of Individual Single-Layer Nb 4 C 3 T x MXene Flakes. *Adv. Electron. Mater.* **2020**, *1901382*, 1901382. [CrossRef]
58. Bauchau, O.A.; Craig, J.I. *Structural Analysis with Application to Aerospace Structures*, 1st ed.; Solid Mechanics and Its Applications; Bauchau, O.A., Craig, J.I., Eds.; Springer: Dordrecht, The Netherlands, 2009; Volume 163, ISBN 978-90-481-2515-9.

Publisher's Note: MDPI stays neutral with regard to jurisdictional claims in published maps and institutional affiliations.

materials

MDPI

Article

Researches and Simulation of Elastic Recovery Phenomena during Roller Burnishing Process of Macro-Asperities of Surface

Agnieszka Kułakowska * and Łukasz Bohdal

Department of Mechanical Engineering, Koszalin University of Technology, Racławicka 15-17 Street, 75-620 Koszalin, Poland; lukasz.bohdal@tu.koszalin.pl
* Correspondence: agnieszka.kulakowska@tu.koszalin.pl

Received: 12 October 2020; Accepted: 18 November 2020; Published: 21 November 2020

Abstract: The paper presents preliminary studies of a new innovative surface treatment method—the process of roller burnishing of macro-irregularities of the surface. As part of the work, the possibility of plastic shaping of the surface macrostructure with indentations (plateau structure), which will show anti-wear properties through appropriate surface shaping and the compressive stress state in the product's top layer, was investigated. The essence of the paper is the analysis of one of the aspects of the application of this processing method, i.e., the influence of the elastic recovery of the product on its technological quality measured by dimensional deviation. The main objective of the work is to develop adequate methods and mathematical models to enable the design of the macro-asperities of the surface burnishing process to maintain the dimensional tolerance of the shaped parts. The results of dependencies of elastic recovery of the asperities and the deviation of height, Δh_t, for sample depths of burnishing were presented. The model tests of the elastic recovery of the model material using the visioplasticity method show that with the increase of the value of the vertical surface asperities, the value of the elastic recovery of the material decreases. The increase of the deviation of the asperities' height causes a decrease in the value of elastic recovery. With the increase of the value of the vertical angle of the surface roughness, the value of the elastic recovery of the material is smaller.

Keywords: numerical analysis; roller burnishing process; roughness; surface state; previous treatment; outline deviations; elastic recovery

1. Introduction

Roller burnishing is used increasingly as a finishing operation [1–6]. It is known that the surface state after previous treatment has significant influence on its quality after roller burnishing [7,8].

The effect of burnishing is to reduce the roughness of the profile while hardening the surface layer of the product and the formation of compressive stress in it [1–3,9]. In the process of burnishing several interacting processes occur simultaneously, phenomena present in the material deformation areas affected on the conditions in the contact zone, and those on some of the burnishing conditions, which also affects its complexity [5,10].

Review of the literature shows that work on the burnishing process has been conducted by many researchers. Several works have investigated the effect of the burnishing process to improve the properties of the parts' mechanical properties, e.g., increased hardness [2], higher wear resistance [6], surface quality [9,11], and increased maximum residual stress in compression. The parameters affecting the surface finish are burnishing force, feed rate, ball material, number of passes, workpiece material, and lubrication [6]. References [1–4] showed that the burnishing force and the number of tool passes are the most predominant of the parameters that have an effect on the surface roughness of the workpiece during the burnishing process.

Determining roller burnishing parameters also causes difficulty in achieving the required quality, because the measurement of process parameters (determining the technological quality such as temperature field, stress field, structural changes, the change in the dimensions under the influence of elastic deformation, etc.) during the burnishing process in the current state technique is impossible.

The process parameters can only be inferred from the properties of the product after the process. As a result of burnishing, there are dimensional changes caused by the following factors:

1. Plastic deformation and the decrease in surface roughness,
2. The concentration and structure squeeze,
3. An axial elongation of the product due to plastic deformation and residual stress generated in the surface layer [6],
4. Moving the material after moving away the burnishing element and increasing the asperities height.

Despite numerous works on burnishing treatment, the problem of springback of the material after unloading (or moving away) the burnishing tool, has been considered in a few studies [12,13]. The problem of elastic recovery is most often occupied by researchers working on pressing an indenter into the material [14–19], and less frequently, rolling [20] or the process of sheet metal forming [21] or cutting [22].

Stilwel and Tabor [18] found that when a spherical indenter is pressed into a plastic material, it creates a cavity, then after it is pulled back, there is an elastic return of the material of both the indenter and the cavity. Shape changes can be described in relation to the Hertz equation for the elastic deformation of a spherical surface. Chodór, in his work [22], using the results from numerical analyses, shows the elastic recovery of the material during the process of micro-cutting and sliding burnishing. Skalski [23] noticed that during the propagation of the plastic area (in the case of pressing a rigid cylinder into a body with a flat surface), a narrow elastic area appears under the cylinder on a plastic substrate. This area does not plasticize even during increasing load.

Despite the large number of works on burnishing, elastic recovery has not been considered. There are works concerning elastic recovery during the scratch test [14,24] but not during roller burnishing. Also, in many papers, modeling and numerical analysis of the process were presented [25–31] but without considering the outline of the surface after previous treatment and material elastic recovery. Reference [32] describe the effects of ball-end hard milling to a variation of the lubrication situation. At the reversal, roller burnishing process parameters on the surface point to micro-contacts at the maximum rolling speed elasto-roughness of the raceways of constant velocity joint shafts. The authors of Reference [33] developed a three-dimensional (3D) numerical finite element model of the ball burnishing process including real surface integrity descriptors resulting from a ball-end-milled AISI 1038 surface in the target workpiece. Specifically, its periodical topological features were used to generate the surface geometry. Different models varying the effect of the coefficient of friction and the direction of application of burnishing passes with regards to the original milling direction were calculated. Reference [34] reports the results of experimental studies on the impact of ball burnishing parameters on the roughness, microstructure, and microhardness of the surface layer of laser-cut C45 steel parts. The distribution of residual stresses generated in the surface layer of these parts were analyzed. The superficial effect of hardening caused after vibration-assisted ball burnishing and its consequences in the tensile behavior of a carbon steel material were studied in Reference [35]. Experiments using different amplitudes and new forces were encouraged to obtain more information about how the material can be modified optimally through vibration-assisted ball burnishing.

Burnishing of the macro-irregularities of a surface is a variation of the typical burnishing process, however it differs in the scope of influence on the surface layer of the product. The research on this new process was undertaken due to the conducted preliminary research, which gave very favorable operational effects of the surface layer shaped in this way. It can be seen that the depth of strengthening the surface layer can be even a few mm, while for a typical burnishing process, only a part of a mm or a few μm. Additionally, the literature shows how to solve the problem of the kinematics of the roughness of the material for the triangular asperities of the surface crushed (burnished) with a flat punch under various friction conditions [36–38]. Kukiełka [10,39,40] stated that the kinematics of the smoothing of the asperities depends only on the apex angle of the rough surface and determined the angle boundaries for which there is a change in the flow mechanism.

The author of Reference [39] qualitatively distinguished three different cases of flow of the surface layer of material. The considerations were carried out with the assumption that the surface asperity after the previous treatments is triangular, symmetrical, and regular. The research shows that in the burnishing process, these cases depend only on the vertical angle θ of the asperity (Figure 1):

- For vertical angles $\theta \leq 80°$, the material core remains undistorted. The indentations of the asperity do not rise, and the deformation of the material occurs only within the unevenness. A gap (discontinuity planes) with the depth 0.5 Rt that separates deep inequalities can be seen. As a result of the material flowing sideways, the surface flattens out.
- For the vertical angles $80° < \theta < 145°$, the zone of plastic deformation includes the material core. Gaps are still visible at the junction of adjacent burrs but of less depth. The asperity valleys rise and total strain in the contact zone of neighboring material overflows.
- For angles $\theta \geq 145°$, as a result of deformation of asperity and the material core, the surface is evened out. This is not at the expense of the material outflows towards the sides of the asperity. There are no planes of material discontinuity in the surface layer. The rise of the indentation is equal to the decrease in the top of the asperity.

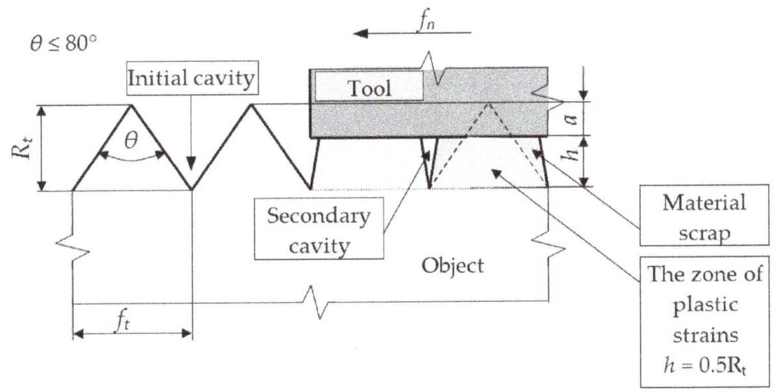

(a)

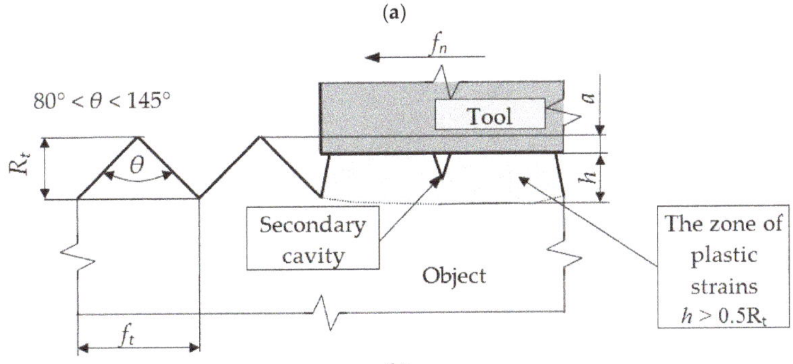

(b)

Figure 1. *Cont.*

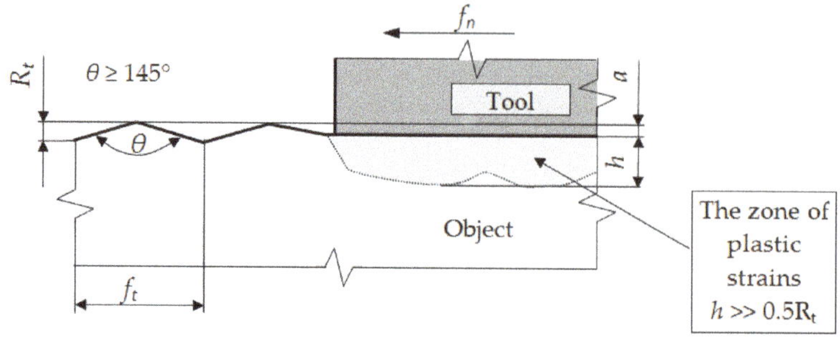

(c)

Figure 1. The influence of the vertical angle of a symmetrical triangle asperity on the depth of strain zone and surface roughness profile after burnishing: $\theta \leq 80°$ (**a**), $80° < \theta < 145°$ (**b**), $\theta \geq 145°$ (**c**) [39].

A significant problem is to ensure the regularity of the outline and the appropriate vertical angle θ of the shaped irregularities in the treatment preceding burnishing. Due to the type of the machined part and its intended use, the value of the feed, f_t, in the preceding machining and the value of the vertical angle θ of the asperity should be selected each time.

In previous paper [12] on the initial stages of the experiment related to elastic recovery issues, we analyzed the influence of the deviations of the height and distance of the asperities after turning on the deviation of the height of the asperities of the surface as well as elastic recovery after burnishing. Research was performed for different burnishing depths. The calculations were performed for apex angles $\theta = 90°$ and $\theta = 120°$. The received results showed that the elastic recovery increases as the burnishing depth increases. Each of the asperities were analyzed separately. The analysis of the state of knowledge shows that none of the researchers analyzed the problem of elastic recovery, which is a novelty and originality of the presented article. In the paper, FEM (Finite Element Method) analysis with using the dynamic explicit method for the roller burnishing process was established. The surface state after turning (as previous treatment) was taken into account. The results of dependencies of elastic recovery of the asperities and the deviation of height, Δh_t, for exemplary burnishing depth is presented. The influence of the surface after previous treatment on the burnished product quality can be analyzed by developed numerical algorithms. It allows for better understanding of the phenomena which occur in the tool–workpiece contact zones. The basis for the development of guidelines for the selection of the conditions of rolling and burnishing processes considering the required technological quality of the product can be formulated. This paper consists of numerical simulations and analysis, model test, and the research on the steel C45, with the aim to understand elastic recovery as well as analysis of elastic recovery of the asperities with different vertical angles.

2. Idea of Elastic Recovery

It is assumed that during the burnishing process, deformations of the material occur in terms of elastic and visco-plastic strains [41]. It is known that visco-plastic strains will remain in the material, while elastic strains change [41–43]. After the burnishing tool has passed, the phenomenon of elastic recovery occurs and visco-plastic deformations remain in the processed material, while the elastic deformations change and settle on the level corresponding to the new state of equilibrium of the object. Thus, the states of displacements, stresses, and strains in the tested object during and after the end of the process differ significantly, because after the process is completed, the nodes are displaced in the elastic range. The value of elastic deformations will decrease by the value $\Delta\phi^{(E)}$, while the elastic deformations will remain at the value $\phi^{(E)}$ in the surface layer of the product. This is shown in Figure 2.

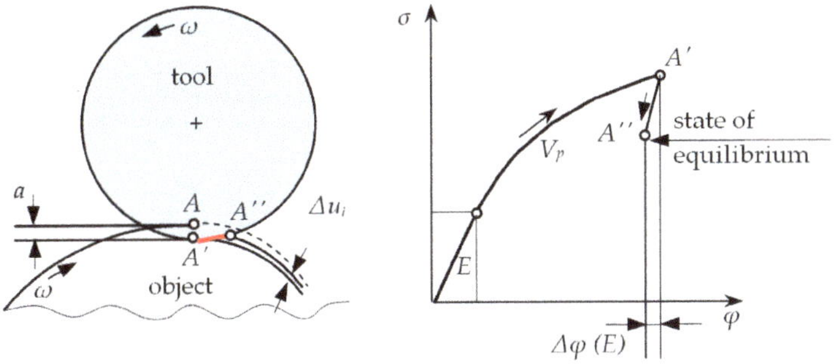

Figure 2. Elastic recovery during the roller burnishing process.

During the process, the burnishing tool is plunged to the depth a. It can be noticed that the contact point of the tool and the workpiece material A, as a result of this depression, moves to the point A'. Then, after the rotation of the workpiece and the burnishing tool, the load is unloaded and point A' is moved to position A''. The radial distance between points A' and A'' is marked as Δu_i and represents the elastic recovery of the material after burnishing, and φ is the actual deformation of the material.

The process of roller burnishing is treated in this study as a process with unknown boundary conditions in the tool–workpiece contact area. It is a geometrical and physical boundary and initial value problem. During experimental research and simulations of the process, special attention was drawn to the elastic strain of the material, which takes place after moving away the burnishing element, called elastic recovery, Δu_i. The state of the surface layer after previous treatment and after roller burnishing are considered together. They take into account two asperities, *A1* and *A2*, after the turning process, which are burnished. Asperity *A1* after turning was characterized by constant height (h) and distance (s). Asperity *A2* possesses different values of the height or distance. The difference between the asperities was called height deviation after turning (Δh_t) and between distances: distance deviation after turning (Δs_t). Similarly, deviations were defined after burnishing and marked as Δh_b and Δs_b. The schematic diagram of elastic recovery in joint analysis of the surface after turning and burnishing is presented in Figure 3.

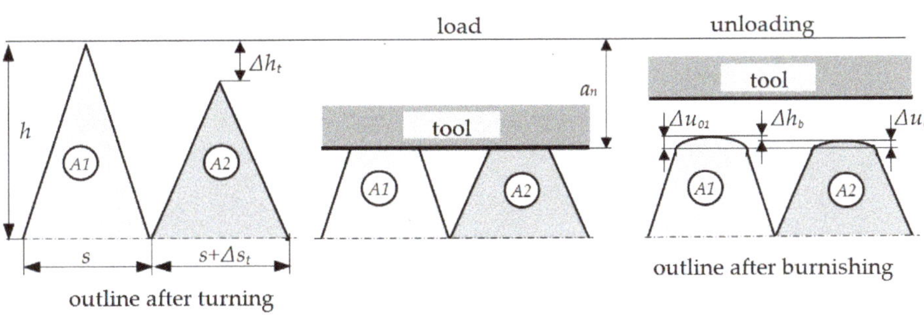

Figure 3. Schematic diagram of the elastic recovery of the material after burnishing: *A1*—first asperity, *A2*—second asperity after previous machining (turning), *h*—asperity's height, Δh_t—height deviation of asperity after turning, Δh_b—asperity's height deviation after burnishing, a_b—burnishing depth, Δu_{01}—elastic recovery of the material of first asperity *A1*, Δu_{02}—elastic recovery of the material of second asperity *A2*.

3. Researches of the Influence of Height and Distance Deviations on the Elastic Recovery

3.1. Experimental Researches—Materials and Methods, Discussion

The experiment was conducted for the following input factors: height deviation (Δh_t) and distance deviation (Δs_t) of the surface asperities after rolling, determined in compliance with Table 1, and the vertical angle of asperities was constant, $\theta = 60°$. The five-level rotatable experiment plan was used to set up the values of the deviations.

Table 1. Experimental plan with real and coded values.

	1	2	3	4	5	6	7	8	9	10	11	12	13
Δh_t	0.01	0.01	0.03	0.03	0.02	0.02	0.04	0	0.02	0.02	0.02	0.02	0.02
Δs_t	0.05	0.15	0.05	0.15	0.2	0	0.1	0.1	0.1	0.1	0.1	0.1	0.1
Δh_t	−1	−1	+1	+1	0	0	+1.414	−1.414	0	0	0	0	0
Δs_t	−1	+1	−1	+1	+1.414	−1.414	0	0	0	0	0	0	0

The values of the input factors were changed in the following ranges: $\Delta h_t = 0.01$–0.04 mm, $\Delta s_t = 0.05$–0.2 mm. The tests were repeated three times for each point of the experiment plan. Tests were performed at room temperature (T = 19 °C). The external surfaces of the samples made of steel C45 were prepared in the turning process on a NEF 520 numerical lathe in "TEPRO" Vacuous Technology Plant in Koszalin, with a VCMT 160402 machining plate. The following machining parameters were used: velocity v_t = 200 m/min and feed f_t = 0.15 mm/rev in rough machining and v_t = 200 m/min, f_t = 0.1 mm/rev in finishing. Six samples were made on the shaft (Figure 4). Each sample included two asperities, which differed in the value of the height and distance deviations, according to the experiment plan. The accuracy of the shaping of the outline was verified on a Werth optoelectronic microscope. Figure 5 shows an example of a shaft with samples prepared in accordance with the experiment plan, Table 1 (first 6 implementations). The values of the height and distance deviations obtained in measurements were not significantly different (at the significance level $\alpha = 0.05$) from those planned on the basis of single factor analysis of variance.

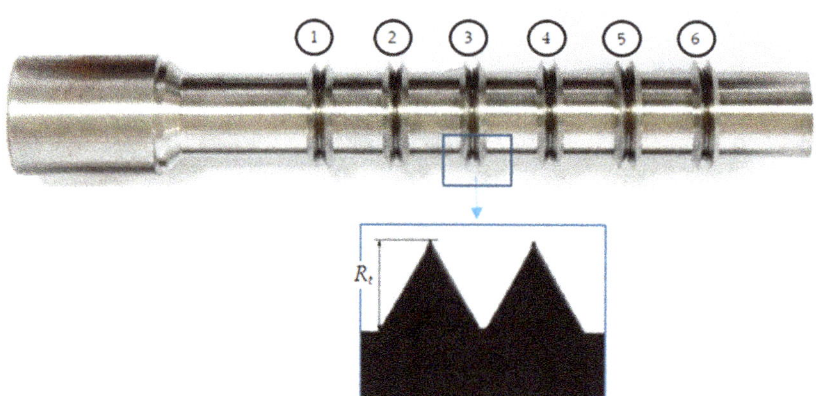

Figure 4. General view of the shaft with the samples after turning process and zoom of the sample (nb 3) after turning process, where R_t is the asperity height.

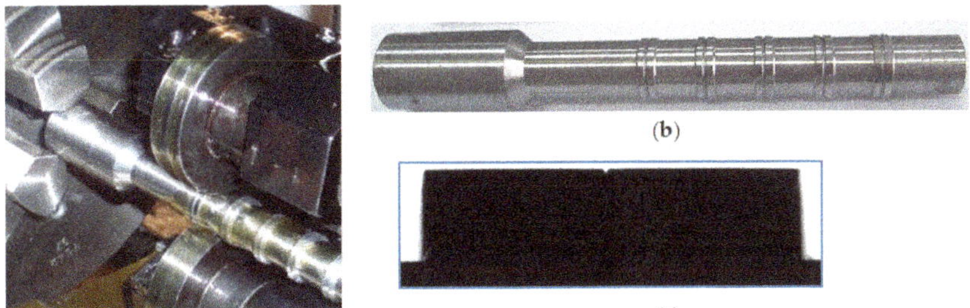

(a) (b) (c)

Figure 5. Burnishing position (**a**), general view of the shaft with the samples after burnishing (**b**), zoom of the asperities after roller burnishing (**c**).

Next, the samples were burnished using a Fette T27 tangent burnishing head with high stiffness to reduce the influence of dimensional changes of the head on results. The burnishing depth for each sample was calculated separately, while assuming that it is $a = 0.5\ R_t$, where R_t is the asperity height without a deviation (constant one).

After the burnishing process, the values of parameters Δu_0 were calculated and profiles with profilometer T8000 were obtained. Exemplary profile of burnished asperities is presented in Figure 6.

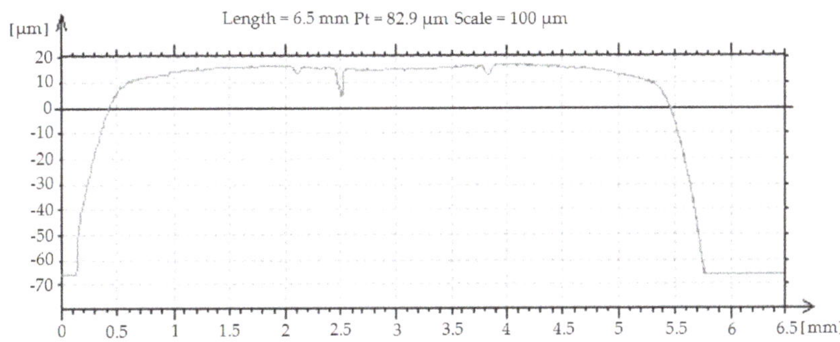

Figure 6. Exemplary profile of the asperities after roller burnishing.

The results of the measurements were developed statistically in the Experiment Planner program developed by Kukiełka L. and Kukiełka S. [44–50]. The following regression equation is obtained of the dependence of the material elastic recovery after burnishing (Δu_0) from the deviations of height (Δh_t) and distance (Δs_t) of asperities after turning, for correlation coefficient $R = 0.985$:

$$\begin{aligned} \Delta u_0 = {} & 0.0007\Delta s_t - 0.013\Delta h_t - 2.423\times 10^{-14}\Delta s_t \Delta h_t + \\ & + 0.0048\Delta {s_t}^2 + 0.12\Delta {h_t}^2 \end{aligned} \quad (1)$$

In order to determine the form of Equation (1), appropriate analytical and experimental tests were carried out, in the following stages: determination of the sets of tested, constants, disturbing and resulting factors, determining the range of variability (research area) of the examined factors, adoption of the class of the mathematical model of the research object, coding of researched factors, implementation of the actual research, experiment plan, the results of the experiment, elimination of results with a gross error, calculating the inter-row variance, checking the homogeneity of variance in the sample, calculating the coefficients as a function of the regression, statistical analysis of the regression function, study of the significance of the multidimensional correlation coefficient, checking the adequacy of the mathematical model, and decoding the regression function. The detailed procedure is given in References [39,41,44]. The graph of the function according to Equation (1) with the correlation coefficient $R = 0.86$ is presented in Figure 7.

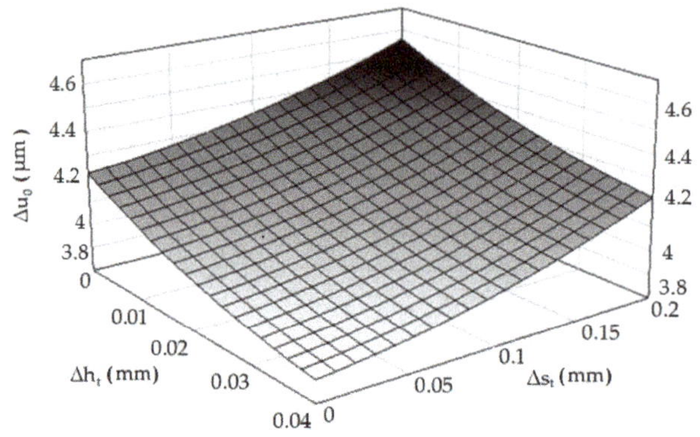

Figure 7. Dependence of the material elastic recovery after burnishing (Δu_0) from the deviations of height (Δh_t) and distance (Δs_t).

It is visible that an increase of the deviation of the asperities' height causes a decrease in the value of elastic recovery, while the opposite situation occurs when the deviation of the distance is analyzed. Knowledge about influence of determined profile parameters on the physical phenomena that occur in the burnishing process is unknown. It is important because they impact on the quality of the formed product's surface layer, tribological properties, and fatigue issues. Many physical phenomena whose observation or measurement is very difficult or impossible take place during processes of burnishing, such as the pressure in contact zones, the friction forces, strain and stresses, elastic recovery, the slip and contact zones, and states of displacement. After passing the burnishing tool, visco-plastic deformations remain in the workpiece, while the elastic deformations change and settle at the level corresponding to the new equilibrium state of the object. Thus, the states of displacements, stresses, and strains in the tested object during and after the end of the process differ significantly, because after the process is completed, the nodes are displaced in the elastic range. In the case of higher asperities, the displacement is greater than in the case of those with a height deviation. The opposite is the case for a distance deviation. These issues are very complex and require further analysis.

3.2. FEM Modeling and Numerical Results

The computer models (2D and 3D) of the process in the real scale were characterized by the possibility of changing the value of the height deviation (Δh_t) and the distance deviation (Δs_t). Exemplary calculations were performed for the apex angle $\theta = 60°$ and the asperities feed $f = 2.7$ mm, and the deviations were changed according to Table 1. Numerical analyses were conducted with the aid of Ansys Structural, LS-Dyna and LS-Prepost programs. In the analyzed case, the object was treated as an elastic body (in terms of reversible deformations) and visco-plastic (in terms of irreversible deformations), while the tool (roller) was treated as a perfectly rigid body. Based on the results obtained, various analyses can be carried out. In the discussed case, only some of the results were presented in order to determine the impact of the burnishing process of the rough surface (with regular triangular asperities) on the value of elastic material return after the burnishing process. Moreover, in incremental material models, it was assumed that the object was made of C45 steel, for which the cumulative value of the yield stress in 2D models is described by the regression equation according to Equation (2):

$$\sigma_p = 976.95\left(0.0667 + \varphi_i^{(\mathrm{VP})}\right)^{0.2052}\left[2.33\times10^{-6}\left(\dot{\varphi}_{\mathrm{i}}^{(\mathrm{VP})}\right)^2 + 8.83\times10^{-5}\dot{\varphi}_{\mathrm{i}}^{(\mathrm{VP})} + 1\right]^{1.337} \tag{2}$$

where: yield stress is σ_p, intensity of the real strain is φ_i, and strain rate is $\dot{\varphi}_i$.

The necessary parameters for this material are Young's modulus E = 210 GPa, Poisson ratio v = 0.29, yield stress R_e = 425 MPa, and hardening modulus E_T = 1024 MPa. In 3D models, object discretization was performed using finite elements of the object with a linear shape function, shell elements (tool), and contact elements (contact of the tool with the object). The nonlinear, anisotropic Coulomb friction model and the coefficient of static friction μ_s = 0.1 and dynamic friction μ_d = 0.05 were adopted. Then, adequate boundary conditions were set—mainly concerning the displacements: tool cavity (up to the value of 0.5 of the roughness height) and the degrees of freedom for the workpiece, as well as the set rotational speed ensuring the required burnishing speed v_b = 0.52 m/s. In this case, the shaft with surface asperities was rotated at n = 477 rpm, then the burnishing element (roll) was moved to the burnished surface. The contact of the roller with the machined surface caused the tool to rotate. In the next step, methods of solving the discrete equation of motion were introduced. For this purpose, in accordance with the adopted algorithm of explicit solution, at each analysis step, an approximation of the column vector of acceleration and velocity increment as a function of the column vector of displacement increment was introduced, all with the following dimension integration and were given the required number of steps, as well as the total calculation time t = 3.5 s. The geometrical model and mesh grid are shown in Figure 8. The 3D FEM model was developed according to Hill plasticity condition and low strain rates. Thermal strains for this model were approximately 2%.

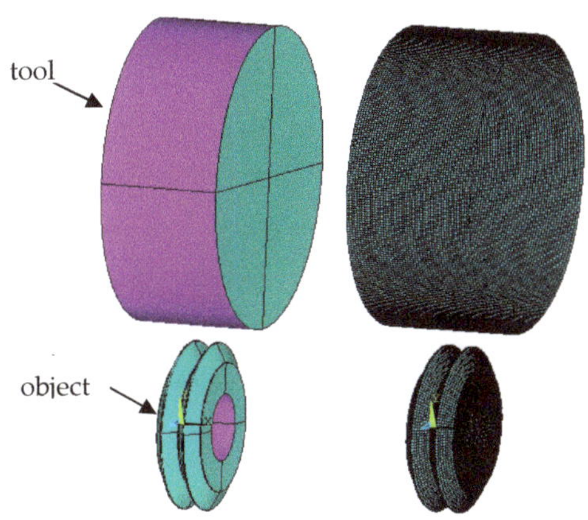

Figure 8. Geometry of the object, tool, and finite element mesh.

The temporary yield stress in the 3D model was described with the aid of the Cowper–Symonds model. A Cowper–Symonds model allows for linear isotropic (β = 1), kinematic (β = 0), or mixed (used in this study) (0 < β < 1) plastic strain hardening, and the effect of the plastic strain velocity is given by the following power relation:

$$\sigma_p = \left[1 + \left(\dot{\varphi}_{\mathrm{i}}^{(p)}/\mathrm{C}\right)\right]^{\mathrm{m}}\left(R_e + \beta E_p \varphi_i^{(p)}\right) \quad (3)$$

where β is the plastic strain hardening parameter, R_e (MPa) is the initial static yield point, $\dot{\varphi}_i^{(p)}$ (s^{-1}) is the plastic strain velocity, C (s^{-1}) is the material parameter defining the effects of the plastic strain velocity, m = 1/P is the material constant defining the sensitivity of the material to the plastic strain velocity, $\varphi_i^{(p)}$ is the plastic strain intensity, and $E_P = \frac{E_T E}{E - E_T}$ is a material parameter dependent on both the plastic strain hardening modulus, $E_T = \partial\sigma_p/\partial\varphi_i^{(p)}$, and Young's modulus, E.

To find the value of material elastic recovery in performed computer simulations, one had to select areas when the tool affected the subject, and after it moved away. Two kinds of analysis were performed. The first, when Δh_t (height deviation) values after the turning process were Δh_t = 0–0.04

mm and Δs_t (distance deviation) values after the turning process were $\Delta s_t = 0$, and the second, when $\Delta s_t = 0$–0.2 mm but $\Delta h_t = 0$. On the basis of computer simulations, the graphs of dependencies of elastic recovery of the first Δu_1 and the second Δu_2 asperity and the deviation of height, Δh_t, for the sample after burnishing were elaborated. The graph of dependencies of elastic recovery Δu_1 and Δu_2 in the asperities *A1* and *A2* from distance deviation Δs_t is presented in Figure 9.

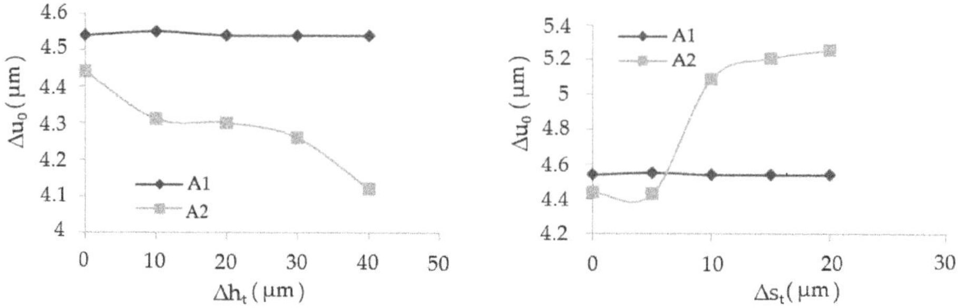

Figure 9. The graphs of dependencies of elastic recovery Δu_1 and Δu_2 accordingly in the asperities *A1* and *A2*, from height deviation, Δh_t (**a**), and dependencies of elastic recovery Δu_1 and Δu_2 accordingly in the asperity *A1* and *A2* from distance deviation, Δs_t (**b**).

It has been observed that together with the increase of the deviation of surface asperities' height, the elastic recovery decreases. Comparing the values of elastic recovery for an asperity with constant geometry *A1* and changeable geometry *A2*, it can be seen that the value of elastic recovery was higher for asperity *A1*.

Computer simulations were also performed to validate the experimental studies according to the experiment plan presented in Table 1. An example of the results obtained from the simulation is shown in Figure 10, including a geometric model with a mesh of finite elements (a), mesh deformation (b), elastic recovery of selected asperity nodes with variable geometry (A) and constant (B).

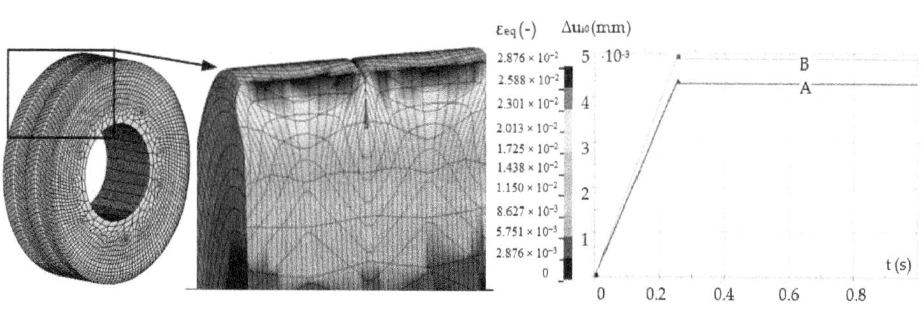

Figure 10. Geometric model with a mesh of finite elements and mesh deformation (**a**), elastic return of selected inequality nodes with variable geometry—A and constant—B (**b**).

On the basis of the obtained results, an Equation (4) was statistically developed, describing the influence of height deviation and distance deviation on the elastic recovery of the material. Multidimensional correlation coefficient $R = 0.8332$.

$$\Delta\hat{u}_0 = 0.0048 - 0.0032\Delta s_t - 0.026\Delta h_t - 3.526 \times 10^{-14}\Delta s_t\Delta h_t + 0.024\Delta s_t^2 + 0.358\Delta h_t^2 \quad (4)$$

As in the case of experimental studies, it was found that an increase of the asperities' height deviation causes a decrease in the value of elastic recovery, while the opposite situation occurs when the distance deviation is analyzed. The results of experimental studies and computer simulations differ slightly from each other. This may result from simplifications and assumptions made when developing the computer model (e.g., refinement of the finite element mesh, shape of elements). Based on Equation (4), a spatial diagram (Figure 11) was prepared.

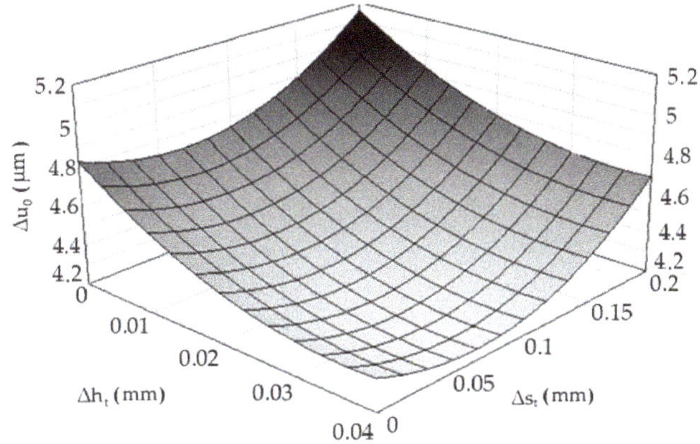

Figure 11. Relation of elastic material recovery after burnishing—$\Delta u_0 = f(\Delta h_t, \Delta s_t)$.

4. Researches of the Phenomenon of Elastic Recovery Depending on the Vertical Angle of the Asperities

4.1. Modelling Researches of the Phenomenon of Elastic Recovery

Experimental studies with the use of a model material were carried out in order to assess the effect of the flow blocking phenomenon through adjacent irregularities on the elastic return of the surface after burnishing. The tests were carried out in accordance with the rheological, geometric, and time scales [4,36,44]. Modeling consists in replacing a real research object with a model object. The real and model objects form a system of equivalent objects whose similarity is determined by mathematical relationships. Then, the modeling results can be translated into a real research object. Physical modeling on substitute material is used in experimental qualitative and quantitative analysis. The most frequently used model material replacing metal is a non-metallic material, such as plasticine, wax, putty, etc. These are materials with much lower plastic flow resistance. Physical modeling on plasticine is used for experimental qualitative and quantitative analysis of the plastic working processes. Qualitative analysis mainly concerns the kinematics of the material flow process, while the physical analysis covers the study of force and energy parameters (force, unit pressure on the contact surface of the tool—model material). In recent years, the technique of visualizing the flow trajectory of material particles of the shaped material, the so-called method of visioplasticity, has been used. It is based on the study of plastic flow kinematics based on the observation and measurements of the coordination grid. Performing experimental tests of visualization on real metal materials is laborious and costly and requires a station with very high-pressure forces. Therefore, often in experiments, the real material is replaced by the so-called model material or numerical analyses are performed [39,42,45]. In order to ensure the condition of rheological similarity, the characteristics for the model material—plasticine—were determined. This is due to its availability, low price, easy workability, plasticity (plasticizing stresses are 100–1000 times lower than the appropriate metal stresses), the possibility of its modification and regeneration, and therefore, multiple use. The material model was developed based on the static compression test. The results of this test are approximate data due to the presence of disturbances in the measurement of the compression force. The factor disturbing the measurement of the compressive force is the friction occurring on the faces of the cylindrical sample.

Under ideal conditions, during a static compression test, the diameter of a cylindrical specimen would change uniformly over its entire height. However, for real conditions, the effect of friction causes the sample material to flow unevenly (a barrel is formed), which is a result of blocking the flow of material on the front contact surfaces of the sample with the substrate and the punch. The static tensile test does not have this disadvantage, however, due to the nature of the material, it is impossible to carry out.

The compression test was carried out on the stand shown in Figure 12a. A cylindrical sample with the initial height h_0 = 50 mm and the initial diameter $d_0 = 2r_0$ = 40 mm was used for the tests. On the basis of the compression test, the dependence of the force as a function of the change of sample height $P = f(\Delta h)$ was developed (Figure 12b). Using the results of the compression test and appropriate dependencies from the works of Dawidenkow and Spiridonowa [39,43,46], a material model was developed in the form of an actual compression diagram (Figure 12c). The obtained material characteristics for the model material are similar to the material characteristics of C 45 steel.

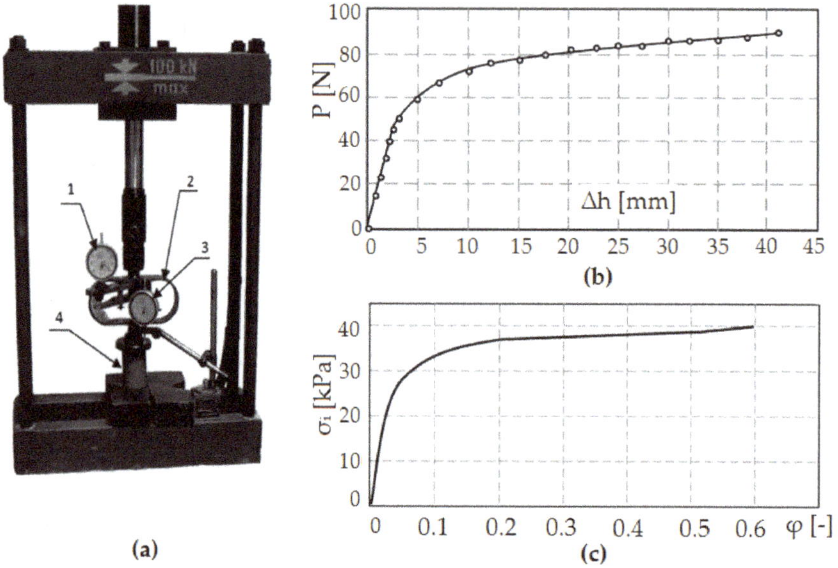

Figure 12. Stand for static compression test on the model material: 1—bow dynamometer, 2—dial displacement sensor, 3—compressed sample, and 4—dial gauge calibrated in force units (**a**), change of the compressive force P as a function of the height change Δh of the cylindrical sample (**b**), the actual compression diagram for the model material (**c**).

The aim of the model research was to determine the relationship between the vertical angle of triangular asperities and the elastic recovery of the model material. The burnishing element in the test was a flat punch mounted on a hydraulic press. The tests allowed for visual assessment of material displacement, numerical determination of the size of deformation, and elastic recovery of the material. It is also possible to determine the magnitude of stresses and the direction of material flow. The samples made of model material (plasticine) were characterized by the vertical angles of asperities from $\theta = 60°$ to $\theta = 150°$, every 15°. Each sample had three asperities with the same vertical angles and was cut along the longer sides to apply a coordination grid. Photos of samples before, during, and after the burnishing process are presented in Table 2.

Table 2. View of the exemplary samples.

	Before Burnishing	During Process	After Burnishing
$\theta = 60°$	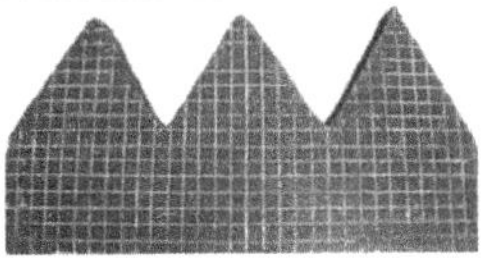	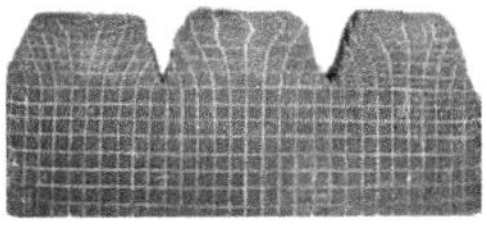	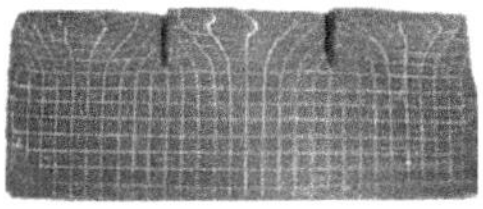
$\theta = 150°$	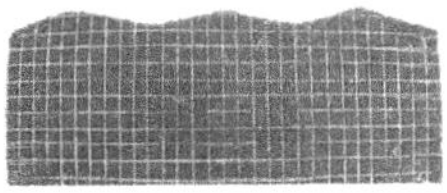	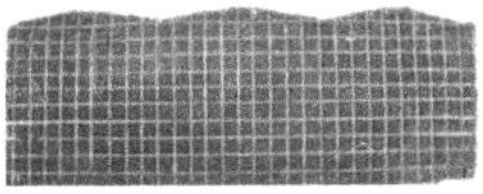	

Then, the samples were placed in the matrix, which task was to properly position the sample in relation to the punch and to ensure appropriate boundary and initial conditions. The stamp was displaced, causing the deformation of the vertices of the triangular asperities and the deformation of the coordination grid. The sample and stamp were covered with a layer of talcum powder to reduce friction between them. Using a coordination grid applied to the samples, it was possible to perform appropriate measurements of the elastic recovery of the material.

The obtained results are presented in the form of a graph in Figure 13. The model tests of the elastic recovery of the model material using the visioplasticity method show that with the increase of the value of the vertical angle of the surface asperities, the value of the elastic recovery of the material decreases. The results were confirmed by numerical analysis. In some areas, the results are very close. The differences may be due to simplifications and assumptions made during creating a computer model. The accuracy of FEM calculations is influenced, among others, by the ratio of the side length of the finite elements used to discretize the object (object and tool), mesh refinement (mainly in areas of strong geometric or physical nonlinearity), and the finite element shape function. Although, in order to determine the significance of the influence of these factors and to determine an effective discrete model for the case of roller burnishing of a rough surface, sensitivity analysis was performed. As a result, acceptable gaps between the results were obtained.

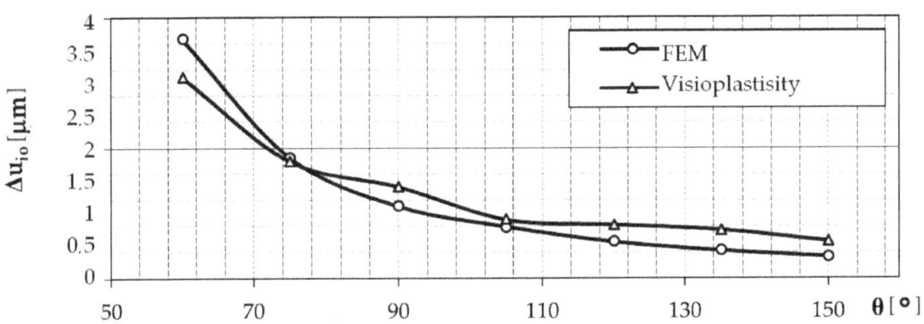

Figure 13. Influence of the vertical angle of surface asperities, θ, on the material elastic recovery Δu_{i0}.

4.2. Computer Simulation of the Phenomenon of Elastic Recovery Depending on the Vertical Angle of the Asperities

Computer simulations were developed for the same values of the asperities' peak angles as in the research on the model material. The calculations were made in two stages. In the first stage, triangular irregularities were burnished, and in the second stage, the burnishing tool was moved away. It was burnished to a depth of a = 0.5 h, where h is the height of the triangular asperity. This approach to the problem made it possible to calculate the value of the material elastic recovery by determining the displacement of the nodal point of the peaks of the asperity. The difference between the node displacement modules at the moment of contact with the burnishing tool and after the tool is fully retracted is the desired result (Figure 14). Table 3 presents selected results from 2D computer simulations before, during, and after the burnishing process, showing displacement maps.

Table 3. Results of sample computer simulations.

	Before Burnishing	During Process	After Burnishing
$\theta = 60°$	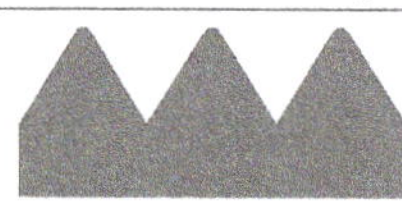	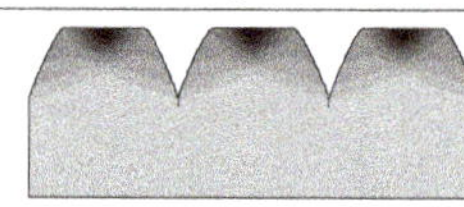	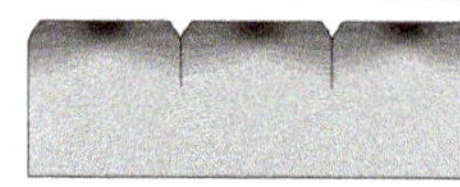
$\theta = 150°$	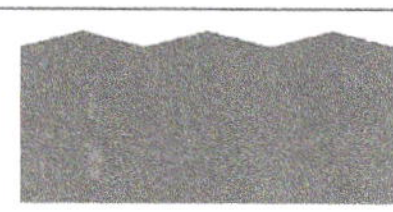	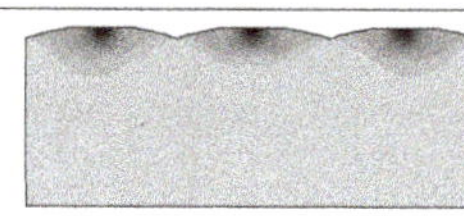	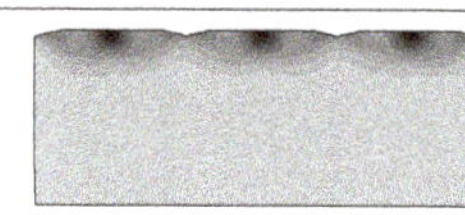

Figures 14 and 15 show diagrams of the displacement of the node lying on the contact surface (the highest point of asperity) along the Y axis.

In the initial phase of the process, the node was slowly displaced due to the impact of the tool on the burnished material. The asperity was deforming. At the moment when the burnishing tool shifted to half of the roughness, the observed node reached the highest value of displacement. The first stage of numerical analysis has been completed. In the second step, when the time exceeded 1 s, the crushing tool was moved back.

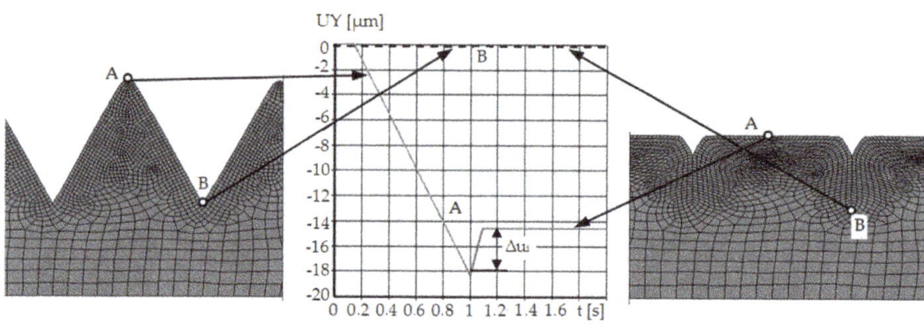

Figure 14. Diagram of displacement along the Y axis of selected nodes of asperity with a vertical angle in the process of crushing with a flat punch: A—top of asperity, B—bottom between asperities.

The phenomenon of elastic recovery of the material is visible and then the stabilization of the position of the node (A) (the peak of the roughness), which has stopped moving along the Y axis. Similar results were obtained for the cases of crushing the remaining asperities with different vertical angles. When analyzing the node located between the asperities (B) (on the bottom), no elastic recovery was found for the case of asperities with a constant vertical angle and height.

The difference between the individual simulation results concerns the displacement value, which is related to the burnishing depth. The burnishing depth resulted from the geometry of the outline, i.e., the vertical angle of the asperities, and thus its height. The distance of the asperity was the same in each case. The elastic recovery also changed its value—the greater the vertical angle of the roughness, the smaller the value of the elastic recovery.

Computer simulations made it possible to observe and quantify the amount of material elastic recovery during the burnishing process. It is possible to check the results at any time during the process. It was confirmed that with the increase of the value of the vertical angle of the surface roughness, the value of the elastic recovery of the material is smaller. Moreover, using the developed numerical model, it is possible to qualitatively and quantitatively determine how much the bottoms between surface asperities rise in the process of their crushing. In cases where the vertical angles of the asperities were equal to $\theta = 60°$ and $\theta = 75°$, the indentations of the asperities (bottom) did not rise, and the material core remained undeformed, and with complete deformation, deformed asperities are visible, separated from each other by gaps (planes of discontinuity) at depth $a = 0.5\ R_t$. In the range of vertical angle $\theta = 90$–$135°$, the plastic deformation zone increased to encompass the material core. The indentations between the asperities (bottom) rise (Figure 15), and with complete deformation, gaps are visible, but of less depth. For the angle $\theta = 150°$, deformations of the roughness and the core of the material occur, and the cavities between the asperities (bottoms) rise completely.

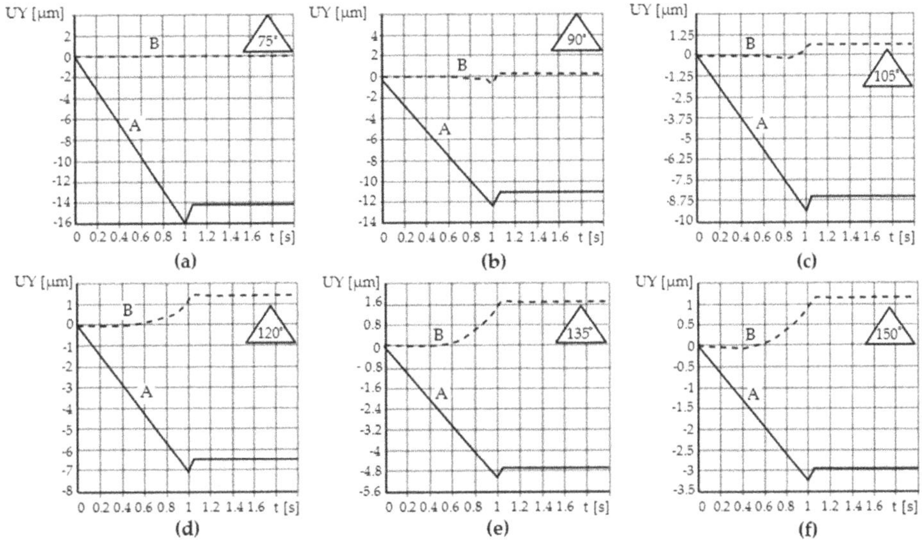

Figure 15. Diagram of displacement along the Y axis of selected asperity nodes with the vertical angle $\theta = 75°$ **(a)**, $\theta = 90°$ **(b)**, $\theta = 105°$ **(c)**, $\theta = 120°$ **(d)**, $\theta = 135°$ **(e)**, $\theta = 150°$ **(f)**, crushed with a flat punch: A—peak of asperity, B—bottom between asperities.

5. Conclusions

The phenomenon of elastic recovery of the material during burnishing is a very complicated issue. The presented results of model tests, simulations, and analyses do not reflect the complexity of the problem. They confirm that this phenomenon occurs and there are methods by which it is possible to try to quantify the dependence of the elastic recovery on, for example, the apex angle of the inequality or deviations of the inequality outline. In the conducted analyses, only the elastic recovery of the burnished material was taken into account. The spring-return occurring in the tool material was omitted, treating the tool as perfectly rigid. It should also be noted that the phenomenon of elastic recovery probably also takes place during the treatments preceding burnishing. It is not possible to continuously measure the elastic recovery during modeling or experimental studies. This possibility is provided by computer simulations using the finite element method. However, the history of the material was not taken into account in the simulations performed. Model tests and computer simulations are a cheaper way than carrying out experimental tests each time. They give a guideline to consider the material elastic recovery when designing the burnishing process, especially in the case of fine machining. Then, when calculating the burnishing depth, its value should be increased by the elastic recovery of the material. Based on the research, the following conclusions can be drawn:

- Knowledge about influence of determined profile parameters on the physical phenomena that occur in the burnishing process is unknown. It is important because they impact on the quality of the formed product's surface layer, tribological properties, and fatigue issues. Many physical phenomena whose observation or measurement is very difficult or impossible take place during processes of burnishing, such as the pressure in contact zones, the friction forces, strain and stresses, elastic recovery, the slip and contact zones, and states of displacement.
- After passing the burnishing tool, visco-plastic deformations remain in the workpiece, while the elastic deformations change and settle at the level corresponding to the new equilibrium state of the object. Thus, the states of displacements, stresses, and strains in the tested object during and after the end of the process differ significantly, because after the process is completed, the nodes are displaced in the elastic range. In the case of higher asperities, the displacement is greater than in the case of those with a height deviation. The opposite is the case for a distance deviation. These issues are very complex and require further analysis.
- The obtained regression equations of the dependence of the material elastic recovery after burnishing (Δu_0) from the deviations of height (Δh_t) and distance (Δs_t) of asperities after turning are reasonable for specific angles of asperities ($\theta = 60°$) for C45 steel.
- The increase of the deviation of the asperities' height causes a decrease in the value of elastic recovery, while the opposite situation occurs when the deviation of the distance was analyzed.
- Comparing the values of elastic recovery for an asperity with constant geometry, *A1*, and changeable geometry, *A2*, it can be seen that the value of elastic recovery was higher for asperity *A1*.
- It was confirmed that with the increase of the value of the vertical angle of the surface roughness, the value of the elastic recovery of the material is smaller. Moreover, using the developed numerical model, it is possible to qualitatively and quantitatively determine how much the bottoms between surface asperities rise in the process of their crushing.

Author Contributions: The individual contributions of the authors: conceptualization, A.K.; methodology, A.K. and Ł.B.; investigation, A.K. and Ł.B.; software, A.K.; data curation, Ł.B.; validation, A.K.; writing—original draft preparation, A.K. and Ł.B.; writing—review and editing, A.K. and Ł.B.; visualization, Ł.B.; supervision, A.K. All authors have read and agreed to the published version of the manuscript.

Funding: The authors gratefully acknowledge the financial support from the Koszalin University of Technology, funding No. 524.02.23, entitled "Analysis of the deformation state of burnished surface with taking into account surface layer state and nonlinearity of the process".

Conflicts of Interest: The authors declare no conflict of interest.

Nomenclature

$\Delta\phi^{(E)}$	increment of logarithmic strain in elastic stage
$\phi^{(E)}$	logarithmic strain in elastic stage
Δu_i	elastic recovery
θ	vertical angle
σ_p	yield stress

References

1. Luca, L.; Neagu-Ventzel, S.; Marinescu, J. Effects of working parameters on surface finish in ball-burnishing of hardened steels. *Precis. Eng.* **2005**, *29*, 253–256. [CrossRef]
2. El-Axir, M.H. An investigation into roller burnishing. *Int. J. Mach. Tools Manuf.* **2000**, *40*, 1603–1617. [CrossRef]
3. Hassan, A.M. The effects of ball- and roller-burnishing on the surface roughness and hardness of some non-ferrous metals. *J. Mater. Process. Technol.* **1997**, *72*, 385–391. [CrossRef]
4. Kułakowska, A. Problems of surface preparation under burnishing rolling in aspect of product quality. *Steel Res. Int.* **2010**, *81*, 218–221.

5. Patyk, R. Theoretical and experimental basis of regular asperities about triangular outline embossing technology. *Steel Res. Int.* **2010**, *81*, 190–193.
6. Przybylski, W. *Technologia Obróbki Nagniataniem*; WNT: Warszawa, Poland, 1987. (In Polish)
7. Kulakowska, A.; Kukielka, L. Numerical analysis and experimental researches of burnishing rolling process with taking into account deviations in the surface asperities outline after previous treatment. *Steel Res. Int.* **2008**, *2*, 42–48.
8. Patyk, R.; Kukielka, L. Optimization of geometrical parameters of regular triangular asperities of surface put to smooth burnishing. *Steel Res. Int.* **2008**, *2*, 642–647.
9. Kulakowska, A. Experimental researches of burnishing rolling process of regular surface asperities prepared in turning process. In Proceedings of the 14th International Conference on Metal Forming, Krakow, Poland, 16–19 September 2012; pp. 127–131.
10. Kukielka, L.; Kukielka, K. Numerical analysis of the physical phenomena in the working zone in the rolling process of the round thread. *WIT Trans. Eng. Sci.* **2007**, *55*, 125–134.
11. Patyk, R. New Method of Technological Surface Layer Shaping of Machine Parts during Duplex Burnishing Rolling Process. In Proceedings of the 14th International Conference on Metal Forming, Krakow, Poland, 16–19 September 2012; pp. 115–118.
12. Kułakowska, A.; Kukiełka, L. *Analiza Numeryczna Wpływu Odchyłek Zarysu Nierówności Powierzchni po Toczeniu na Odchyłki Zarysu Nierówności Powierzchni po Nagniataniu Powierzchniowym*; Projektowanie Procesów Technologicznych; Komisja Budowy Maszyn PAN: Poznań, Poland, 2006; pp. 171–178. ISBN 978-83-903808-7. (In Polish)
13. Li, F.L.; Xia, W.; Zhou, Z.Y.; Zhao, J.; Tang, Z.Q. Analytical prediction and experimental verification of surface roughness during the burnishing process. *Int. J. Mach. Tools Manuf.* **2012**, *62*, 67–75. [CrossRef]
14. Bucaille, J.L.; Felder, E.; Hochstetter, G. Mechanical analysis of the scratch test on elastic and perfectly plastic materials with the three-dimensional finite element modeling. *Wear* **2001**, *249*, 422–432. [CrossRef]
15. Jardret, V.; Zahouani, H.; Loubet, J.L.; Mathia, T.G. Understanding and quantification of elastic and plastic deformation during a scratch test. *Wear* **1998**, *218*, 8–14. [CrossRef]
16. Meza, J.H.; Moré Farias, M.-C.; Martinz de Souza, R.; Riaño, L.J.C. Using the ratio, maximum load over unload stiffness squared, Pm/Su2, on the evaluation of machine stiffness and area function of blunt indenters on depth-sensing indentation equipment. *Mater. Res.* **2007**, *10*, 437–447. [CrossRef]
17. Pelletier, H. Predictive model to estimate the stress-strain curves of bulk metals using nanoindentation. *Tribol. Int.* **2006**, *39*, 593–606. [CrossRef]
18. Stilwell, A.; Tabor, D. Elastic Recovery of Conical Indentations. *Proc. Phys. Soc.* **1961**, *78*. [CrossRef]
19. Zajkowska, A.; Derpeński, Ł.; Seweryn, A. Metody wyznaczania rzeczywistej krzywej umocnienia materiału, Cz. II. Test wciskania kulistego wgłębnika. *Acta Mech. Autom.* **2011**, *5*, 143–147. (In Polish)
20. Lee, J.D. A large-strain elastic-plastic finite element analysis of rolling process. *Comput. Met. Appl. Mech. Eng.* **1998**, *161*, 315–347. [CrossRef]
21. Zang, S.L.; Liang, J.; Guo, C. A constitutive model for spring-back prediction in which the change of Young's modulus with plastic deformation is considered. *Int. J. Mach. Tools Manuf.* **2007**, *47*, 1791–1797. [CrossRef]
22. Chodór, J. Modelowanie i Analiza Przemieszczania Materiału Obrabianego Podczas Mikroskrawania i Nagniatania Ślizgowego z Uwzględnieniem Nieliniowości. Ph.D. Thesis, Politechnika Koszalińska, Koszalin, Poland, 2011. (In Polish).
23. Skalski, K. *Analiza Zagadnienia Kontaktowego Ciała Sprężysto-Plastycznego (na Podstawie Zasad Wariacyjnych i Metod Elementów Skończonych)*; Wydawnictwa Politechniki Warszawskiej: Warszawa, Poland, 1979. (In Polish)
24. Gauthier, C.; Lafaye, S.; Schirrer, R. Elastic recovery of a scratch in a polymeric surface: Experiments and analysis. *Tribol. Int.* **2001**, *34*, 469–479. [CrossRef]
25. Yang, H.; Guo, L.; Zhan, M.; Sun, Z. Research on the influence of material properties on cold ring rolling processes by 3D-FE numerical simulation. *J. Mater. Process. Technol.* **2006**, *177*, 634–638. [CrossRef]
26. Schulze, V.; Bleicher, F.; Groche, P.; Guo, Y.B.; Pyun, Y.S. Surface modification by machine hammer peening and burnishing. *CIRP Ann.* **2016**, *65*, 809–832. [CrossRef]
27. Janczewski, Ł.; Toboła, D.; Brostow, W.; Czechowski, K.; Kot, M.; Zagórski, K. Effects of ball burnishing on surface properties of low density polyethylene. *Tribol. Int.* **2016**, *93*, 36–42. [CrossRef]
28. Nestler, A.; Schubert, A. Roller burnishing of particle reinforced aluminium matrix composites. *Metals* **2018**, *8*, 95. [CrossRef]

29. Sánchez Egea, A.J.; Rodríguez, A.; Celentano, D.; Calleja, A.; López de Lacalle, L.N. Joining metrics enhancement when combining FSW and ball-burnishing in a 2050 aluminium alloy. *Surf. Coat. Technol.* **2019**, *367*, 327–335. [CrossRef]
30. Juanjuan, H.; Wei, Z.; Guangchun, W. Investigation of influence factors on surface roughness of micro-scale features. *Precis. Eng.* **2019**, *56*, 524–552.
31. Kuznetsov, P.; Tarasov, S.Y.; Dmitriev, A.I. Nanostructuring burnishing and subsurface shear instability. *J. Mater. Process. Technol.* **2015**, *217*, 327–335. [CrossRef]
32. Denkena, B.; Grove, T.; Maiss, O. Surface texturing of rolling elements by hard ball-end milling and burnishing. *Int. J. Adv. Manuf. Technol.* **2017**, *93*, 3713–3721. [CrossRef]
33. Amini, C.; Jerez-Mesa, R.; Travieso-Rodriguez, J.A.; Llumà, J.; Estevez-Urra, A. Finite element analysis of ball burnishing on ball-end milled surfaces considering their original topology and residual stress. *Metals* **2020**, *10*, 638. [CrossRef]
34. Skoczylas, A.; Zaleski, K. Selected properties of the surface layer of C45 steel parts subjected to laser cutting and ball burnishing. *Materials* **2020**, *13*, 3429. [CrossRef]
35. Llumà, J.; Gómez-Gras, G.; Jerez-Mesa, R.; Rue-Mascarell, J.; Travieso-Rodriguez, J.A. Mechanical strengthening in S235JR steel sheets through vibration-assisted ball burnishing. *Metals* **2020**, *10*, 1010. [CrossRef]
36. Hill, R. *The Mathematical Theory of Plasticity*; Oxford University Press: Oxford, UK, 1986.
37. Szczepiński, W. *Mechanika Plastycznego Płynięcia*; Małe Monografie; PWN: Warszawa, Poland, 1978. (In Polish)
38. Johnson, W. *Impact Strength of Materials*; Edwards Arnolds: London, UK, 1972.
39. Kukiełka, L. *Teoretyczne i Doświadczalne Podstawy Powierzchniowego Nagniatania Tocznego z Elektrokontaktowym Nagrzewaniem*; Monografia Wydziału Mechanicznego Wyższej Szkoły Inżynierskiej w Koszalinie no. 47; Wyższa Szkoła Inżynierska w Koszalinie: Koszalin, Poland, 1994. (In Polish)
40. Kukiełka, K.; Kukiełka, L.; Olender, J.; Kuchta, R. *Numerical Analysis of State of Deformation and Stress in Process of Trapezoidal Rolling*; Zeszyty Naukowe Wydziału Mechanicznego no. 33; Politechnika Koszalińska: Koszalin, Poland, 2004; pp. 167–178.
41. Kulakowska, A. *Modelowanie, Analiza i Prognozowanie Cech i Wyników Procesu Nagniatania Naporowego Tocznego Warstwy Wierzchniej o Zdeterminowanej Strukturze Geometrycznej Powierzchni*; Monografia no. 307; Politechnika Koszalińska: Koszalin, Poland, 2016. (In Polish)
42. Kowalczyk, L. *Modelowanie Fizykalne Procesów Obróbki Plastycznej*; Instytutu Technologii Eksploatacji: Radom, Poland, 1995. (In Polish)
43. Dawidenkow, N.N. *Izbrannyje Trudy*; Naukowa Dumka: Kijew, Ukraine, 1981; Volume 1. (In Ukrainian)
44. Kukiełka, L. *Podstawy Badań Inżynierskich*; Politechnika Koszalińska: Koszalin, Poland; PWN: Warszawa, Poland, 2002. (In Polish)
45. Patyk, R. Kształtowanie Regularnych Symetrycznych Nierówności Trójkątnych w Procesie Nagniatania Naporowego tocznego Wałków Stalowych. Ph.D. Thesis, Koszalin University of Technology, Koszalin, Poland, 2006.
46. Kukiełka, L.; Geleta, K.; Kukiełka, K. Modelling of initial and boundary problems with geometrical and physical nonlinearity and its application in burnishing processes. In Proceedings of the 14th International Conference Metal Forming, Krakow, Poland, 16–19 September 2012; Wiley-VCH Verlag GmbH & Co.KGaA: Weinheim, Germany, 2012; pp. 1375–1378, ISBN 978-3-514-00797-0.
47. Kukiełka, L. New damping models of metallic materials and its application in non-linear dynamical cold processes of metal forming. In Proceedings of the 13th International Conference Metal Forming 2010, Toyohashi, Japan, 19–22 September 2010; Verlag Stahleisen GmbH: Düsseldorf, Germany, 2010; pp. 1482–1485, ISBN 978-3-514-00774-1.
48. Kukiełka, L.; Geleta, K.; Kukiełka, K. Modelling and analysis of nonlinear physical phenomena in the burnishing rolling operation with electrical current. In Proceedings of the 14th International Conference Metal Forming, Krakow, Poland, 16–19 September 2012; Wiley-VCH Verlag GmbH & Co.KGaA: Weinheim, Germany, 2012; pp. 1379–1382, ISBN 978-3-514-00797-0.

49. Kukiełka, L.; Kukiełka, S. *Zastosowanie Programu Experiment Planner 1.0 do Wspomagania Eksperymentu w Technologii Nagniatania, Obróbka Kształtująca i Powierzchniowa Nagniataniem*; Zeszyty Naukowe Wydziału Mechanicznego nr 34; Wydawnictwo Uczelniane Politechniki Koszalińskiej: Koszalin, Poland, 2004; pp. 131–140. (In Polish)
50. Kukielka, L. Mathematical modelling and numerical simulation of non-linear deformation of the asperity in the burnishing cold rolling operation. In *Computational Methods in Contact Mechanics V, Proceedings of the 5th International Conference on Computational Methods in Contact Mechanics, Seville, Spain, 7 June 2001*; Book Series: Computational and Experimental Methods; WIT Press: Ashurst, UK, 2001; pp. 317–326.

Publisher's Note: MDPI stays neutral with regard to jurisdictional claims in published maps and institutional affiliations.

materials

Article

Tensile Deformation Behavior of Typical Porous Laminate Structure at Different Temperatures

Ping Wang, Ye-Da Lian * and Zhi-Xun Wen

School of Mechanics, Civil Engineering and Architecture, Northwestern Polytechnical University, Xi'an 710072, China; wangpingping@mail.nwpu.edu.cn (P.W.); zxwen@nwpu.edu.cn (Z.-X.W.)

* Correspondence: lianyeda@nwpu.edu.cn

Received: 8 November 2020; Accepted: 25 November 2020; Published: 26 November 2020

Abstract: In this study, the Ni-Cr-W superalloy GH3230 is used as the test material. According to the actual structure of the flame tube, a porous laminate structure specimen is designed. The structure consists of impact holes, overflow holes and pin fins. High-temperature tensile tests at 650 °C, 750 °C and 850 °C were carried out to study the high-temperature mechanical properties and fracture mechanism of the specimens of porous laminate structure, and the strain nephogram of the specimens were obtained by digital image correlation (DIC) technique. Due to the large number and dense arrangement of overflow holes, an obvious hole interference effect can be found from the strain nephogram. The stress concentration around the pore and the interference between the pores provide priority places and paths for the initiation and propagation of microcracks. The test found that the microcracks of the porous laminate structure first occurred around the hole, the overflow surface fractured first, after which the impact surface fractured. The strength of the alloy exhibits a significant temperature sensitivity to temperature. From 650 °C to 750 °C, the ultimate strength (σ_b) and yield strength ($\sigma_{0.2}$) decrease slightly, but they decrease significantly at 850 °C. The microstructure of the fracture surface shows that all microcracks occur at the interface between the matrix and the carbides but that the fracture mode of the specimens gradually changes from intergranular fracture to transgranular fracture as the temperature increases. Due to the pinning effect of the intracrystalline diffusive solute atoms on the dislocations, the stress-strain curves of the high-temperature tensile tests at 650 °C and 750 °C showed zigzag characteristic fluctuations during the strengthening stage.

Keywords: digital image correlation technology; Ni-Cr-W superalloy; microstructure; fracture mechanism

1. Introduction

In general, in order to make superalloys suitable for long-term use under critical temperature and stress conditions, a large number of alloying elements such as chromium, molybdenum, boron and C are added to the alloys [1,2]. Among them, the Ni-Cr-W superalloy is a kind of nickel based deformation superalloy with solid solution strengthening and carbide dispersion strengthening. A large amount of refractory elements such as W and Cr are added to the alloy to improve the strength of the matrix, and a small amount of C element is added to form dispersed carbides to hinder the grain growth and strengthen the grain boundary [3–5]. Therefore, it has excellent fatigue creep mechanical properties, corrosion resistance and oxidation resistance, and is widely used in gas turbine and aero-engine components, such as blades, flame tubes, etc. [6–9].

At present, some research has been done on the microstructure and mechanical properties of Ni-Cr-W superalloys. Han y et al. [10] studied the effect of carbides formed by grain boundary element segregation on the mechanical properties and microstructure of Ni-Cr-W superalloys. It is found that, with the increase of the heat exposure time, the morphology of M_6C carbide changes from grain to strip, which reduces the mechanical properties of the material. Hu and Tang et al. [11,12] studied the precipitation behavior of $M_{23}C_6$-type carbides at the grain boundary of a Ni-Cr-W superalloy and their effect on the mechanical properties. The results show that the large interfacial energy at the large angle grain boundary is conducive to the formation of $M_{23}C_6$-type carbides and that the decrease of the tensile strength and yield strength is mainly caused by the fracture of the $M_{23}C_6$-type carbide. Hao et al. [13] studied the effect of the C element content on the microstructure and fracture properties of Ni-Cr-W-Fe superalloys and found that a high C element content was prone to forming $M_{23}C_6$ carbides, which accelerated the crack along the grain boundary extension. MQ et al. [14] studied the effect of grain refinement on the microstructure and mechanical properties of Ni-Cr-W superalloys. Bai et al. [15] studied the effect of temperature on the tensile mechanical properties of Ni-Cr-W superalloys. In summary, although there have been some studies on the mechanical properties of this type of superalloy, most research work has not considered the impact of the real structure in engineering practice on the material properties, especially the high-temperature mechanical properties. With the increase of the turbine inlet temperature and the increase of engine efficiency, higher requirements are put on the operating temperature of hot components such as blades and combustion chambers. At present, film-cooling hole cooling methods are mostly used. However, the existence of dense vent membrane holes makes the structural stress state complex, and cracks are more likely to occur around the holes [16,17]. Therefore, it is necessary to understand the mechanical properties of the film-cooling holes' structure.

In this paper, according to the actual structure of the combustion chamber, the porous laminate structure was designed with a Ni-Cr-W superalloy GH3230, and its high temperature tensile mechanical properties were studied at different temperatures. The surface strain field of the structure was obtained by Digital image correlation (DIC) technology.

2. Experimental Material and Process

2.1. Material and Specimens

The deformed superalloy GH3230 was used in the present experiment. The experimental GH3230 superalloy materials are all from Beijing Institute of Aerial Materials (Beijing, China). The nominal chemical composition is given in Table 1. The ingot was prepared by vacuum induction furnace + electroslag remelting, and then the microstructure was homogenized by standard heat treatment (1310 °C/1 h + 1320 °C/2 h + 1330 °C/2 h + 1340 °C/4 h, 1120 °C/4 h (AC), 870 °C/32 h (AC) (AC: air cooling)). Finally, the original material was prepared by a cold rolling process. The test alloy is a Ni-Cr-based solid solution strengthened deformation superalloy, which is an isotropic polycrystalline material. After being grinded, polished and electrochemically etched in a solution of 95 mL HCL + 10 mL CH_3COOH (by using a voltage of 4 V and corrosion time of 10 s), the grain distribution and microstructure of the specimen were observed through a scanning electron microscope (SEM), and the results are shown in Figure 1. The microstructure of the alloy is mainly composed of γ matrix, M_6C carbide and $M_{23}C_6$ carbide (Figure 1a). Granular M_6C carbide is the primary carbide, which is uniformly distributed in the alloy matrix (Figure 1c). The microstructure of $M_{23}C_6$ carbide is linear or lamellar, mainly distributing at the grain boundary (Figure 1d). The carbide is enriched in some regions of the grain boundary (Figure 1b). According to the results of the EDS analysis, M_6C carbide mainly contains Ni, W, Co, W, etc., while $M_{23}C_6$ type carbide is mainly enriched in Ni, W, Cr, etc., as shown in Figure 1e,f.

Table 1. Chemical composition of the GH3230 alloy (wt.%).

Element	C	Cr	Ni	Co	W	Mo	Al	Ti
Mass fraction/%	0.05–0.15	20.00–24.00	balance	≤5.00	13.00–15.00	1.00–3.00	0.20–0.50	≤0.10
Element	**Fe**	**La**	**B**	**Si**	**Mn**	**S**	**P**	**Co**
Mass fraction/%	≤3.00	0.005–0.05	≤0.015	0.25–0.75	0.30–1.00	≤0.015	≤0.05	≤0.50

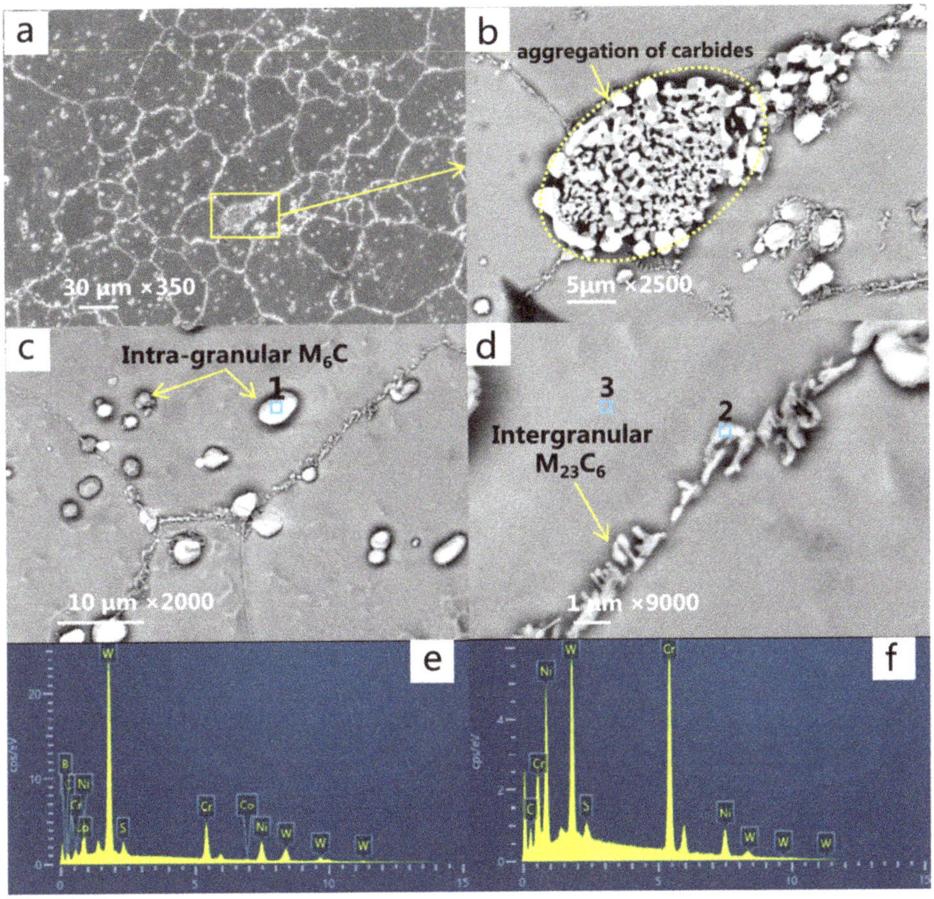

Figure 1. Scanning electron microscope (SEM) images of the microstructure of the GH3230 superalloy: (**a**) microstructure and morphology; (**b**) segregation and aggregation of intergranular carbides; (**c**) morphology and distribution of M_6C-type carbides in grains; (**d**) morphology and distribution of $M_{23}C_6$ type between the grains; (**e**,**f**) the results of the EDS analysis at positions 1 and 2 in the figure, respectively.

In order to simulate the structure of the flame tube in actual engineering and in combination with its structural characteristics, a double-layer thin wall plate structure is designed. The structure is composed of pin fins and two kinds of film-cooling holes: overflow holes and impact holes. The number and arrangement of the two kinds of holes are different. In order to eliminate the influence of the recast layer on crack initiation around the holes, picosecond laser drilling technology is used for punching. The diameter of the overflow hole is 1.5 mm, the arrangement is a triangle symmetry, and the plate thickness is 0.5 mm. The hole diameter of the impact hole is 0.9 mm, the arrangement mode is a rhombic distribution, and the plate thickness is 0.8 mm. The pin fin connecting the two layers is a cylinder with a diameter of 0.15 mm and a height of 0.25 mm. The actual picture and detailed size of the specimen are shown in Figure 2.

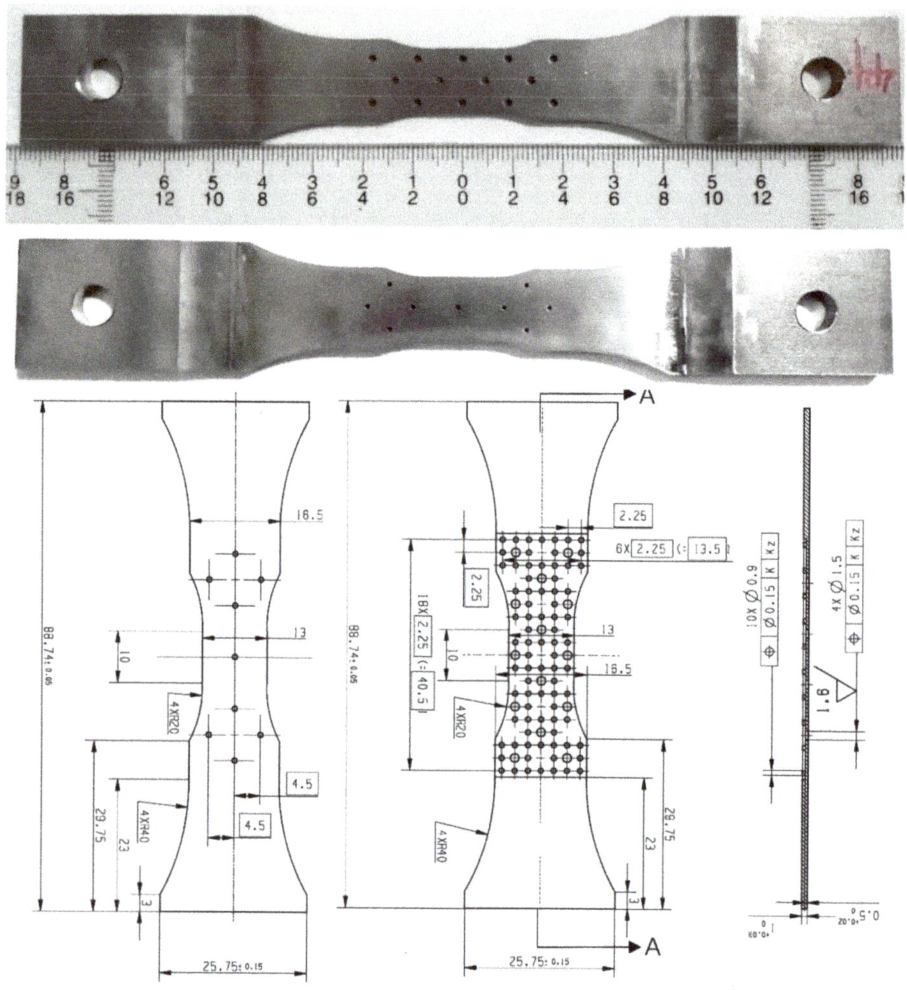

Figure 2. Physical picture and geometry schematic of the specimens (unit: mm).

2.2. Experimental Procedure

In order to study the mechanical behavior and microstructure evolution of porous laminates under a high-temperature, uniaxial tensile condition, the experiments at 650 °C, 750 °C and 850 °C were carried out on a high-temperature electronic creep testing machine (RDL100, Changchun Machinery Experimental Equipment Co., Ltd., Changchun, China) according to the actual working conditions, and each condition was tested twice. The test process is controlled by displacement, and the tensile rate is 0.2 mm/min. During the test, a K-type thermocouple was attached to the upper, middle and lower sections of the specimen to monitor the temperature in real time. In order to study the interference effect between film-cooling holes and the thermal strength of the structure, digital image correlation (DIC) (XTDIC Version 8.2.1, Xi'an Jiaotong University, Xi'an, China) was used to measure the strain field on the surface of the specimen. Before the test, high-temperature paint was sprayed on the surface of the test piece to make the speckle area. During the whole experiment, a high-temperature lens was used to collect images every 10 s. The images are processed by the DIC system, and the changes of the gray value in the speckle region are calculated. The strain data of the specimen surface is obtained, and the strain nephogram is finally outputted. The schematic diagram of the test process and the preformed speckle pattern of the test piece surface are shown in Figure 3. After the test, the yield strength, ultimate tensile strength and elongation were obtained. In order to observe the special microstructure of the grain boundary, carbides and matrix, a metallographic analysis was carried out by grinding, polishing and electrochemical corrosion of the test alloy. The electrochemical corrosion was carried out in 95 mL HCl (w% ≥ 99.5%) + 10 mL CH3COOH (w% ≥ 99.5%) with a voltage of 4 V and corrosion time of 10 s. The microstructure and microcracks at the fracture surface were observed by optical microscope (OM) (VHX-6000, KEYENCE, Osaka, Japan) and scanning electron microscope (SEM) (ZEISS Gemini300, Carl Zeiss AG, Oberkochen, Germany). The high temperature failure mechanism is explained.

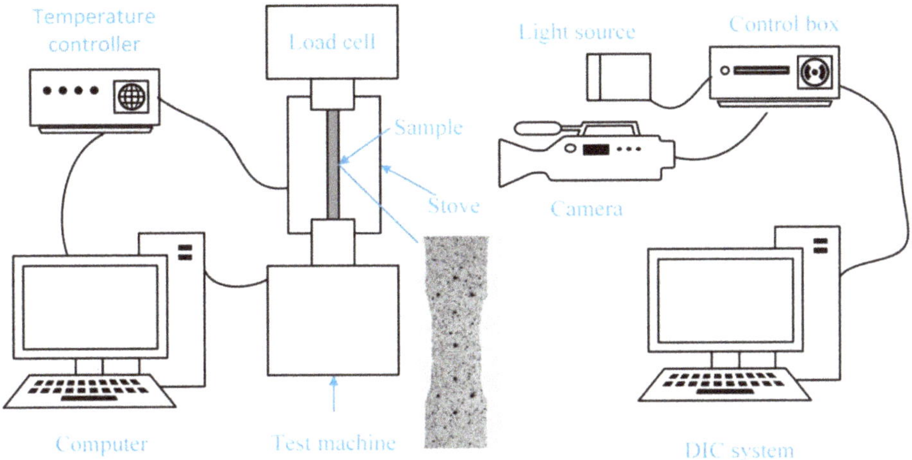

Figure 3. Schematic diagram of the test process and preformed speckle pattern on the surface of the specimens.

3. Experimental Results and Discussion

3.1. Tensile Mechanical Behavior at High Temperature

Since the cross-sectional area of the specimen varies along the tensile direction, the minimum cross-sectional area is used to calculate the stress, and the strain is calculated based on the distance between the two most distant holes along the tensile direction of the specimen. The high-temperature tensile stress-strain curves are shown at different temperatures in Figure 4. The specific high-temperature mechanical parameters of the tested alloy are shown in Table 2. The results show that the specimen has no obvious yield stage under all temperature conditions, so that the stress value corresponding to the 0.2% plastic deformation of the test piece is used as its yield strength ($\sigma_{0.2}$).

From the stress-strain curve, it can be seen that the mechanical properties of the specimen at 650 °C are the most excellent, with the best values for the yield strength ($\sigma_{0.2}$) and ultimate strength (σ_b) but with a lower elongation. The mechanical properties at 750 °C are slightly lower than those at 650 °C, but at 850 °C the yield strength ($\sigma_{0.2}$) and ultimate strength (σ_b) of the material are significantly reduced, and there is no strengthening stage. Compared with 650 °C and 750 °C, the yield strength ($\sigma_{0.2}$) and ultimate strength (σ_b) at 850 °C are reduced by 20.6% and 52.4%, respectively; however, the elongation (δ) is 1.18 and 1.06 times that at 650 °C and 750 °C, respectively. The variation trends of $\sigma_{0.2}$, σ_b and δ at different temperatures are shown in Figure 4b. It can be inferred that the temperature has no significant effect on the elastic deformation ability of materials. In addition, the stress-strain curves show obvious zigzag fluctuations in the strengthening stage at 650 °C and 750 °C. This phenomenon is usually related to the portevin Le Chatelier effect based on dynamic strain aging (DSA), that is, the pinning effect of diffused solute atoms on dislocations [18–20].

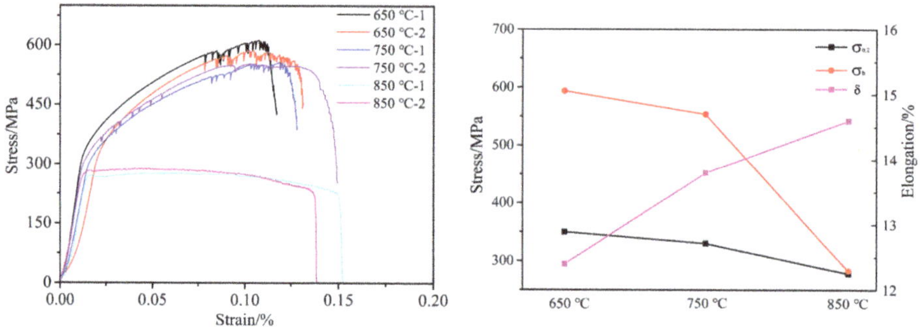

Figure 4. Tensile test results for different temperatures: 650 °C/750 °C/850 °C. (**a**) stress-strain curve of the high-temperature tensile experiment; (**b**) the change trend of high-temperature tensile mechanical properties parameters ($\sigma_{0.2}$, σ_b and δ) with the test temperature.

Table 2. Mechanical characteristics of the alloys determined at different testing temperatures.

Temperature	Yield Stress (MPa)	Ultimate Stress (MPa)	Average Yield Stress ($\sigma_{0.2}$) (MPa)	Average Ultimate Stress (σ_b) (MPa)	Average Elongation (δ/%)
650 °C	368.99	582.60	350.0	593.8	12.4
	331.08	605.08			
750 °C	332.57	555.16	330.2	553.9	13.8
	327.80	552.68			
850 °C	271.98	277.34	278.0	282.9	14.6
	284.09	288.37			

The studies in [15,21] found that below 800 °C the yield strength ($\sigma_{0.2}$) and ultimate tensile strength (σ_b) of the Ni-Cr-W deformed superalloy decreased slightly as the temperature increased, but that above 800 °C the ultimate tensile strength (σ_b) dropped drastically, which is consistent with the high-temperature tensile results in our study. Therefore, being the same kinds of material, the micromechanism should be the same. At a high temperature, the main reason for this phenomenon is the existence of the secondary phase $M_{23}C_6$ at the grain boundary, which reduces the strength of the grain boundary. As a result, cracks initiate at the junction of $M_{23}C_6$ and the grain boundary, and then propagate, resulting in alloy failure. In addition, the high dislocation density and cross slip of the internal structure of the alloy at a high temperature are also responsible for the phenomenon. Therefore, this kind of material shows a certain sensitivity to temperature. When the temperature is too high, this will lead to the rapid deterioration of its thermal strength.

3.2. Mode of Crack Propagation and Strain Field around Holes

Macroscopic fracture paths are similar at three temperatures. From Figure 5a–c, it can be found that both sides of the test pieces are broken from the middle part of the test piece and that the initial cracks are generated at the middle film-cooling holes. However, due to the differences in the size and arrangement of the impact holes and overflow holes, the crack propagation paths on both sides of the specimen are different. Cracks on the surface of the overflow holes are first generated on the middle overflow hole and expand outward along the direction perpendicular to the loading axis. When the cracks develop toward the edge of the specimen, the edge breaks. The inwardly expanding cracks are in a zigzag shape, which first expands along the direction perpendicular to the load axis, then expands along the 45° direction connecting the two holes, and then returns to the direction perpendicular to the load axis. The cracks on the plate with impact holes are first generated on the middle hole before expanding along the direction perpendicular to the loading axis until they extend to the edge, breaking the entire impact surface. The "zigzag" fracture mode of the plate with overflow holes is mainly caused by the serious stress concentration around the hole at the beginning, resulting in crack initiation. Then, due to the interference effect between the holes, the stress level in the direction 45° to the loading direction is higher, and the crack propagates along the 45° direction. When the crack propagates for a certain distance, the axial stress of the specimen begins to dominate the fracture process, and the crack will propagate along the path that is perpendicular to the loading axis until the plate with overflow holes breaks. Due to the small diameter and sparse layout of the impact holes, the interference effect between the holes is not obvious, causing the cracks to expand along the direction perpendicular to the loading axis until the plate with impact holes fractures. The micromorphology of the overflow hole after the tensile test at 850 °C is shown in Figure 5d–f. It can be seen from the figure that the circular overflow hole has been elongated into an ellipse. Compared with the original test piece, the edge of the overflow hole parallel to the loading direction was complete, but a large number of microcracks were generated at the edge of the hole perpendicular to the direction of the loading axis. Furthermore, the crack initiation position was at the grain boundary. This is mainly due to the machining defects around the pores and the presence of intergranular carbides, which provide a place for microcracks and the nucleation of micropores [13].

The images taken by the high-temperature camera were processed by the DIC system, and the strain nephograms at each temperature were obtained, as shown in Figure 6. Through the strain field nephogram, it can be found that there are obvious hole interference effects between the overflow holes under the three temperature conditions, which shows that there is an obvious high strain band area between the adjacent holes. The results show that the banded zone is 45°, −45° and 90° to the tensile direction and that the maximum strain occurs around the hole where the strain band intersects. It can be seen in Figure 6 that when $\sigma = \sigma_{0.2}$, the strain bands are clear. When the ultimate strength is reached, the width of the strain band increases and the high strain area becomes larger. Among these areas, the high strain region between the two big holes in the middle expands the fastest, which makes the middle part of the specimen become a high strain area. Due to the rapid propagation of cracks, the stress drops sharply, but the strain increases slowly. From the strain nephogram, it can also be found that the cracks first occur at the edge of the two holes in the middle before extending to both ends. The propagation of the cracks leads to a sharp increase in the strain in the region between the two holes, which is obviously higher than that in the other regions. When the cracks between the two holes are connected, the plate with overflow holes breaks, and then, 2 to 3 s later, the plate with impact holes fractures, and the whole specimen finally breaks.

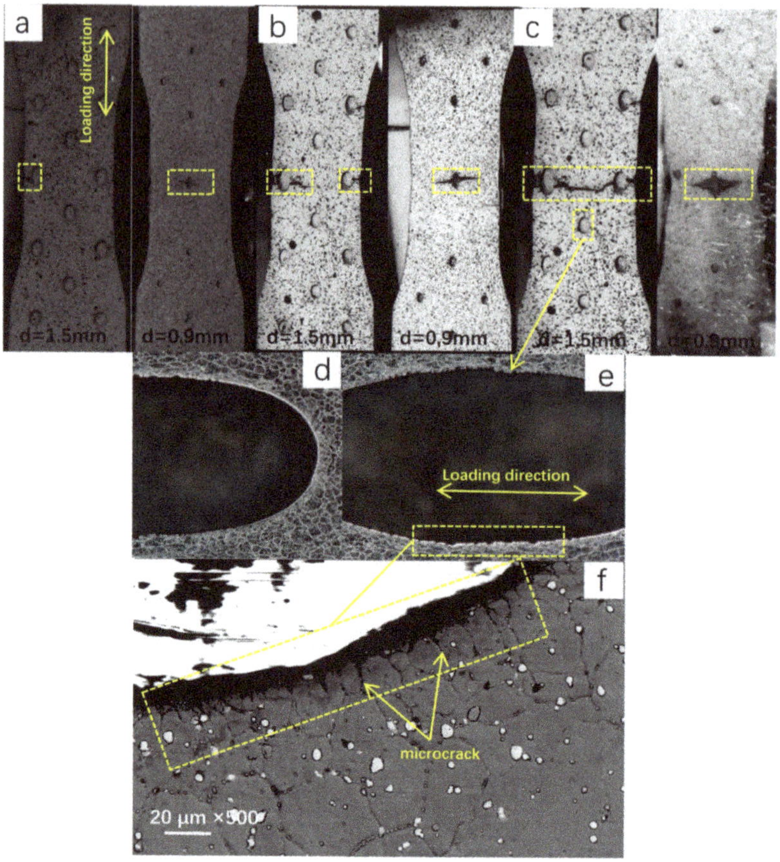

Figure 5. Images of test pieces during testing: (**a**) 650 °C; (**b**) 750 °C; (**c**) 850 °C; (**d**–**f**) microcracks around the hole at 850 °C.

By observing the strain nephogram of the plate with impact holes, it is found that the hole interference effect also exists between the impact holes but that it is not as obvious as that between the overflow holes. This is related to the size, density and layout of the impact holes. The strain nephogram has obvious differences at different temperatures. At 750 °C, because the distance between the impact holes arranged in a diamond shape at the upper and lower ends of the specimen is small, strain bands of ±45° and 90° to the tensile direction are formed between the holes when yielding. However, at 650 °C, due to the low temperature and high strength of the material, there is no obvious difference in the strain of each part of the whole face, while at 850 °C the strain of the whole surface is relatively large, and there is also no obvious difference in strain due to the low strength of the material. Because the isolated impact hole is far away from other impact holes, there is no strain band between the isolated hole and other holes under the three temperature conditions. However, the strain around the isolated hole is the greatest. As time increases, the strain in the middle part is obviously higher than for the other parts. The crack propagates from the periphery of the hole along the high strain region to both ends. The fracture mode shown by the strain nephogram is consistent with the actual fracture mode of the specimen in the test.

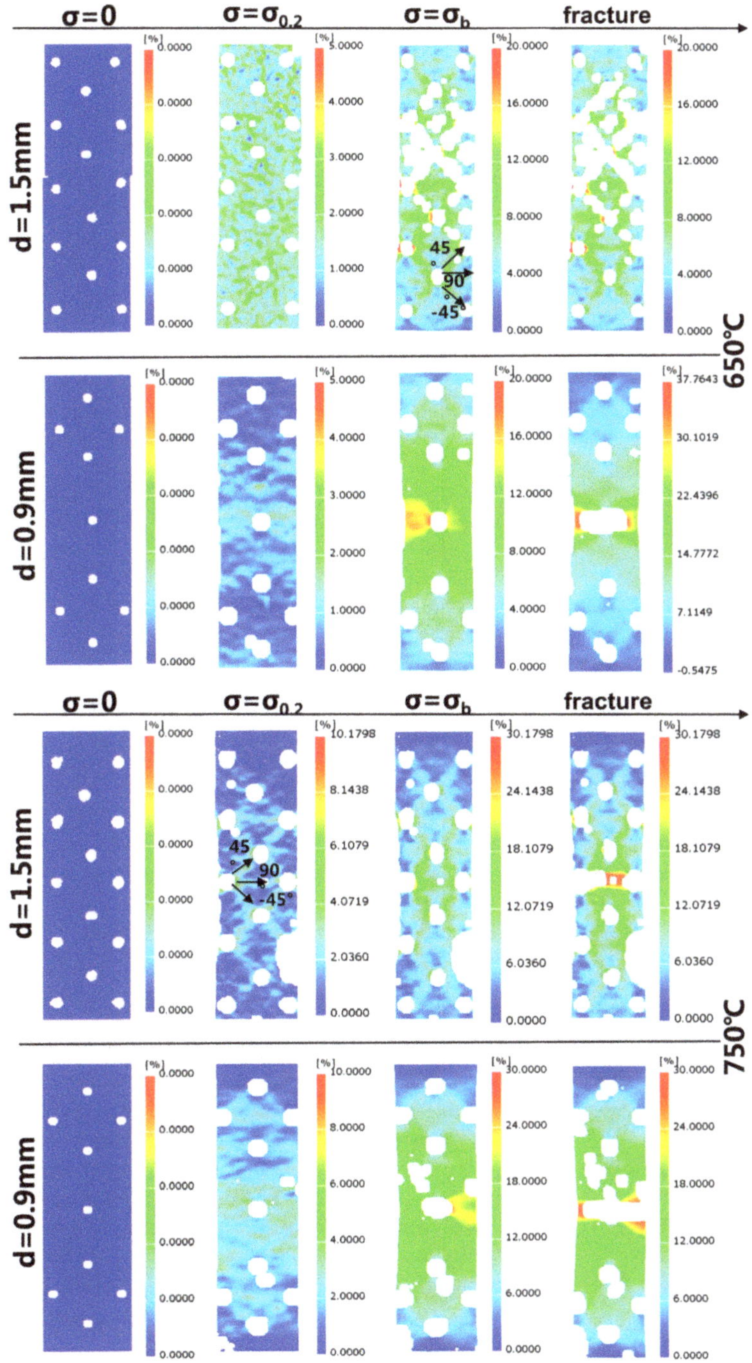

Figure 6. *Cont.*

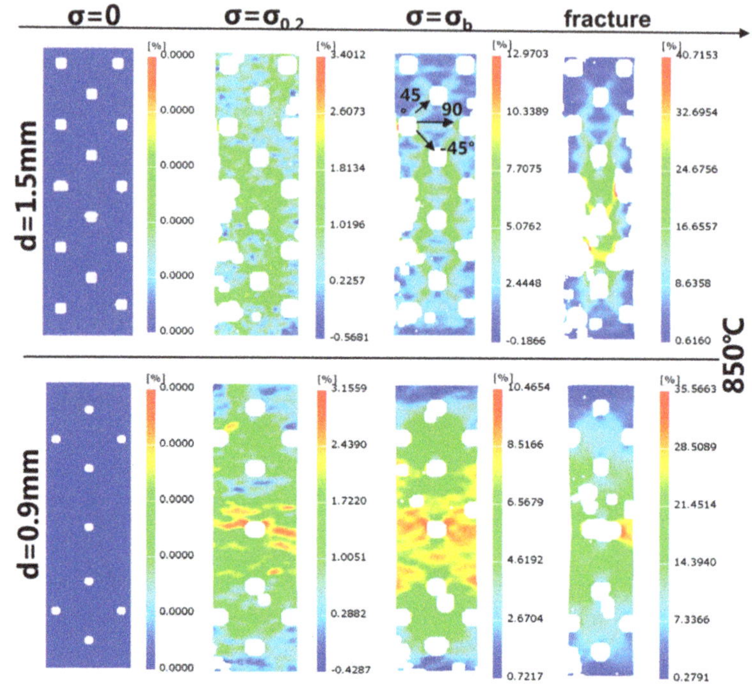

Figure 6. DIC strain nephograms of the specimen plates with overflow holes (d = 1.5 mm) and impact holes (d = 0.9 mm) at three different stress state stages ($\sigma = 0/\sigma = \sigma_{0.2}/\sigma = \sigma_b$) and, before fracture, under three temperature conditions.

3.3. Fracture Analysis

Figure 7b,e,g shows the fracture morphology seen under the optical microscope under different temperature conditions. It can be seen that the macromorphology corresponds to the fracture mode described in Section 3.2. Under the interference effect between the holes, a crack propagation path appears along the strain band on the plate with overflow holes, while there is no obvious hole interference effect on the plate with impact holes. The fracture morphologies at 650 °C and 750 °C are similar, with the fracture being straight and the Diameter Shrinkage phenomenon not being obvious. However, the fracture surface at 850 °C shows an obvious Diameter Shrinkage phenomenon, and the fracture deformation is serious, which indicates that the plasticity of the material is significantly improved at a high temperature. The fracture surface is observed after electrochemical corrosion, as shown in Figure 7c,f,i, where it can be clearly seen that the fracture at 650 °C is a typical intergranular fracture, the fracture at 750 °C shows the mixed fracture characteristics of an intergranular fracture and transgranular fracture, and the fracture at 850 °C is almost all transgranular fracture. Figure 8 shows the microstructure of the fracture surface and fracture side under SEM. The fracture mode can also be identified by Figure 8a–f, which is consistent with the results observed by optical microscope. In addition, by comparing the microstructure of the fractured test pieces with the microstructure of the material before the test, it can be found that, after the high-temperature tensile test, the amount of carbide on the alloy surface increases when compared with the original structure, but the size decreases slightly. At all temperature conditions, there are no significant differences between the carbides near the fracture surface and those around the holes, but the grains near the fracture surface are obviously elongated. Figure 8g–i shows the micromorphology of the fracture surface after the high-temperature tensile test. It can be seen that all the fractures are ductile dimple fractures, and a large number of second-phase particles are uniformly distributed on the fracture surface. The EDS analysis confirmed that the second-phase particles were granular M_6C carbides. At the same temperature, the grain size of the plate with overflow holes is larger than that of the plate with impact holes. This is due to the large number and dense arrangement of overflow holes, which makes the stress on the plate with overflow holes larger under the same loading conditions.

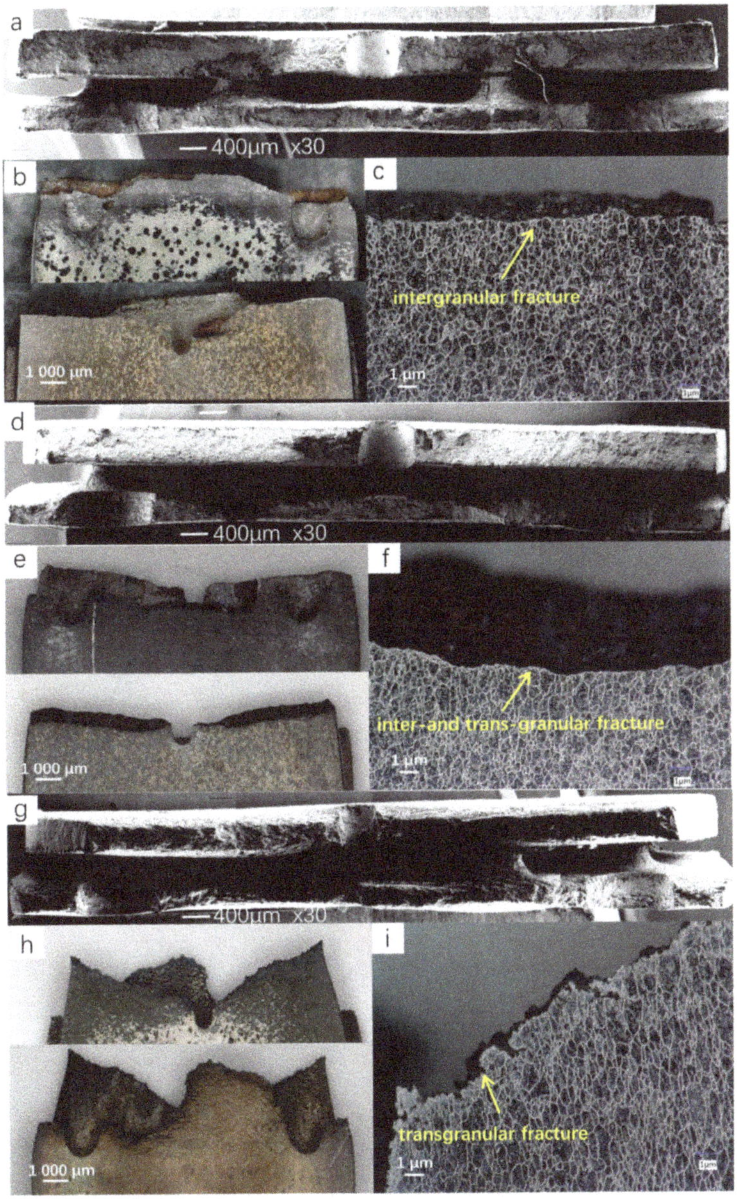

Figure 7. Fracture morphology and side morphology after electrochemical corrosion. (**a**–**c**) 650 °C; (**d**–**f**) 750 °C and (**g**–**i**) 850 °C.

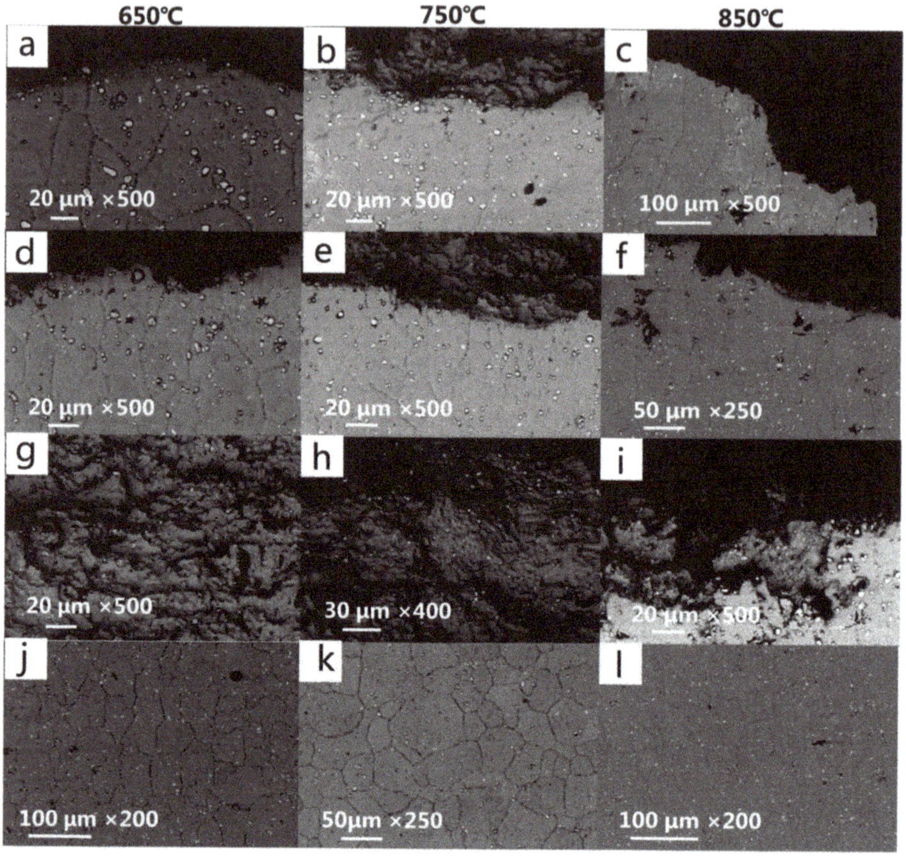

Figure 8. Fracture morphology at 650 °C/750 °C/850 °C. (**a–c**) Inclined fracture view of the surface of the plate with overflow holes and (**d–f**) the surface of the plate with impact holes; (**g–i**) fracture surface morphology and (**j–l**) microstructure near the fracture surface.

3.4. High-Temperature Fracture Mechanism of Porous Laminates

The unique structure, which can simulate actual engineering, has a significant influence on the mechanical properties of the test pieces. The effective cross-sectional area is reduced by 73.2–84.6% due to the lamellar and porous structure, resulting in a significant loss of tensile mechanical properties. According to the strain field measured by DIC in Section 3.2, it can be seen that there is an obvious stress concentration area around the holes. Moreover, due to the large diameter and dense arrangement of the overflow holes, there is a significant pore interference effect between the overflow holes under the three temperature conditions. It is shown that there are obvious high strain banded regions between adjacent holes. The direction of the strain band is ±45° and 90° to the tensile direction, and the distribution is symmetrical. The strain around the holes where the deformation bands intersect is the largest, which provides a place for the initiation of cracks on the surface with overflow holes. When the initial microcrack is generated around the holes, the propagation direction of the microcrack is consistent with that of the interference band. When the crack propagates for a certain distance, the effective cross-sectional area decreases, and the stress on the surface with impact holes increases rapidly. The cracks are produced at both ends of the impact hole in the center of the test piece and propagate rapidly to both ends. Then, the cracks on both sides extend outward at the same time until they penetrate the whole test piece, resulting in its fracture failure. The pin fins between the laminates did not bear a direct axial force and did not undergo significant deformation, but they still played a certain role in the strength of the structure itself, which was mainly manifested by bonding the plate with overflow holes and impact holes in order to improve the stiffness of the laminate structure.

According to the high-temperature tensile stress-strain curve of the alloy in Figure 4, it can be seen that the test alloy shows a significant sensitivity to temperature. The ultimate strength (σ_b) and yield strength ($\sigma_{0.2}$) of the specimens are higher at 650 °C and 750 °C but decrease significantly at 850 °C, while the elongation (δ) increases significantly. When the temperature exceeds a certain critical value, the thermal strength of the material will deteriorate rapidly, and the fracture mode of the material will change from intergranular fracture to transgranular fracture with the increase of the test temperature. Under the three temperature conditions, the crack is generated at the grain boundary around the homogeneous hole, as shown in Figure 5f. The reason for this is that the strength of $M_{23}C_6$ carbide at the grain boundary is high, so that it is difficult for dislocation to cut or bypass it, resulting in the dislocation stacking at the $M_{23}C_6$-type carbide and forming a dislocation nail. However, when the stress reaches a certain value, the local stress concentration occurs in the $\gamma/M_{23}C_6$ bonding area, which makes the cracks appear in the $\gamma/M_{23}C_6$ bonding area [20]. From the previous article, one can see that the test material is mainly composed of the matrix phase and two kinds of carbide phases, belonging to the typical carbide dispersion strengthened superalloy. Therefore, the strength of carbides plays an important role in the mechanical properties of the alloy, especially $M_{23}C_6$ carbide, which can strengthen the grain boundary. At a lower temperature (650 °C), the strength of the matrix phase is greater than that of the grain boundary, so that the crack will propagate along the grain boundary of the interference band between the holes, resulting in an intergranular fracture of the alloy. In addition, after the strengthening stage, the stress-strain curve appears to zigzag due to the cross slip of matrix deformation in the crystal and the pinning effect of diffusion solute atoms on the dislocation motion. With the increase of temperature, both the grain boundary strength and matrix phase strength decrease, but the decrease rate of the matrix strength is faster than that of the grain boundary [22]; when the temperature reaches 850 °C, the strength of the grain boundary will be greater than that of the matrix, and the crack will propagate along the inter-hole strain band. Because the strength of the matrix phase is greatly reduced, the dislocation can easily pass through the M_6C-type carbide by the climbing mechanism and Orowan bypass mechanism [23]. The crack will propagate through the grain until the specimen breaks. At the same time, due to the decrease of the material strength, the plasticity of the specimen will increase significantly, resulting in a significant increase in the elongation (δ) of the test piece, and the polygonal grains will be elongated into long strips during the tensile test. The microstructure evolution and fracture mechanism of the alloy under high-temperature tensile testing are drawn for different temperatures according to the experimental phenomena, as shown in Figure 9.

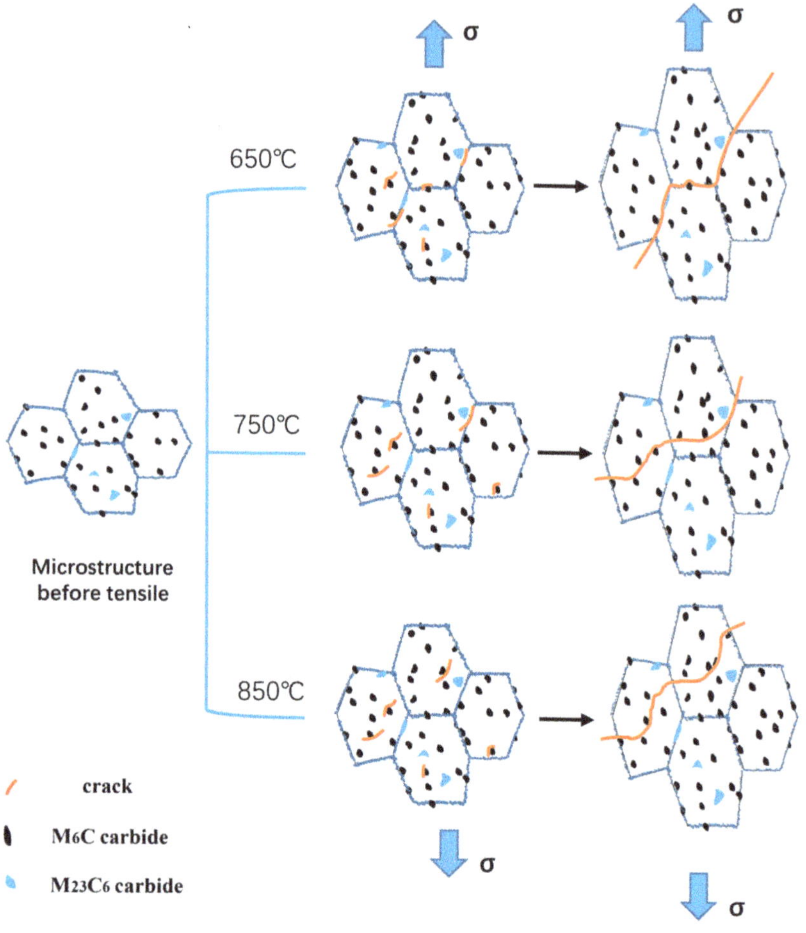

Figure 9. The schematic of the microstructure evolution and fracture mechanism of the tested alloy at different temperatures.

4. Conclusions

In this study, the Ni-Cr-W superalloy GH3230 was used as the experimental material, and porous laminate specimens were designed in combination with an actual engineering structure. High-temperature tensile tests at 650 °C, 750 °C and 850 °C were carried out. The high temperature mechanical properties and fracture mechanism of deformed superalloy porous laminate specimens were studied by means of digital image correlation (DIC), an ultra depth of field optical microscope, a scanning electron microscope (SEM) and EDS element analysis technology. The main conclusions are as follows:

1. Through the strain field results measured by DIC, it is found that an obvious stress concentration area is formed around the pores of the film-cooling holes, Moreover, due to the large diameter and dense arrangement of the overflow holes, there is a significant pore interference effect between the overflow holes under three temperature conditions. It is shown that there are obvious high strain banded regions between adjacent holes. The direction of the strain band is ±45° and 90° to

the tensile direction, and the distribution is symmetrical. On the contrary, no obvious interference effect was found on the side with impact holes, and the crack initiation was mainly due to the stress concentration around the holes. The pin fins between the laminates have no obvious effect on the structural strength, and their function is mainly to bond the laminates on both sides.

2. Because the strength of the matrix phase and carbide phase decreases with the increase of temperature and because the rate of strength reduction differs, the strength of the tested alloy shows a significant temperature sensitivity. The ultimate strength (σ_b) and yield strength ($\sigma_{0.2}$) decrease slightly from 650 °C to 750 °C but decrease significantly at 850 °C. The plasticity of the material increases significantly with the increase of temperature. In addition, the stress-strain curves of the tensile tests at 650 °C and 750 °C show characteristic zigzag fluctuations in the strengthening stage due to the pinning effect of the diffusive solute atoms on the dislocation.
3. All the initial microcracks are generated at the grain boundary around the hole perpendicular to the loading direction, before extending macroscopically along the interference band. The interference effect provides a preferential path for the crack growth. On a microlevel, the fracture mode of the specimen changes from intergranular fracture to transgranular fracture with the increase of temperature.

Author Contributions: Design study, Y.-D.L. and Z.-X.W.; Literature search, P.W.; Formal analysis, P.W.; Investigation, Y.-D.L. and P.W.; Data curation, P.W.; Writing—Original draft preparation, Y.-D.L. and P.W.; Writing—Review and editing, Y.-D.L.; Funding acquisition, Z.-X.W. All authors have read and agreed to the published version of the manuscript.

Funding: This research was funded by the National Natural Science Foundation of China (Project No.: 51875462), the Natural Science Basic Research Plan in Shaanxi Province of China (Project No.: 2020JC-16), the National Science Technology Major Project of China (Project No.: 2017-IV-0003-0040, 2017-V-0003-0052) and Taicang Scientific Research Institute Innovation leading special plan (Project No.: tc2019dyds11).

Conflicts of Interest: The authors declare no conflict of interest.

References

1. Zhao, Y.; Zhang, J.; Song, F. Effect of trace boron on microstructural evolution and high temperature creep performance in Re-contianing single crystal superalloys. *Prog. Nat. Sci.* **2020**, *30*, 371–381. [CrossRef]
2. Wen, Z.; Zhang, D.; Li, S. Anisotropic creep damage and fracture mechanism of nickel-base single crystal superalloy under multiaxial stress. *J. Alloys Compd.* **2017**, *692*, 301–312. [CrossRef]
3. Wei, C.N.; Bor, H.Y.; Chang, L. The effects of carbon content on the microstructure and elevated temperature tensile strength of a nickel-base superalloy. *Mater. Eng. A* **2010**, *527*, 3741–3747. [CrossRef]
4. Pollock, T.M.; Tin, S. Nickel-Based Superalloys for Advanced Turbine Engines: Chemistry, Microstructure and Properties. *J. Propuls. Power* **2006**, *22*, 361–374. [CrossRef]
5. Sheng, L.Y.; Du, B.N.; Lai, C.; Guo, J.T.; Xi, T.F. Influence of Tantalum addition on microstructure and mechanical properties of the nial-based eutectic alloy. *Strength Mater.* **2017**, *49*, 109–117. [CrossRef]
6. Lu, Y.L.; Liaw, P.K.; Wang, G.Y.; Benson, M.L.; Thompson, S.A.; Blust, J.W.; Browning, P.F.; Bhattacharya, A.K.; Aurrecoechea, J.M.; Klarstrom, D.L. Fracture modes of HAYNES (R) 230 (R) alloy during fatigue-crack-growth at room and elevated temperatures. *Mater. Eng. A* **2005**, *397*, 122–131. [CrossRef]
7. Liu, Y.; Hu, R.; Li, J.; Kou, H.; Li, H.; Chang, H.; Fu, H. Characterization of hot deformation behavior of haynes230 by using processing maps. *J. Mater. Process. Technol.* **2009**, *209*, 4020–4026. [CrossRef]
8. Du, B.; Hu, Z.; Sheng, L.; Cui, C.; Yang, J.; Zheng, Y.; Sun, X. Tensile, creep behavior and microstructure evolution of an as-cast ni-based k417g polycrystalline superalloy. *J. Mater. Technol.* **2018**, *34*, 95–106. [CrossRef]
9. Yang, C.; Xu, Y.; Nie, H.; Xiao, X.; Jia, G.; Shen, Z. Effects of heat treatments on the microstructure and mechanical properties of Rene 80. *Mater. Des.* **2013**, *43*, 66–73. [CrossRef]
10. Han, Y.; Xue, X.; Zhang, T.; Hu, R.; Li, J. Grain boundary character correlated carbide precipitation and mechanical properties of Ni-20Cr-18W-1Mo superalloy. *Mater. Eng. A* **2016**, *667*, 391–401. [CrossRef]
11. Hu, R.; Bai, G.; Li, J.; Zhang, J.; Zhang, T.; Fu, H. Precipitation behavior of grain boundary M23C6 and its effect on tensile properties of Ni–Cr–W based superalloy. *Mater. Eng. A* **2012**, *548*, 83–88. [CrossRef]

12. Tang, B.; Jiang, L.; Hu, R.; Li, Q. Correlation between grain boundary misorientation and M23C6 precipitation behaviors in a wrought Ni-based superalloy. *Mater. Charact.* **2013**, *78*, 144–150. [CrossRef]
13. Ou, M.; Hao, X.; Ma, Y.; Liu, R.; Zhang, L.; Liu, K. Effect of carbon on the microstructure and stress rupture properties of a new Ni-Cr-W-Fe alloy for advanced ultra-supercritical power plants. *J. Alloys Compd.* **2018**, *732*, 107–115. [CrossRef]
14. Gao, Z.; Guo, W.; Zhang, C.; Tan, J. Development of fine-grained structure in Ni-Cr-W based superalloy and its effect on the mechanical properties. *Mater. Sci. Eng. A* **2017**, *682*, 156–163. [CrossRef]
15. Bai, G.; Li, J.; Hu, R.; Tang, Z.; Xue, X.; Fu, H. Effect of temperature on tensile behavior of Ni–Cr–W based superalloy. *Mater. Eng. A* **2011**, *528*, 1974–1978. [CrossRef]
16. Wen, Z.; Pei, H.; Yang, H.; Wu, Y.; Yue, Z. A combined CP theory and TCD for predicting fatigue lifetime in single-crystal superalloy plates with film cooling holes. *Int. J. Fatigue* **2018**, *111*, 243–255. [CrossRef]
17. Zhang, Y.; Wen, Z.; Pei, H.; Wang, J.; Li, Z.; Yue, Z. Equivalent method of evaluating mechanical properties of perforated Ni-based single crystal plates using artificial neural networks. *Comput. Methods Appl. Mech. Eng.* **2020**, *360*, 112–725. [CrossRef]
18. Mulford, R.A.; Kocks, U.F. New observations on the mechanisms of dynamic strain aging and of jerky flow. *Acta Metall.* **1979**, *27*, 1125–1134. [CrossRef]
19. Beukel, A.V.D. Theory of effect of dynamic strain aging on mechanical-properties. *Phys. Status Solidi* **1975**, *30*, 197–206. [CrossRef]
20. Hrutkay, K.; Kaoumi, D. Tensile deformation behavior of a nickel based superalloy at different temperatures. *Mater. Eng. A* **2014**, *599*, 196–203. [CrossRef]
21. Bai, G.; Li, J.; Hu, R.; Xue, X.; Ma, J.; Hu, S.; Fu, H. Microstructure and high temperature tensile deformation behavior of Ni-Cr-W Superalloy. *Rare Met. Mater. Eng.* **2011**, *40*, 1300–1304.
22. Lian, Z.W.; Yu, J.J.; Sun, X.F.; Guan, H.R.; Hu, Z.Q. Temperature dependence of tensile behavior of Ni-based superalloy M951. *Mater. Sci. Eng. A* **2008**, *489*, 227–233. [CrossRef]
23. Betten, J.; Tsang, N. *Creep Mechanics*; Springer Science & Business Media: Berlin, Germany, 2003; Volume 56, p. 69.

materials

MDPI

Article

Investigation of the Impact of Load on the Magnetic Field Strength of the Crane by the Magnetic Metal Memory Technique

Agnieszka Koson-Schab * and Janusz Szpytko

Faculty of Mechanical Engineering and Robotics, AGH University of Science and Technology, A. Mickiewicza Av. 30, PL30-059 Krakow, Poland; szpytko@agh.edu.pl

* Correspondence: koson@agh.edu.pl

Received: 24 October 2020; Accepted: 4 December 2020; Published: 6 December 2020

Abstract: The paper deals with the problem of applicability of the metal magnetic memory (MMM) technique in the crane structural inspection and monitoring. The MMM method does not require the external magnetization of a structure that results in reduction of downtime of maintenance operations. Measurement of the intensity of the self-magnetic leakage signal can be an alternative to other non-destructive methods used for inspection of a large crane's structure and equipment. However, the complexity of the residual magnetization effect in the MMM technique is the problem with its application. Thus, the magnetic flux leakage behavior on the crane girder surface under different measurements and the crane's load conditions is analyzed based on the results obtained during experiments carried out on the overhead traveling crane.

Keywords: metal magnetic memory; magnetic field non-destructive evaluation; diagnostics; gantry crane

1. Introduction

Currently, there is a growing interest in obtaining information from the research of operating structures in terms of their degradation processes for the purpose of improving their construction, manufacturing, and operation. Construction safety requirements are related to the development of new diagnostic methods enabling the detection and localization of structural degradation.

Structures of large-scale industrial transport devices, subjected to loads resulting from the type of their work, are exposed to high stressors. These stressors affect the fatigue strength of the structure as well as damage the material structure. In the interest of safety, you should perform frequent diagnostic tests of these devices.

Nowadays, technical diagnostics increasingly use non-invasive diagnostic methods for diagnosing or monitoring (continuous control) without interrupting the normal operation of the object [1–3]. One of the least invasive methods of diagnosis is magnetic metal memory (MMM). This is used for inspection of welds, structural elements such as loaded cyclically on rotors, gears, steel ropes, and much more. The MMM method is one of the fastest and cheapest for detecting and assessing early caused damage based on the natural stress-induced magnetic flux leakage (MFL) [4,5].

The MMM method, based on detecting the impact of changes in the residual magnetization of the examined object on its magnetic flux leakage, is increasingly used in diagnostics. It allows indicating places or dangerous areas, cracks, or other defects of elements made of ferromagnetic materials [6,7].

The MMM method is based on the magneto elasticity Villari effect. Local disturbances in the earth's magnetic field are measured at the surface of ferromagnetic materials. Ferromagnetic materials in the presence of the geomagnetic field are subjected to stress, which causes the induction of a magnetic field [8–11].

The analysis of the test results consists of comparing the normal and tangential components of the magnetic flux leakage along the measuring path.

It is based on the known phenomenon of magnetic flux leakage of material with different magnetic permeability. Changes in the sign and leakage magnetic field *Hp* value are indications on the basis of which the assessment of stress and strain in the material of the tested element is made, define *Hp* by the Formula (1) [12]:

$$H_p = \frac{1}{\mu_0} B_p \quad (1)$$

where:

B_0 is the vertical value of the magnetic induction component [T];
μ_0 is vacuum magnetic permeability $4\pi \times 10^{-7}$ [H/m].

Areas of stress concentration are detected in which processes leading to a reduction in material strength can occur [11,13–15].

The MMM method can detect the residual magnetic field due to mechanical stress and geomagnetic field and can be performed without artificial magnetization. The magnetometer using the MMM method is based on magnetoresistance effects. It is characterized by a small volume, high sensitivity, and reliability. This makes it easier to detect the fatigue of the tested elements [2,16].

It is widely accepted that the magnetic state of ferromagnetic material may be irreversibly altered by mechanical loading due to magneto-elastic effects [17]. On the macroscopic scale, there are close relationships between the mechanical and magnetic properties of ferromagnetic material. Under the influence of mechanical stress in the ferromagnetic material, deformation occurs and changes the internal magnetic field. In general, the MMM is an after effect which is directly correlated with the distribution of natural magnetic fields and the accumulated stress-strain which has been experienced by the ferrous or paramagnetic material.

The heterogeneity of the structure, mechanical properties, and the magneto-plastic effect are the result of the leakage magnetic field, which will lead to an irreversible increase in the magnetic induction under a weak magnetic field. The ferromagnetic element is subjected to the external stresses in a certain direction [18].

The MMM signals are a weak magnetic signal, which consists of vulnerable factors such as lift-off height, chemical composition, the specimen shape, the notch shape, the heat-treating technology, the initial magnetization state of ferromagnetic materials, and environmental magnetic field [10,14,19,20].

The objective of the paper is to determine the experimental impact of the structure load change on the level of the own magnetic field, using the MMM method and measuring the magnetic field strength in the elastic deformation zone, for the purpose of inferring changes in the technical condition of the structure, on the example of the girder of the overhead crane.

2. Crane Girder Bridge Testing Approach with Use MMM Technique

The aim of the experiment was to determine the impact of the load change on the level of self-magnetic leakage field (SMLF)of a double girder bridge crane in terms of analyzing the possibility of using its residual magnetic flux.

As a result of stresses and plastic deformations, changes in the structure of the material caused by fatigue, local irregularities are observed as local changes in the degree of the material magnetization.

The value measured in the test is the value of the selected component of the magnetic field strength *H* measured near the diagnosed object, represented by Formula (2):

$$H = \frac{B}{\mu_r \cdot \mu_0}, \quad (2)$$

where [21]:

H is magnetic field strength [A/m],

B is magnetic induction [T],
μ_r is relative magnetic permeability,
μ_0 is vacuum magnetic permeability $4\pi \times 10^{-7}$ [H/m].

After analyzing the available literature on the research of objects using the MMM method, our own test plan was developed to reduce the cost and time of the experiment. The research plan cover: object model, the duration of individual experiments, identifying confounding factors and constants. Figure 1 shows the test object model.

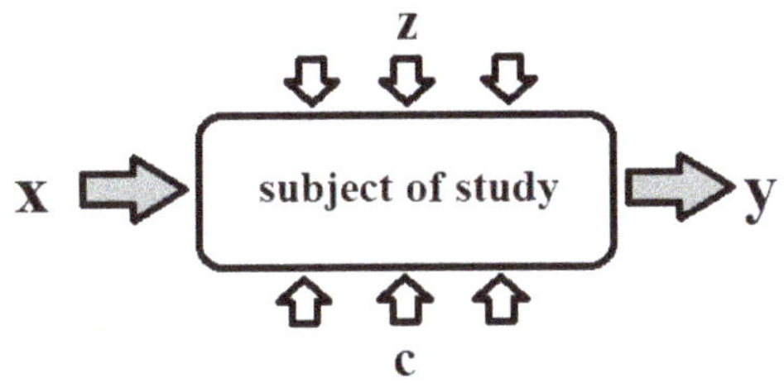

Figure 1. Block diagram of the research subject.

The model research object (the girder of a double-girder overhead crane) can be represented by the function defined by Equation (3):

$$y(t) = f\ (x, z, c, t), \quad (3)$$

where:

y—output values: magnetic field strength H in the tested girder section in [A/m];
x—input values: $x = \{x_1, x_2\}$;
x_1—beam load time in [s],
$x_{/1}$—beam deflection in [mm],
c—solid factors: $c = \{c_{/1}, c_{/2}, \ldots\ c_i\}$;
c_1—the high of raising in [mm], c_2—load lifting speed in [m/s], c_3—ambient temperature of crane in [°C], c_4—crane girder load in [kg], c_5—the place on the girder where the load was applied in [m], c_i—other that did not affect the results of the tests carried out,
z—interference: $z = \{z_1, z_2, \ldots\ z_i\}$;
z_1—measurement errors, z_2—magnetic field disturbances caused by the device's operation in [A/m], z_3—material structure, z_i—factors unnoticed affecting the measurement.

The SMFL signal measured using the MMM method is influenced by static and dynamic loads, movements of driving and lifting mechanisms, and transient states of the drive systems.

The frequency inventers used in the crane drive system generate the external disturbances that affect the measurement signal. This drawback was eliminated by carrying out the measurement experiments when crane motors and frequency inverters were turned out. The other exogenous conditions, such as temperature (if the measurement is performed in the temperature range −20–60 °C), have minor impact on the measurement results.

Research assumptions:

The experiments were carried out on the double-girder overhead traveling crane with hoisting capacity $\boldsymbol{m}$ = 1000 kg, a span of the girders $\boldsymbol{l}$ = 8 m, and trolley wheelbase $\boldsymbol{p}$ = 1 m. Figures 2 and 3 present the crane subjected to examine, and girder beam cross-section, respectively. The girder is made of S235JR structural non-alloy steel [22]. The beam, supported on both sides at its ends, was subjected to elastic deformation, giving an elastic force in its center of 1000 kg, resulting in deflection of the beam f = 12.17 mm. The loading conditions of the tested girder are shown in Figures 3 and 4.

Figure 2. The tested double-girder overhead traveling crane with a trolley and a hoist.

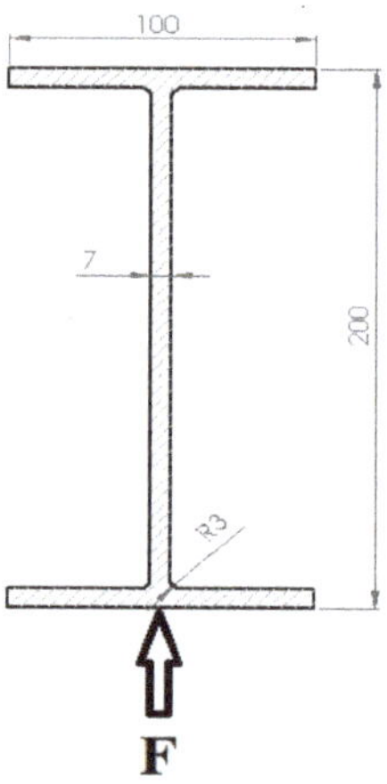

Figure 3. Crane's girder cross-section, where: F is tested plane.

The r = 1.4 m length section between points G and H (Figure 4) of the crane's girder bottom surface was selected for examination using (where G is the starting point and x is the direction of the measurement). Points G and H are appointed symmetrically to the center of the crane beam.

The SMFL signal was measured using the Tester Stress Concentration TSC-4M-16, which is equipped with a scanning device. The scanning device is a four-wheeled cart with the flux-gate transducers, and the encoder for sensing position of the cart moved along a measuring element (Figure 5). The magnetometers installed in the scanning device allow to measure Hx and Hy distribution of the self-magnetic leakage field signal along the surface of an inspected structure.

The measurement device allows to measure the magnetic field intensity in the range ±2000 A/m. The basic relative error of H_p measurement is 5%, and the sensor movement accuracy is 0.9 mm. The magnetometer TSC-4M-16 is equipped with the software for data acquisition and analysis.

The measurements of the magnetic field strength in the designated spar section were carried out according to our own test plan.

During the inspection, the trolley with a winch was set in the middle of the crane's girder. The load was placed in the center of the trolley. Figure 4 presents the dimensions and position of the trolley with a load during the experiments. The crane's girder, where: $\boldsymbol{l}$ is crane span, $\frac{l}{2}$ is the distance of the suspended load from the beginning of the girder, $\boldsymbol{m}$ is load suspended on a rope, $\boldsymbol{k}$ is load-lifting height measured from the ground, $\boldsymbol{r}$ is the examined section between points G and H of the girder. In order to eliminate interference magnetic field strength, the tests were performed with the driving and lifting mechanisms turned off.

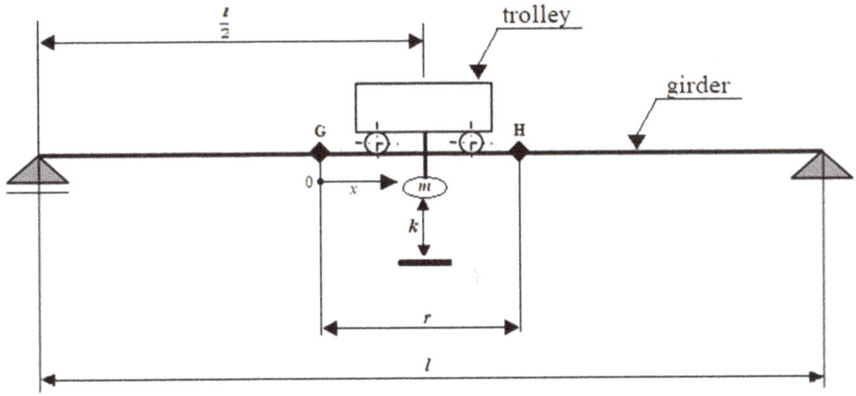

Figure 4. Experimental setup.

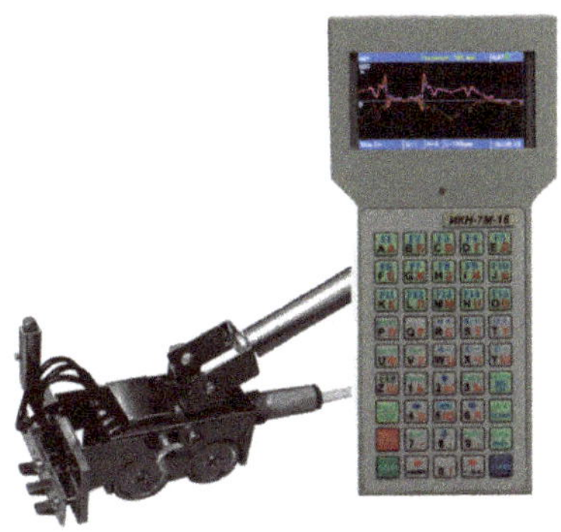

Figure 5. Tester of Stress Concentration TSC-4M-16 and scanning 4-channel sensor.

All measurements of the magnetic field strength were carried out in the same conditions:

- the trolley with the load was placed in the middle of the spar span,
- constant ambient temperature $T = 22\ ^\circ\mathrm{C}$,
- during the measurement, all devices disturbing the magnetic field are turned off,
- the same section of the spar's lower surface was always tested.

The research was divided into two stages.

During the first stage, the intensity of the residual magnetic field in the spar loaded for 4 h was tested to the following scenario repeated every 24 h for 4 days, the results are presented in Table 1

- measurement variant A—first measurement with the crane load $\boldsymbol{m} = 0$ kg, $\boldsymbol{k} = 0$ cm, $\boldsymbol{t} = 0$ s,
- measurement variant B—the experiment carried out immediately after the load was applied, $\boldsymbol{m} = 1000$ kg, $\boldsymbol{k} = 10$ cm, $\boldsymbol{t} = 0$ s,
- measurement variant C—the experiment for the crane maximum load $\boldsymbol{m} = 1000$ kg, $\boldsymbol{k} = 10$ cm, $\boldsymbol{t} = 14{,}400$ s (4 h) carried out four hours after variant B, performed after 4 h of loading of the crane girders with a load of $\boldsymbol{m} = 1000$ kg,
- measurement variant D—immediately after the measurements in variant C, the load is removed from the girder, $\boldsymbol{m} = 0$ kg, $\boldsymbol{k} = 0$ cm, $\boldsymbol{t} = 0$ s.

During the second stage of the research, measurements of the magnetic field strength were carried out after 10 loads of 1000 kg for 5 s:

- measurement variant E—The first measurements were made according to the measurement variant A, where m = 0 kg, t = 0 s, in Table 2, these measurements were called the measurement cycle 0. Then, the load m = 1000 kg was raised to $\boldsymbol{k}$ = 10 cm, after $\boldsymbol{t}$ = 5 s the load was abandoned ($\boldsymbol{m}$ = 0 kg, $\boldsymbol{t}$ = 0 s). This operation was repeated 10 times. The movement mechanisms were turned off, then the magnetic field strength was measured (measurement cycle 10). Six cycles were performed, after each, the magnetic field intensity in a given beam section was examined.

Four repeated measurements were carried out for each case A, B, A, B, C, D and E.

3. Results and Discussion

The results of the measurements magnetic field strength were analyzed. During the analysis of the measurements, the focus was on the graphs based on the value of the magnetic field strength at a given section of the tested object. The results of measurement experiments carried out for each case A, B, C, D, and E are presented in Table 1 in the form of maximum, minimum, and average values of tangential Hx and normal Hy component of SMFL are listed as determined as the mean values of four measurements repeated for each case. Maximum, minimum, and average values of magnetic field strength are listed as Hx and Hy in Table 1.

Table 1. Minimum, maximum, and average values of tangential and normal component of the magnetic signals for experiments A–D.

Measurement Day	A Sample	Measuremet Variant	Hx_{min}	Hx_{max}	Hx_{avg}	Hy_{min}	Hy_{max}	Hy_{avg}
1	1	A	56.6	624.3	360.7	143.4	244.5	176.2
	2	B	79.3	549.8	351.1	129.6	202.8	161.1
	3	C	81	532	334.2	105.6	191.9	137.9
	4	D	3.3	592	331.8	71.8	228.5	173.2
2	5	A	17.8	527.6	306.2	134.8	231.5	166.9
	6	B	32	612	197.9	131.8	196.6	16.3
	7	C	76.6	537.6	158.2	99.9	203.8	141.2
	8	D	−38.9	574.3	205.7	137.3	234.9	172.2
3	9	A	12.1	508.8	318.4	117	191.1	142.2
	10	B	154.4	586.5	391	117.1	192.1	160.5
	11	C	161	590	390.2	135.4	192.1	156.9
	12	D	122.1	643.1	4115	136.6	195	156.4
4	13	A	23.3	644.3	409.9	117	191.5	152.9
	14	B	121	639.8	4184	139.6	213.8	175.6
	15	C	164.4	583.1	397.5	121.6	188.3	150.2
	16	D	126.6	615.4	415.6	141.6	224.1	170

Graphs were made for Hx min and max and Hy min and max to compare the maximum, and minimum values of the magnetic field strength in the tested section of the crane girder during the experiment. These graphs are shown in Figure 6.

The minimal values of the tangential component x the strength of the magnetic field in the tested girder, after raising the load, reach higher values. With the passage of time under load, Hx_{min} values are increasing. After unloading the crane, the magnetic field strength in the tested section of the Hx_{min} girder decreases.

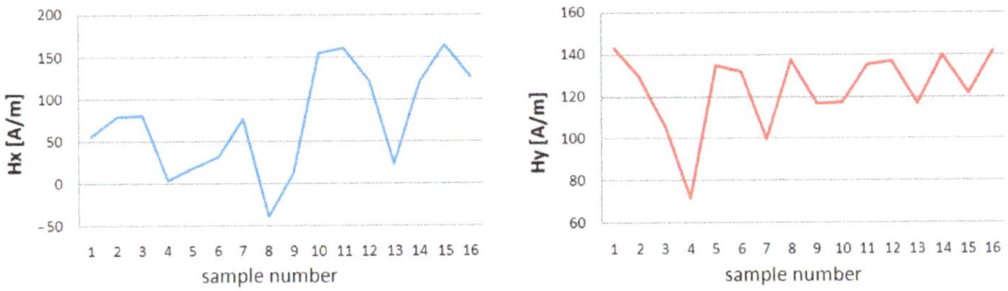

Figure 6. Minimum of tangential *Hx* and normal component *Hy* after repeatedly applying variants A, B, C, D, for 4 days long.

The lowest *Hx* value is obtained after several hours of relaxing the crane girder (Figure 7). The minimum normal component y of the magnetic field strength for each loading and unloading of the girder reacts with a change in value. The more measurement days, the smaller the range of change.

The same happens with max *Hx* and *Hy* values. Each operation related to the load causes a change in the value of the magnetic field strength for both components.

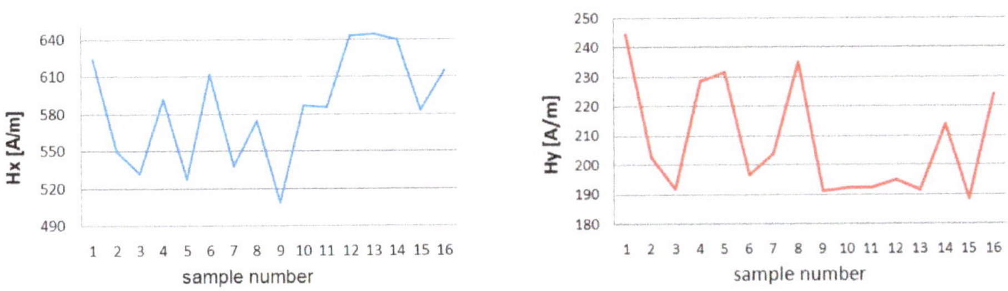

Figure 7. Maximum of tangential *Hx* and normal component *Hy* after repeatedly applying variants A, B, C, D, for 4 days long.

The maximum values of the magnetic field strength in the tested spar cross-section, after loading with the mass m = 1000 kg, decrease their value. After the load time t = 4 h, the measurement shows a decrease in the H_{max} value for the x and y components. However, after removing the load where m = 0, H_{max} reaches higher values.

Figures 8–11 show graphs of tangential *Hx* and normal *Hy* component of the magnetic field, for samples 1, 2 and 15, 16. On their basis, it is possible to notice the characteristics of changes in magnetic field strength waveforms *Hx* and *Hy* at the beginning of the experiment, then immediately after the first loading of the device, and before the last unloading and after the whole experiment.

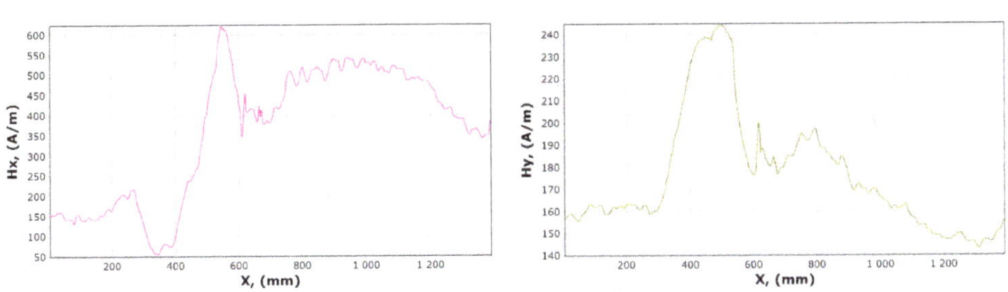

Figure 8. Graphs of tangential *Hx* and normal *Hy* component of the magnetic field, for sample 1.

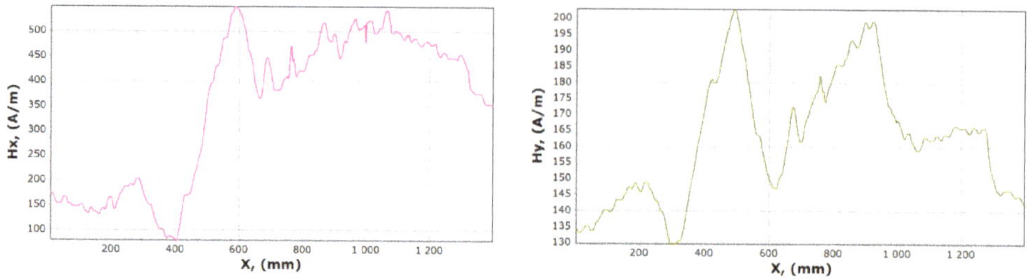

Figure 9. Graphs of tangential *Hx* and normal *Hy* component of the magnetic field, for sample 2.

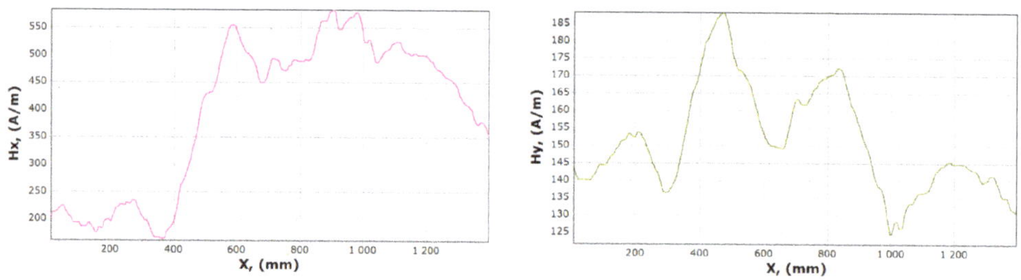

Figure 10. Graphs of tangential *Hx* and normal *Hy* component of magnetic field, for sample 15.

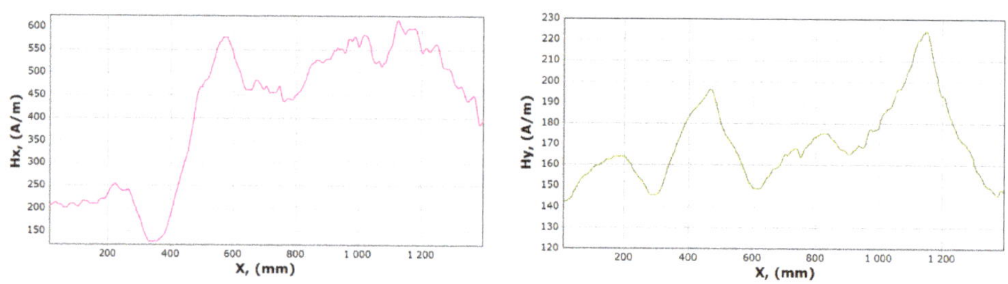

Figure 11. Graphs of tangential *Hx* and normal *Hy* component of the magnetic field, for sample 16.

After comparing the charts of the *Hx* and *Hy* components, we notice slight differences in the course of the chart for *Hx*. The diagrams, regardless of the load and its length, are similar due to the curve. For *x* in the range 340–370 mm for unloaded girder, Hx_{min} is about 50 A/m. After the last experiment, unloading the girder, this value is about 120 A/m. Whereas Hx_{max} is in the range 550–600 mm is reduced from 620 to 570 A/m. The *Hy* changes its max and min and the course of the curve during each experiment. The graph shows tendencies to large changes in magnetic field strength for *Hy*. It can be stated that the magnetic field in the element subjected to stress reduces its amplitude vibration.

The next stage of the research included measurement variant E. The influence of dynamic loads on the intensity of the girder's own magnetic field on the section determined in accordance with Figure 3 was examined.

After unloading the device magnetometric measurements were carried out. A total of 6 series of 10 loads and unloading of the girder were made. The measurement results for one magnetometer are presented in Table 2. The maximum, minimum, and average values of *Hx* and *Hy* are shown in Figures 12 and 13.

Table 2. Minimum, maximum, and average values of tangential and normal component of the magnetic signals for experiments E.

A Sample	Load Cycle	Hx_{min}	Hx_{max}	Hx_{avg}	Hy_{min}	Hy_{max}	Hy_{avg}
17	0	255.4	670.9	534.7	134.1	190.9	165
18	10	263.3	656.5	486.8	125.9	182.1	151.8
19	20	275.4	643.1	487.5	136.1	171.4	156.4
20	30	272	679.8	496.3	124.1	181.1	157.4
21	40	272.1	652	487.2	125.9	172.9	152.6
22	50	276.9	660.9	508.6	139.9	183.5	162.4
23	60	261	717.5	514	141	199.4	165.1

After the first series of short loads on the crane girder, a change in the value of the magnetic field in the measured plane of the beam is noticeable. The minimum Hx values increase from 255 to 275 A/m, while the Hy decreases from 134 to 126 A/m. The second series of loads causes a slight decrease in the value of H for x. For y, H_{min} achieves a value close to the measurement before the beam is loaded. The results of measurements of successive series show the independence of the magnetic field strength from the number of series made. The maximum values of H for x and y show a similar graph. After the first 10 loads, H_{max} decreases, the next series increases the level of the magnetic field in the beam plane being tested for both components. In the third series, you can see a drop in value. After the last 6 trial, H_{max} reaches the highest values Hx_{max} = 719 A/m and Hy_{max} = 200 A/m.

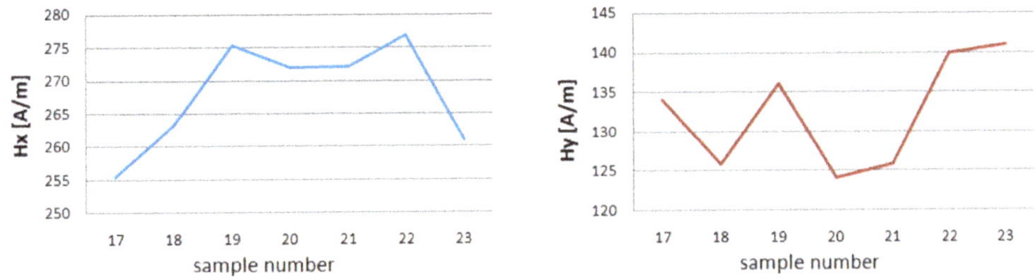

Figure 12. Minimum of tangential component Hx and Hy for experiment E.

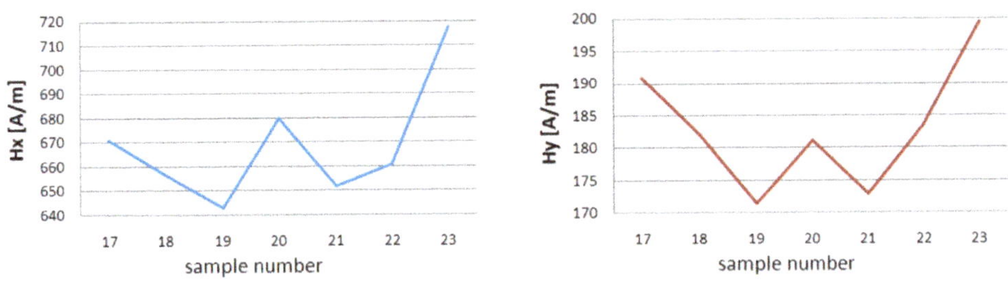

Figure 13. Maximum of tangential component Hx and Hy for experiment E.

Examples of magnetometric charts made on the basis of the first measurement taken without a girder load, and the last one is shown in Figures 14 and 15.

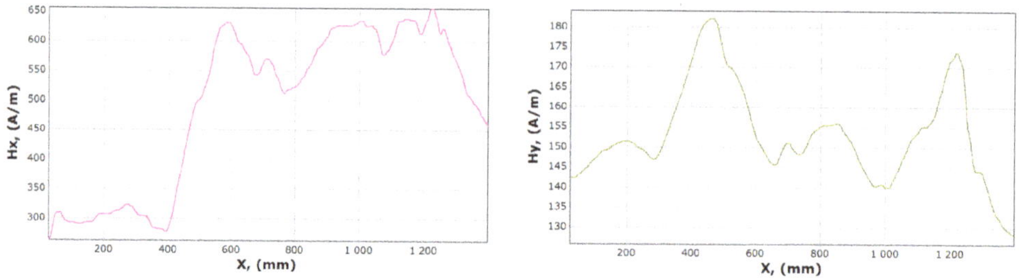

Figure 14. Graphs of tangential *Hx* and normal *Hy* component of the magnetic field, for sample 17.

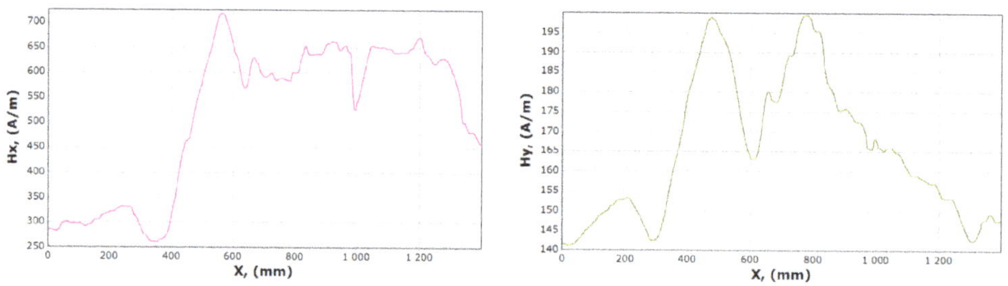

Figure 15. Graphs of tangential *Hx* and normal *Hy* component of the magnetic field, for sample 23.

The curve illustrating the *x* component before and after the experiment involving a similar course, when the curve illustrating the component *Hy* contains a significantly different character. The minimum *Hx* reaches in the range of 350–400 mm while the maximum in the range 550–600 mm. In the case of the *y* component, it is impossible to clearly determine the fixed place where it reaches its maximum and minimum.

4. Conclusions

The paper addresses the problem of the crane structure inspection by applying the MMM method.

The crane girder was the subject of inspections of the effect of load on changes in tangential and normal components of the magnetic field using MMM. When analyzing the results of measurement experiments, it can be seen that with each load change, the magnetic field strength changes. These changes are visible in Figures 5 and 6. A similar situation occurs when the load is applied dynamically. After each series of measurements, the change in *Hx* and *Hy* measurements is visible (Figures 11 and 12).

Based on the research, it can be concluded that short-range, high-value forces act between atoms of a given material with elastic properties. During elastic deformation of such material, interatomic distances and angles between atomic bonds change. As a result, it is associated with large energy changes because of the energy of elasticity increases. Elastic deformations are small and these small deformations are accompanied by relatively high forces. After the force causing the elastic deformation disappears, the material returns to its previous dimensions, the material atoms will take their previous positions again.

Microplastic deformations by dislocation development may cause changes in the magnetic flux leakage. However, according to our best knowledge, the relation between the microplastic deformation and magnetic flux leakage is not reported in the literature.

The tests may support the safety assessment of engineering structures subjected to loading. The knowledge of the influence of the load on the level of the magnetic field intensity allows us to predict the degradation time of the device structure. The use of the MMM method for monitoring can be helpful in the early detection of structural damage. This will result in increased work safety.

The next stage of the research will be to determine the intensity of the magnetic field during beam loads causing plastic deformation. These works are aimed at obtaining as much information as possible about the behavior of the magnetic field strength in the girders of the gantries working under load. Further research may be directed towards the challenging problem to obtain the damage indicator parameter and predict the absolute fatigue life of a crane's structure. It requires analysis of magnetic flux leakage behavior under various load conditions during elastic and plastic deformations. Future work may consider to carry out more experiments, to collect more data and use data driven techniques to identify the influence of elastic load and plastic deformation on the magnetic flux leakage signal and their coupling effects.

Author Contributions: Conceptualization, A.K.-S. and J.S.; methodology, J.S.; software, A.K.-S.; validation, J.S.; formal analysis, A.K.-S.; investigation, A.K.-S.; resources, A.K.-S., data curation, A.K.-S.; writing—original draft preparation, A.K.-S. and J.S.; writing—review and editing, A.K.-S., visualization, A.K.-S.; supervision, A.K.-S. and J.S.; project administration, J.S.; funding acquisition, J.S. All authors have read and agreed to the published version of the manuscript.

Funding: The work has been financially supported by the Polish Ministry of Education and Science (Polish Ministry of Science and Higher Education).

Conflicts of Interest: The authors declare no conflict of interest.

References

1. Bi, Z.; Kong, L. The research on force-magnetic effect of high-speed train based on metal magnetic memory method. *Proc. Chin. Intell. Syst. Conf.* **2018**, *1*, 21–32.
2. Hu, Z.; Fan, J.; Wu, S.; Dai, H.; Liu, S. Characteristics of Metal Magnetic Memory Testing of 35CrMo Steel during Fatigue Loading. *Metals* **2018**, *8*, 119. [CrossRef]
3. Ariffin, A.; Ahmad I, M. Detection of cracked position due to cyclic loading for ferromagnetic materials based on magnetic memory method. *J. Technol.* **2015**, *75*, 67–70. [CrossRef]
4. Roskosz, M.; Bieniek, M. Evaluation of residual stress in ferromagnetic steels based on residual magnetic field measurements. *NDT E Int.* **2012**, *45*, 55–62. [CrossRef]
5. Dubov, A.; Kolokolnikov, S. The metal magnetic memory method application for online monitoring of damage development in steel pipes and welded joints specimens. *Weld. World* **2013**, *57*, 123–136. [CrossRef]
6. Kosoń-Schab, A.; Szpytko, J. Magnetic metal memory in the assessment of the technical condition of crane girders for the needs of safety. *J. Konbin* **2019**, *49*, 49–73. [CrossRef]
7. Kosoń-Schab, A.; Smoczek, J.; Szpytko, J. Magnetic memory inspection of an overhead crane girder—Experimental verification. *J. Kones* **2019**, *26*, 69–76. [CrossRef]
8. Moonesan, M.; Kashefi, M. Effect of sample initial magnetic field on the metal magnetic memory NDT result. *J. Magn. Magn. Mater.* **2018**, *460*, 285–291. [CrossRef]
9. Chongchong, L.; Lihong, D.; Haidou, W.; Guolu, L.; Binshi, X. Metal magnetic memory technique used to predict the fatigue crack propagate on behavior of 0.45% C steel. *J. Magn. Magn. Mater.* **2016**, *405*, 150–157. [CrossRef]
10. Kolokolnikov, S.; Dubov, A.A.; Steklov, O. Assessment of welded joints stress–strain state in homogeneity before and after post weld heat treatment based on the metal magnetic memory method. *Weld. World* **2016**, *60*, 665–672. [CrossRef]
11. Dubov, A.A. Principle features of metal magnetic memory method and inspection tools as compared to known magnetic NDT methods. In Proceedings of the 16th Annual Conference on Nondestructive Testing, Montreal, QC, Canada, 30 August–3 September 2004.
12. Juraszek, J. Residual magnetic field for identification of damage in steel wire rope. *Arch. Min. Sci.* **2019**, *64*, 79–92.
13. Dubov, A.A. Energy Diagnostics is a physical basis of the metal *magnetic* memory method. In Proceedings of the 11th European Conference on Non-Destructive Testing, Prague, Czech Republic, 6–11 October 2014.
14. Huang, S.; Wang, S. Metal Magnetic Memory Testing. In *New Technologies in Electromagnetic Non-Destructive Testing*; Springer Series in Measurement Science and Technology, Tsinghua University Press and Springer Science+Business Media: Singapore, 2016; pp. 169–183.

15. Chen, L.; Que, P.W.; Jin, T. A giant-magnetoresistance sensor for magnetic-flux-leakage nondestructive testing of a pipe line. *Russ. J. Nondes. Test.* **2005**, *41*, 462–465. [CrossRef]
16. Sonntag, N.; Skrotzki, B.; Stegemann, R.; Löwe, P.; Kreutzbruck, M. The role of surface topography on deformation-induced magnetization under inhomogeneous elastic-plastic deformation. *Materials* **2018**, *11*, 1518. [CrossRef] [PubMed]
17. Dubov, A.A. Detection of metallurgical and production defects in engineering components using metal magnetic memory. *Metallurgist* **2015**, *59*, 164–167. [CrossRef]
18. Ren, S.K.; Ren, X.Z. Studies on laws of stress-magnetization based on magnetic memory testing technique. *J. Magn. Magn. Mater.* **2018**, *449*, 165–171. [CrossRef]
19. Shui, G.S.; Li, C.W.; Yao, K. Non-destructive evaluation of the damage of ferromagnetic steel using metal magnetic memory and nonlinear ultrasonic method. *Int. J. Appl. Electromagn. Mech.* **2015**, *47*, 1023–1038. [CrossRef]
20. Shi, P.; Jin, K.; Zheng, X. A magneto mechanical model for the magnetic memory method. *Int. J. Mech. Sci.* **2017**, *124*, 229–241. [CrossRef]
21. Vlasov, V.T.; Dubov, A.A. *Physical Bases of the Metal Magnetic Memory Method*; ZAO "TISSO": Moscow, Russia, 2004; pp. 6–86.
22. *EN 10025-2: 2019 Hot Rolled Products of Structural Steels—Part 2: Technical Delivery Conditions for Non-Alloy Structural Steels*; The European Committee for Standardization: Brussels, Belgium, 2019.

Publisher's Note: MDPI stays neutral with regard to jurisdictional claims in published maps and institutional affiliations.

materials

MDPI

Article

Single-Sensor Vibration-Scanning Method for Assessing the Mechanical Properties of 3D Printed Elements

Ryszard Buchalik * and Grzegorz Nowak *

Department of Power Engineering and Turbomachinery, Silesian University of Technology, Konarskiego 18, 44-100 Gliwice, Poland
* Correspondence: ryszard.buchalik@polsl.pl (R.B.); grzegorz.nowak@polsl.pl (G.N.)

Abstract: This paper considers issues related to the assessment of the mechanical properties of elements made with 3D printing technology. To enable experimental testing, an automated test stand was built to perform amplitude and phase angle measurements of any point of the specimen. A contactless, optical measurement method was selected, as it is especially adequate when it comes to elements with small dimensions and masses. One innovative element of the test stand is the original method of phase angle measurement using a single vibration sensor fitted with a system forcing and ensuring full measurement synchronization and dynamic state repeatability. Additionally, numerical models of tested objects were produced and simulations of their oscillations were performed. Based on that, the properties of the tested material (PLA) were considered, with a special focus on the density, elastic modulus, and damping. The analyses were conducted for a few elements with different dimensions at different vibration frequencies.

Keywords: 3D printing; additive manufacturing; material properties; phase shift measurement; vibration scanning; resonance; vibration shapes

Citation: Buchalik, R.; Nowak, G. Single-Sensor Vibration-Scanning Method for Assessing the Mechanical Properties of 3D Printed Elements. *Materials* **2021**, *14*, 1072. https://doi.org/10.3390/ma14051072

Academic Editor: Giovanni Bruno

Received: 27 January 2021
Accepted: 19 February 2021
Published: 25 February 2021

1. Introduction

In recent years, there has been a huge increase in interest in Additive Manufacturing (AM) as a modern technology for the production of a wide range of elements. The method's advantages are the relatively fast production, the possibility of creating complex shapes and joining different materials, and the ease of obtaining the desired color or finish [1]. The use of varied internal structures in this technology makes it possible to shape the properties of produced elements, which is extensively presented in [2]. Gibson et al. [3] describe the application of bulk metallic glasses in AM, which behave like thermoplastics. Considering the recent development of 3D printing technology and the increasing potential of the technology's application in many different fields, it is important to establish the mechanical properties of elements made in this way [4,5]. There are works describing certain features and mechanical (material) properties of such materials [6–12], including product documentation of the raw material supplied by the manufacturer. However, most of them primarily concentrate on testing the material in steady states and on slow changes in essential parameters, such as the load. They offer very scarce information on the material behavior under fast-changing loads (a few dozen Hz and more), i.e., vibration. In this area, work focuses on the impact of geometric discontinuities (e.g., cracks) on the dynamic state of the element [13,14]. The technical data supplied by the manufacturers of devices, on the one hand, and filament producers, on the other hand, are not always confirmed by experimental testing results. Selected examples from the literature focus on the impact of the infill and the printing direction [15,16], taking account of fatigue [17] and creep [18,19] processes, the printing system vibrations [20] and damping [21]. Testing materials made with Additive Manufacturing technology is onerous due to the substantial impact of a great number of factors. The finished product parameters vary, depending on a number of parameters of the production process. The produced object's mechanical parameters depend, in a complex way, on the temperature of the nozzle depositing the material, the method of the material feeding, the shape and diameter of the nozzle and the nozzle thermal conductivity, the ambient temperature, the cooling method and intensity, the layer thickness, the printing speed, the material storage conditions (the history of temperature, time, and humidity), the printer accuracy, and the whole printed element cooling rate. Another key component of the 3D printing system that has an essential impact on the final product parameters is the program creating the extruder movement trajectory [22]. In the case of fused filament fabrication technology, an enormous number of combinations of printing parameters are used on the shell and in the infill. The parameters of the first layer (in contact with the base platform) also differ from the other layers. An element made with this technology is usually characterized by strongly anisotropic properties, which is mainly due to the existence of layers and contact between them [15,16]. An essential part of the production process is to determine the spatial orientation according to which the object will be made. Material cooling also generates internal stresses and may result in bending of the finished product [23].

The behavior of a mechanical element when exposed to vibrations and the possibility of resonant frequency excitation within it can lead to serious problems in the design and operation of machines [24]. There are studies based on the vibration analysis of elements made with 3D printing technology [25–31]. Some of them focus on determining damping inside the material if oscillating motion occurs and make use of comparative analysis. An exhaustive theoretical description of the phenomenon is presented in [26]. Gietl et al. [21] presented the way in which the impact of the medium (air resistance) is eliminated from measurements using a fiber-optic sensor. Efforts aiming to establish material parameters seem especially important when it comes to design and prediction of the occurrence of potentially harmful resonances of elements made with 3D technology. They will make it possible to simulate this type of system at the prototype-making stage and avoid many operating problems, e.g., excessive wear, noise, fatigue-related damage, etc. It is worth mentioning that very technologically advanced elements, even including turbine blades, can also be successfully produced with AM technology [32]. Due to the progress in the field of searching for materials with high vibration-isolating capabilities [33], it is also important to search for new methods for determining the material's elasticity and ability to dissipate the energy (damping) of its vibrations (damping), characterized by simplicity and the possibility of application in various conditions. The description of material parameters is additionally complicated in the case of structures reinforced with fibers made of different materials and embedded within the base material [34]. Damping or elasticity calculations can be very complex for specimens composed of many different material layers characterized by different properties [35] and different joints or infill [36]. The destruction of material due to vibration shapes, among other factors, can be taken into account [37].

In connection with the above-described aspects of AM technology and its expected development, vibration testing methods should be widely used. Typically, vibration testing involves the use of a sensor rigidly and directly mounted on the tested object. The only more widespread type of non-contact vibration sensor is the inductive sensor; however, its use is practically only possible in the case of materials showing magnetic properties, with a significantly different electrical and magnetic permeability from a vacuum. The only rational choice in the case of AM materials with a relatively low density that are light, small, and non-magnetic seems to be the optical measurement method. Due to the significant technological progress and reduction of the costs of photosensitive and laser elements, it seems reasonable to develop optical methods for vibration measurement.

There are several basic methods of optical vibration measurement, which allow various physical quantities to be measured. The first may be displacement-triangulation sensors in many arrangements [38,39] or an analysis of the image of the object on which the appropriate pattern is displayed at defined angles [40–42]. Due to the differential-integral dependence, the mechanical quantities can be mutually converted. However, it should be remembered that the numerical accuracy of the result changes, depending on the frequency of vibrations.

The aim of this paper is to consider issues related to the mechanical properties of objects made with 3D printing technology. An analysis of selected mechanical parameters is presented based on dynamic testing and the use of resonance frequencies. Experiments were performed to this end using a novel method of vibration measurement on a purpose-made measuring stand. The analysis covers the vibration amplitude and the phase shift by means of an unconventional method using only one displacement measuring sensor. Experimental data collected in this way can be coupled with a sophisticated numerical method [43] to determine the material properties and their uncertainty. The method proposed below can be applied to specimens made from different kinds of materials, even those which are naturally non-isotropic and non-uniform [44].

2. Materials and Methods

A purpose-designed test stand was constructed to perform measurements of dynamic states of specimens made with 3D printing technology.

2.1. Test Stand

The test stand (Figures 1 and 2) is composed of a vibration exciter and a measuring system; the whole stand is computer-controlled. The measuring system's fundamental element is a triangulation displacement sensor. Its operating principle is the emission of a laser beam illuminating the investigated point of the specimen; the beam is then subjected to diffusive reflection. The reflected signal returns to the sensor at an angle, depending on the reflection point distance from the sensor, which is recorded by a light-sensitive matrix (one-dimensional image sensor) placed in the instrument and fitted with an appropriate optical system (lenses). The optical sensor returns a digital string of data representing the signal sampled with the frequency of 48 kHz and resolution reaching a few micrometers in the operating range of 10 mm. The test stand also includes a robotic arm to which the above-mentioned sensor is fixed. The arm enables the full control of motion in three spatial directions.

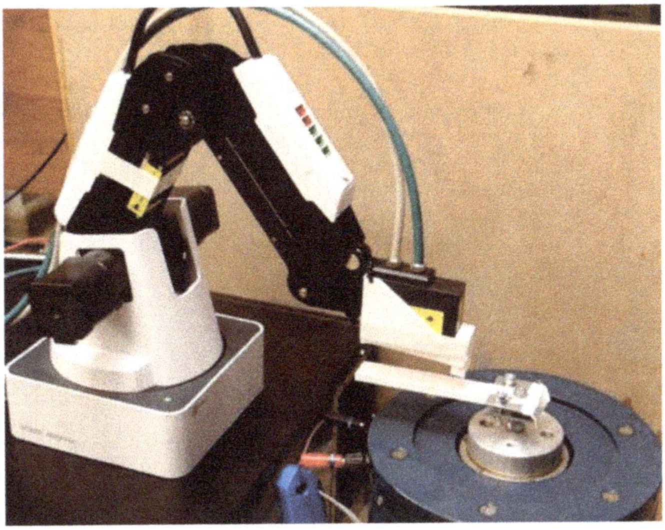

Figure 1. Test stand general view.

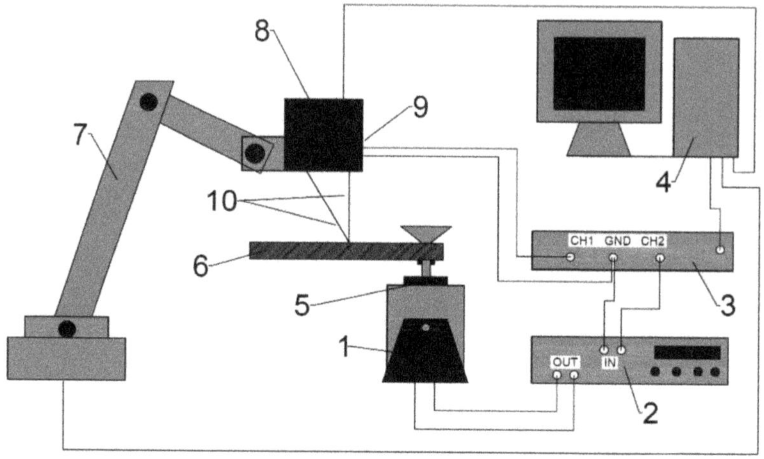

Figure 2. Diagram of the test stand: 1—vibration exciter; 2—amplifier; 3—digital-to-analog converter; 4—computer; 5—mounting; 6—specimen under investigation; 7—robotic arm; 8—optical sensor (distance measurement); 9—synchronizing signal connection; and 10—laser beam.

The analysed specimen is fixed to the electromagnetic core of the vibration exciter powered by an analogue amplifier. The signal is generated using the first channel of the computer sound card. The second channel is used to generate a rectangular waveform (with the frequency of 48 kHz), which is the synchronizing signal for the optical sensor. In this case, each ascending slope triggers a recording of the observed point distance from the sensor. The entire system is operated, programmed, and controlled in the LabView® environment, including the readout and processing of measurement quantities (LabView® Sound and Vibration Toolkit). After the measurement points and the range of tested frequencies are determined, the measurement is fully automatic. The stand's basic operating mode is vibration scanning of the specimen. This process requires, firstly, an array containing tested frequencies (usually, a range of values with monotonic, arithmetic, or geometric progression is determined) and, secondly, the coordinates defining all points of measurement (packets of three numbers—x,y,z). The number of individual measurements is the product of the number of geometrical locations and the number of scanned frequencies. Such a single measurement consists of placing the sensor in a set location and starting to send a signal with a set frequency to the input of the electromagnetic exciter amplifier, which is thus set in motion. At the same time, the exciter core acts as the tested specimen support. The motion of the support is thus the base excitation with an amplitude and frequency set by the user. After the exciter is activated, it is necessary to bring the oscillating system to a quasi-steady state, which, as demonstrated by the tests, takes a few seconds, depending on the vibration frequency. At this stage, the synchronizing signal is flat—no data are collected. Then, the next stage of collecting data from the optical displacement sensor begins, at which point the synchronizing signal starts to oscillate at 48 kHz. The data collection time should be long enough to capture many full oscillation periods (being the effect of the input function frequency), including potential occurrence of acoustic beats. An appropriately long measurement time, i.e., a period producing an adequately large set of measuring data, enables correct data processing and, first of all, makes it possible to improve the accuracy by averaging data for many periods. Based on the testing results, the duration of the data acquisition stage adopted in this study was one second. After the measurements are completed, the signal forcing the exciter motion and the sensor synchronization is turned off.

The number of collected data defining the measurement point distance from the sensor should be equal to the product of the sampling frequency (48 kHz) and the measuring time. The obtained set of measurement results is then processed. First, the number of collected data is checked. It is also checked whether each data value is included within the measuring range and whether there are no transmission errors. In the next step, the data are filtered using a band-pass filter to capture the one and only frequency-dependent component of the currently tested quantity (equal to the input function supplied to the vibration exciter). Fourier analysis is then conducted and the main frequency is determined, together with its amplitude and phase. It is also confirmed whether it is equal to the input function frequency (with a small error margin taking account of the numerical character of the testing and the finite nature of the data set).

In addition, methods allowing compensation of the elasticity of the specimen holder may be used [45]. The proposed arrangement of the test stand may also be employed for imaging and determining significant characteristics for nonlinear elements [46]. After minor modifications, it can also be applied to investigate complicated models made on a small scale, especially damping in the structure [47].

2.2. Phase Angle Determination Procedure

The most innovative part of the constructed test stand is the possibility of measuring the phase angle on a test rig fitted with a single measuring sensor. The process makes use of the mutual relative repeatability of the input function signal and of the signal synchronizing the measuring sensor (and also of the signal triggering data acquisition) described above in Section 2.1. Data collection always begins precisely at the same instant of the vibration exciter cycle defined by the base excitation waveform, i.e., the measurement always begins at the same moment of the cycle duration (t_0), at a set value of the input function frequency. Consequently, resultant signals (the time-dependent phase shift curve) collected at two points, or more specifically, their relative phase shift, represent the real phase shift between oscillating motion at these points at a set frequency. The idea of measurement procedure realization is illustrated in Figure 3. The top and bottom waveforms represent the measurement results for two different points on the specimen. The blue line marks the history of the actual time-dependent displacement of the tested point on the specimen. The red line illustrates the synchronizing signal, the beginning of which is synchronized with the input function—it always appears at the same moment of the cycle, and consequently, at the same position of the exciter core, and hence, at the same position of the entire specimen. The red markings denote collected measurement data, based on which, after they are connected, the phase shift θ can be determined. The quantity on the vertical axis in Figure 3 can be expressed as displacement, velocity, or acceleration, which is irrelevant for the phase angle measurement idea.

The phase angle determination method consists of setting the measuring sensor so that the measuring light beam falls onto the tested element. A two-channel signal is generated next, lasting long enough for the element vibrations to stabilize (usually from a fraction of a second to a few seconds). The signal from one of the channels consists of a sinusoidal waveform starting from zero; the other channel consists of a zero curve (no signal) for a set, precisely measured over time (vibration stabilization). A signal is then generated with a frequency from several to a few dozen kHz, determining the data sampling frequency, i.e., the accuracy of frequency and the phase shift measurement by the measuring sensor. The sampling frequency depends on the maximum frequency of the measuring sensor operation—the higher the sampling frequency, the more accurate the history of oscillations and thus phase angle determination accuracy. The procedure is repeated, varying the measuring sensor location so that the light beam falls onto another point on the tested element.

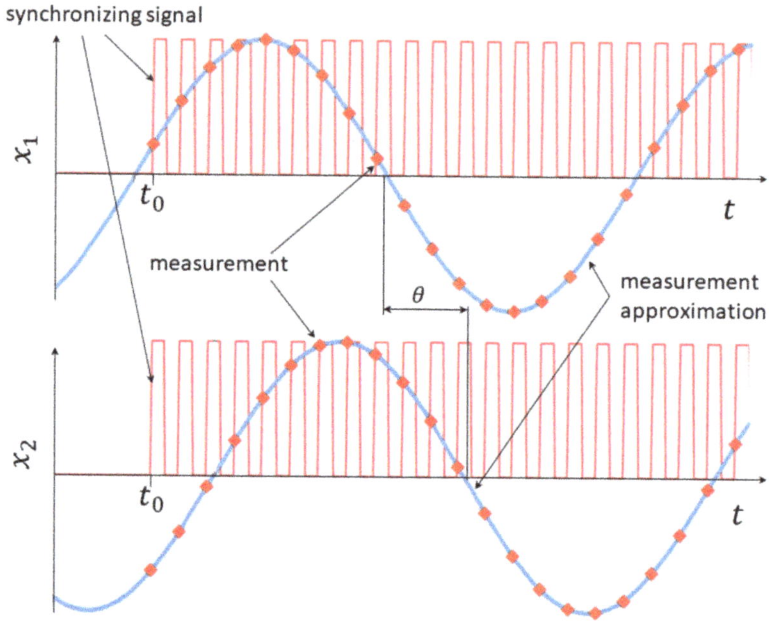

Figure 3. Schematic diagram of the phase shift measurement method.

The measurements result in two one-dimensional tables with numerical values, each of which includes recorded measurements. Plotting their (approximated) history in a chart as a function of time makes it possible to draw a comparison between signals from the two points and determine the phase angle shift between them (θ) Therefore, the essence of the measurement is the use (by controlling the exciter) of fully repeatable base excitation precisely synchronized in time with data acquisition. The key component here is the tested process repeatability—the same history of the input function force is applied to the same system, which results in identical specimen behavior. The measuring sensor synchronizing signal (the moment of the start of the measurement) always appears at the same instant of the period of the exciter vibrations. This makes it possible to subsequently collect data for any number of measuring sensor locations, with a separate measurement being performed for each of these locations. Combining the data collected in this way enables determination of their relative phase shift characteristics. It is assumed that the material properties do not change during the test.

3. Beam Vibration Model

The tests presented in the paper were carried out for a prismatic cantilever beam with a uniformly distributed mass fixed on one end and subjected to dynamic input functions (the Euler–Bernoulli model). Because, during the testing, the oscillating motion measurements are only possible in one (vertical) direction (due to the test bench design) and the beam in question is symmetric, the vibrations may be considered one-dimensional. Only oscillations occurring in this vertical direction are taken into account; torsional forms are ignored, even though they are also present, resulting in slight displacements in the analysed direction. Additionally, oscillations due to deviation from the assumed ideal and symmetrical dimensions of the system are also ignored. Under these assumptions, the analysed specimen reference geometry can be simplified. The analysis only covers the change in vibration values along the specimen length, because a change of the measurement coordinate in other spatial directions should not affect the results—the system can be considered one-dimensional. The oscillating motion direction is perpendicular to the "length" coordinate.

The natural vibration form is determined using a model constructed based on Equation (1). In the determination of the vibration frequency of a prismatic cantilever beam with a constant density along the beam's entire length, the basic assumption is equilibrium between internal forces (the left side of the equation) and inertia (the right side of the equation) for each element along the beam length (dx):

$$\frac{\partial V_x}{\partial x}dx = m\frac{\partial^2 y}{\partial t^2}dx, \tag{1}$$

where V_x denotes the shearing force; m is the linear mass (mass per length); and x and y are axial and lateral coordinates, respectively.

Considering the relation between the bending moment M_x and the shearing force for the prismatic beam,

$$\frac{dM_x}{dx} = V_x \tag{2}$$

and taking account of the beam deflection equation

$$\frac{M_x}{EI} = \frac{\partial^2 y}{\partial x^2}, \tag{3}$$

the fourth-order differential equation with respect to time (t) is obtained (4):

$$\frac{\partial^2 y}{\partial x^2} + \frac{EI}{m}\frac{\partial^4 y}{\partial t^4} = 0, \tag{4}$$

where E and I are the elastic modulus and the second moment of the area, respectively. In terms of the considerations presented herein, the essential solutions are only those corresponding to standing waves, i.e., cases where natural vibration forms are created. The solution of Equation (4) can therefore be presented as a function of separable variables—a product of two functions where one depends on time and the other only depends on space.

Consequently, vibration forms can be expressed using relation (5), which is a general solution of differential Equation (4) [48]:

$$X(x) = C_1\sin(\alpha x) + C_2\cos(\alpha x) + C_3\sinh(\alpha x) + C_4\cosh(\alpha x), \tag{5}$$

where X denotes deflection at axial coordinate x, and $\alpha = \sqrt[4]{\frac{m\omega^2}{EI}}$. Constants C_n are determined using boundary conditions. Since one end of the beam is fixed ($x = 0$), no motion is possible there (neither translation nor rotation occurs).

Moreover, due to free motion of the other end, the forces acting at this point ($x = l$) are zero, l is the length of the beam. The solution expressing frequencies in this situation is

$$f = \frac{\beta}{l^2}\sqrt{\frac{EIl}{m}}, \tag{6}$$

where $\beta = 0.560$ and $\beta = 3.507$ for the first and second bending form under analysis, respectively. The analysed forms of the prismatic cantilever beam natural vibration are presented in Figure 4.

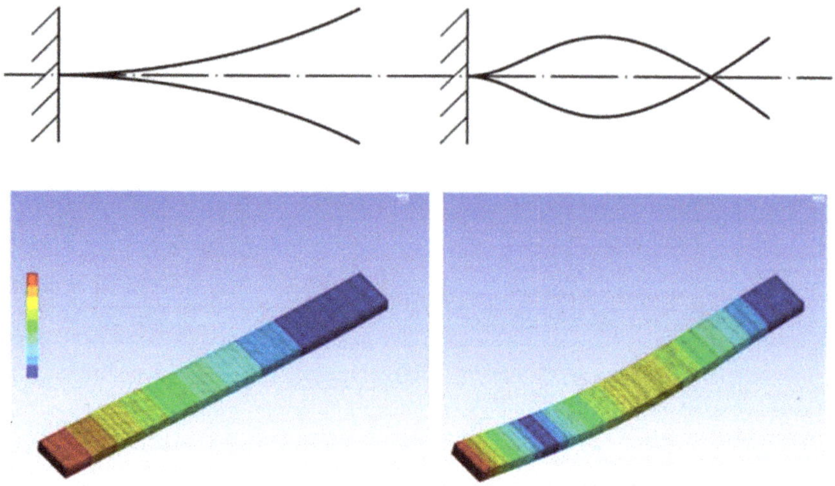

Figure 4. First (**left**) and second (**right**) vibration form in the direction perpendicular to the beam; top: schematic diagram of vibration forms; bottom: FEM results (blue- smallest amplitude, red biggest amplitude).

4. Experimental Testing

The tests were performed on specimens in the form of a prismatic cantilever beam with a rectangular cross-section made with Additive Manufacturing technology. The specimens were printed using the Makerbot Replicator 2X device with the following settings: Infill of 100% (printing degree: 45°, moved with every layer by 90°) and two shells. The nozzle temperature was 210 °C, and the printing proceeded horizontally (the shortest side upwards). The Basf Innofil 3D PLA WHITE filament was used. Several specimens with different dimensions in the form of rectangular prisms were made for the purpose of testing. The ones selected for the measurements were as follows: I—197 × 20 × 6 mm; II—197 × 20 × 4 mm; and III—117 × 20 × 4 mm. The outermost 19-mm part of the specimen was slotted into the fixing system made of two flat steel elements. The specimen was clamped in between using screws and the whole specimen was attached to the electromagnetic exciter core. The fixing system (cf. Figure 2) was a compact structure with a much higher stiffness than the tested specimens. The first step was to establish the specimen's true density using laboratory pan scales. The calculations were performed by adopting the specimen dimensions defined in the geometry source file. The actual measurement of geometry indicated slightly higher values due to the surface texture being an effect of the manufacturing technology (excessive extrusion, material swelling, and excessive looseness). The filament's technical data sheet [49] lists selected data characterizing its raw state and the printed product material. The density is given as 1.26 g/cm^3. The measurement results are listed in Table 1. The determined density value resulted from the specimen's geometrical dimensions and mass.

The specimen refers to the tested beam part protruding from the fixture. The first to be tested was the 197 × 20 × 6 mm specimen. It is hereinafter referred to as specimen I. To make a rough estimate of its dynamic characteristic, the specimen was modeled using the Ansys Mechanical package. The model only comprises the part protruding from the support, i.e., the tested dimensions are 178 × 20 × 6 mm. The beam was fixed on the 6 × 20 front plane to obtain the cantilever beam model. The established value of density (1.24 g/cm^3) and the elastic modulus value read from the material specification sheet data (2852 MPa) were applied. Because the geometrical dimensions and mass were measured with the accuracy of 0.1 mm and 0.01 g, respectively, the density determination uncertainty was estimated at the level of 0.03 g/cm^3.

Table 1. Specimen data.

Element	Length mm	Width mm	Thickness mm	Mass g	Density g/cm^3
I	197.0	20.0	6.0	29.16	1.24
II	197.0	20.0	4.0	19.55	1.24
III	117.0	20.0	4.0	11.55	1.24
				avg. ->	1.24

4.1. Determination of Resonance Frequencies Using the Displacement Amplitude

Following the procedure described in Section 2.1 above, the tested specimens' displacement amplitude was measured in frequency ranges determined through numerical simulations. The obtained values of frequency (numerical simulation) for subsequent vibration modes are listed in Table 2. The vibration forms for which the motion of every point only occurs in the direction possible to record due to the test bench design (vertical) are marked with an asterisk (*). This direction remains the same, irrespective of changes in the horizontal coordinate normal to the specimen axis. The selected vibration forms therefore represent motion that can be defined using a single coordinate and are a solution of the Euler–Lagrange equation. Mode 2 (Table 2) represents transverse vibration in the second plane (left-right, not analysed); modes 4, 6, and 8 are related to the rotation of cross-sections (specimen torsion); and mode 9 involves simple compression and axial tension of the specimen (longitudinal wave mode).

Table 2. Vibration modes of specimen I.

Mode	1	2	3	4	5	6	7	8	9
Freq., Hz	46.6 *	153.5	290.6 *	698.2	807.4 *	911.3	1565.0 *	2102.9	2134.4

In the case of modes used for further analysis (modes 1 and 3 from Table 2), the selection of the Poisson ratio, and consequently, of the shear stiffness (the Kirchhoff modulus), for the analysed geometry does not have an impact on the change in the frequency or form of these natural vibrations (assuming the defined Young modulus).

In the next step, the specimen was attached to the vibration exciter and the above-described (Section 2) vibration scanning was carried out. The location of measurement points on the tested system's geometry is presented in Figure 5. Measuring frequencies were in the range of 43–48 Hz with the scanning step of 0.1 Hz and 284–300 Hz with the scanning step of 0.2 Hz. An ambient temperature was maintained at the level of 23 ± 1 °C; air from the surroundings was gently blown onto the specimen by a fan.

Measurements were performed at 10 points located 20 mm apart along the specimen's center line (points 1–10) and at two points on the element holding the specimen (Point 11 and 12). Point 10 on the specimen and Point 11 on the fixture were as close to each other as possible, and close to the fixing element edge (about 2 mm). The measurements performed in the fixing system enabled an indirect assessment of stiffness through observations of the vibration phase angle between Point 11 and 12.

Table 3 presents the results of the displacement amplitude measurement for all 12 points under analysis. The coordinates of the points defined in relation to the fixing screw (D) are given in the second row of the table. It can be seen that the first resonance frequency is included in the investigated range. However, a deeper analysis of the results indicates that precise determination of the resonant frequency is debatable. For the investigated vibration form (mode 1), the highest vibration amplitude occurs at the specimen's far-end (which is the most distant from the support (Point 1)). The simplest pattern of finding the resonance frequency value (by finding the maximum amplitude in this place) gives the result of 45.5 Hz.

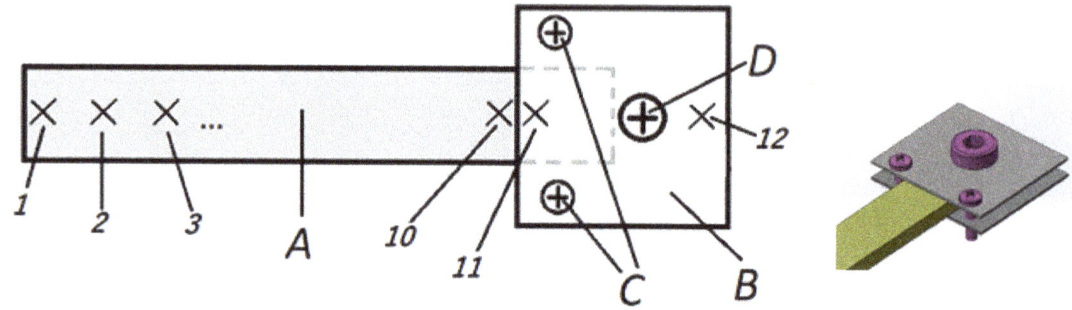

Figure 5. Diagram of the measurement: A—tested specimen; B—fixing metal element; C—fixing screws; D—exciter fixing screw; and 1–12—points of measurement.

Table 3. Measurement results of specimen I.

pt. Number →	**1**	**2**	**3**	**4**	**5**	**6**	**7**	**8**	**9**	**10**	**11**	**12**
position x mm →	194	174	154	134	114	94	74	54	34	24	19	−9
Freq., Hz ↓						Amplitude, mm ↓						
44.2	1.116	0.954	0.787	0.628	0.489	0.354	0.238	0.139	0.073	0.050	0.040	0.012
44.3	1.170	1.007	0.826	0.657	0.511	0.372	0.249	0.143	0.073	0.050	0.039	0.010
44.4	1.240	1.056	0.873	0.694	0.543	0.391	0.260	0.147	0.074	0.049	0.038	0.008
44.5	1.307	1.119	0.921	0.732	0.565	0.411	0.272	0.153	0.075	0.048	0.037	0.007
44.6	1.380	1.182	0.968	0.769	0.596	0.432	0.285	0.158	0.075	0.048	0.036	0.009
44.7	1.475	1.252	1.026	0.817	0.629	0.453	0.298	0.164	0.075	0.046	0.033	0.012
44.8	1.555	1.329	1.091	0.858	0.665	0.475	0.312	0.168	0.075	0.045	0.031	0.016
44.9	1.638	1.399	1.148	0.908	0.699	0.497	0.324	0.173	0.073	0.043	0.028	0.022
45.0	1.720	1.461	1.196	0.949	0.729	0.518	0.336	0.177	0.074	0.040	0.025	0.027
45.1	1.789	1.527	1.248	0.987	0.753	0.535	0.344	0.180	0.068	0.036	0.021	0.034
45.2	1.844	1.571	1.289	1.019	0.775	0.548	0.350	0.181	0.064	0.033	0.018	0.040
45.3	1.878	1.603	1.311	1.040	0.786	0.555	0.353	0.179	0.061	0.028	0.015	0.046
45.4	1.896	1.615	1.326	1.051	0.792	0.556	0.352	0.176	0.057	0.023	0.013	0.051
45.5	1.893	1.613	1.326	1.045	0.793	0.552	0.347	0.171	0.052	0.019	0.013	0.056
45.6	1.869	1.595	1.304	1.036	0.777	0.542	0.338	0.164	0.046	0.015	0.014	0.060
45.7	1.828	1.561	1.283	1.008	0.757	0.524	0.325	0.156	0.041	0.013	0.017	0.064
45.8	1.764	1.506	1.237	0.977	0.731	0.504	0.310	0.146	0.035	0.011	0.020	0.066
45.9	1.689	1.442	1.187	0.936	0.697	0.481	0.293	0.134	0.030	0.011	0.024	0.068
46.0	1.601	1.372	1.130	0.891	0.660	0.452	0.274	0.124	0.023	0.013	0.027	0.069
46.1	1.515	1.295	1.065	0.840	0.623	0.425	0.255	0.113	0.018	0.015	0.029	0.069
46.2	1.427	1.219	1.004	0.790	0.586	0.399	0.236	0.102	0.014	0.017	0.031	0.069

Alternative approaches, i.e., finding the maximum vibration ratio between Point 1 and Point 12, 11, and 10, result in the respective values of 44.5, 45.5, and 45.8 Hz (marked). The curves illustrating the vibration forms for these three values are presented in Figure 6.

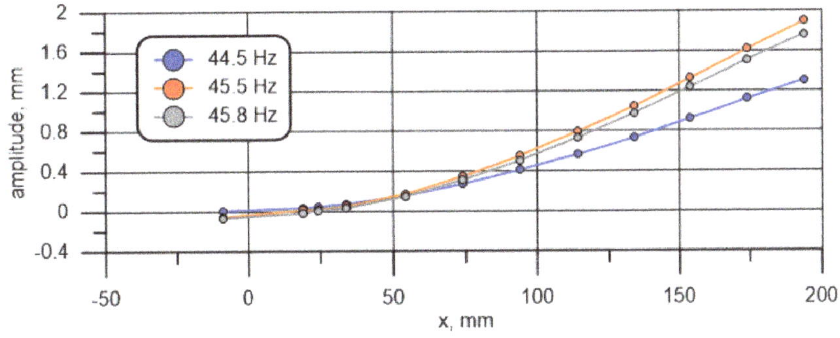

Figure 6. Amplitude at frequencies close to the first resonance frequency as a function of location.

It can be seen that the shape of the forms changes; the nodal point is shifted towards the specimen center with a rise in the base excitation frequency. Negative values represent a shift of 180° in relation to the other points (the value is rounded down or up to 0° (360°) or 180°). Because, under the adopted theoretical assumptions and using the constructed numerical model only the specimen's protruding part is simulated, the most appropriate approach in this situation is to adopt a frequency for which the waveform node is located on the interface between the specimen's tested part and the support as the natural frequency of mode 1. Considering the discrete character of the measurements, it is assumed that the point is located half way between measuring points 10 and 11. It follows from Table 3 that the lowest amplitudes for these two points, which at the same time result in the occurrence of the minimum of the approximating function between them (built on all 12 points of measurement), correspond to 45.6 Hz. This value is adopted as the first resonance frequency adequate for further consideration.

A similar procedure can be applied for the next resonance amplitude. The value determined in this way is 290.6 Hz if the criterion is the maximum amplitude of the end, which is the most distant point from the fixing screw, and 296.2 Hz if the waveform node is assumed on the specimen and fixture interface.

Figure 7 illustrates the natural vibration forms for these frequencies. It should be noted that for the two frequencies, the point for which the approximated amplitude is zero, i.e., the point corresponding to the specimen cross-section that does not move vertically, but only rotates, is located at a position close to $x = 158$ mm.

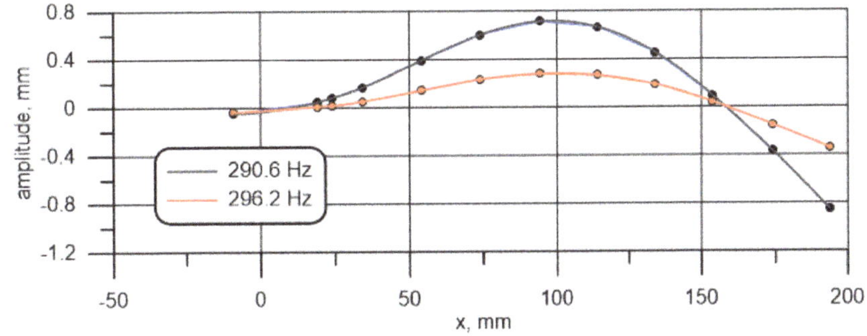

Figure 7. Amplitude at frequencies close to the second resonance frequency as a function of location.

The difference between the node location for the two frequencies (Figure 7) is less than 0.5 mm and, considering the measurement accuracy, is of no practical significance. A higher vibration amplitude for the lower frequency does not necessarily mean a different location of the excited wave nodal point. It may just be due to the way in which energy is supplied to the oscillating system or characteristics of the exciter and the amplifier system. The value of 290.6 Hz was adopted for further consideration.

Figure 8 presents the curves illustrating changes in the beam end amplitude, depending on the frequency in the resonance area of the first two bending modes. It can be observed that the curve representing the second mode is more asymmetric in relation to the extremum. This problem may be due to an unknown relation between the exciting force and frequency. It is unknown because, firstly, the characteristic of the sound card amplification, the amplifier, and the exciter is not perfectly linear, and secondly, because the specimen mass and motion affect the motion of the exciter. This characteristic has been corrected to a certain extent, but only for an exciter without a specimen attached to it. If the specimen is attached, especially if there is motion near resonance, additional inertia forces are introduced into the system. In turn, this has a significant impact on the motion of the exciter core and the forces generated within it. Modeling this phenomenon would require knowledge on the exciter's internal structure (core geometry, number of cores, magnetic and electromagnetic field distribution, etc.) and the amplifier's response to induced currents. The same phenomenon has, among others, an impact on the shape of the curves in Figure 8 and the amplitudes in Figures 6 and 7.

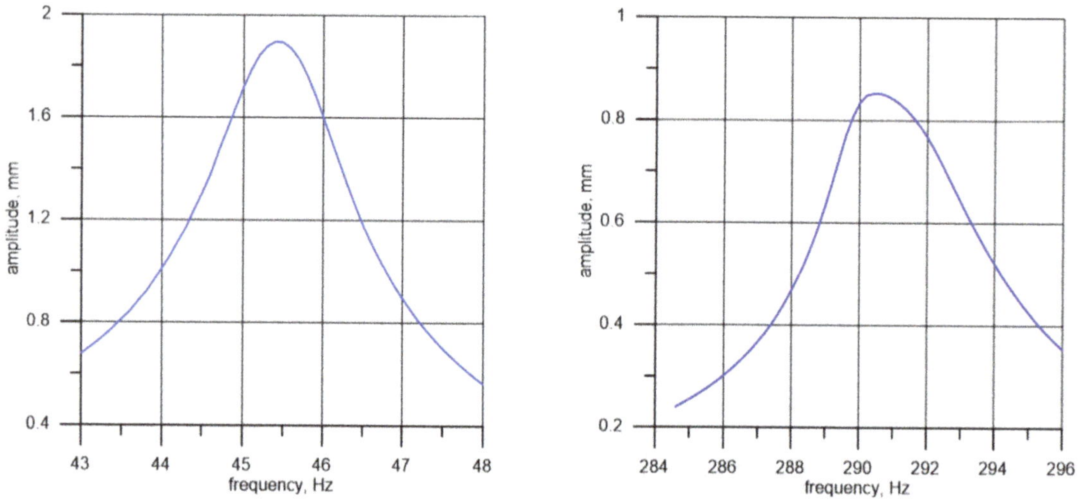

Figure 8. Amplitude depending on the vibration frequency for the first (**left**) and second (**right**) vibration form of specimen I.

Apart from the measurements mentioned above, tests on the analysed element free vibration were also performed. For this purpose, the beam was statically bent by about 3.5 mm at its free end. The end was then released and the specimen motion was recorded using the optical displacement sensor employed previously. The sensor was placed at Point 1 (cf. Figure 5). The recorded changes in the vibration signal for the 197 × 20 × 6 mm beam are illustrated in Figure 9. The Fourier analysis performed for this data set indicates the frequency value of 44.77 Hz.

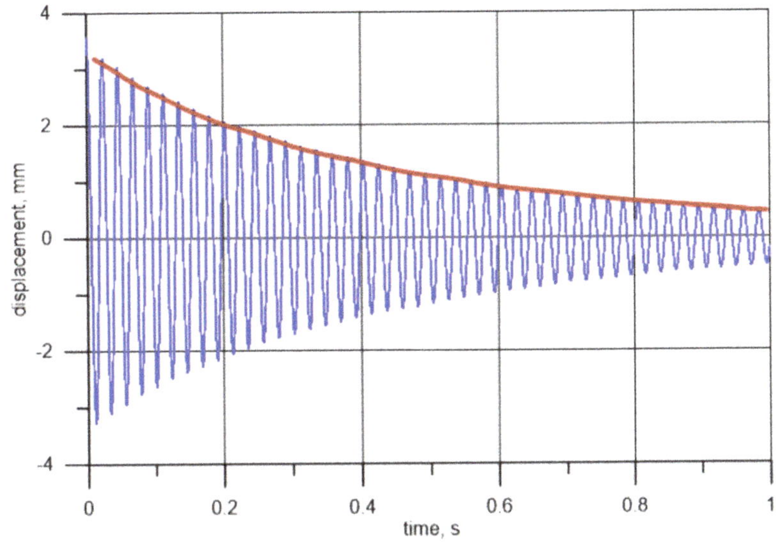

Figure 9. Displacement (distance) depending on selected time points for free vibrations with a decreasing amplitude for specimen I.

4.2. Evaluation of Vibration Damping

The measurement results presented in Figure 9 were used to evaluate vibration damping of the tested specimens. To this end, an algorithm that detects subsequent maxima of the history of changes in values of displacement in subsequent periods and then calculates the ratio between energy (V) in the oscillating element in a given cycle and in the cycle preceding it was created. Based on that and using slight transformations, it was possible to determine the loss factor in relation to a single cycle.

$$\eta = \frac{\Delta V}{V} = \frac{V_{i-1} - V_i}{V_{i-1}} \tag{7}$$

It was assumed that stored energy is proportional to the square of the amplitude, as in the case of a harmonic oscillator. The obtained results are shown in Figure 10 (left). The same results, but calculated on a different time scale and taking account of the division into values established using minima and maxima (for subsequent periods), are presented in Figure 10 (right). The average value of η for the first second of the measurements totals 8.4% for the function maxima shown in Figure 9, and 8.2% for the minima. With time, and as the amplitude decreases, the relative error, and consequently the spread of points around the average value, increases. For this reason, measurements should be limited to the initial phase of oscillations right after they are induced. Figure 10 illustrates changes in the relative energy drop in subsequent cycles.

The presented procedure is equivalent to matching a curve described by equation $Ae^{-\zeta\omega_n t}$ to measurement data [50] and establishing a specific damping ratio ζ as a loss factor per radian:

$$\zeta = \frac{\eta}{4\pi}. \tag{8}$$

Factor 4 results from the fact that, at resonance, the loss factor is twice the loss damping ratio. Loss factor values lower than zero contradict the laws of physics. If such a value is found, it is an effect of measuring errors and indicates that the apparatus has an inadequate accuracy. This is particularly visible and significant at small amplitudes, where the relative error becomes unacceptably large. The red line in Figure 10 represents a straight line matched to measuring data. It can be seen that as the displacement amplitude decreases, so does the loss factor, which is due to the material's smaller strains. Additionally, the aerodynamic interaction between the oscillating element and the environment decreases substantially.

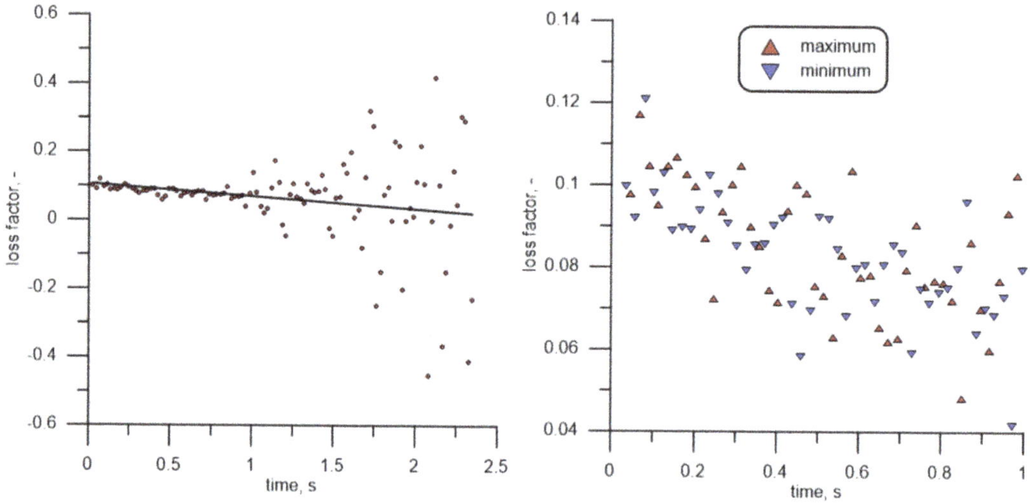

Figure 10. Loss factor depending on time (**left**) and a detailed view of the loss factor, depending on the time (**right**).

Figure 9 also shows the displacement amplitude history of Point 1 (red curve). It can be seen that at a free-end displacement of the order of 3.5 mm, after about 1 s, the amplitude falls below 1 mm. This explains the rising relative error mentioned above. The presented methods deliver loss at every one period, contrary to matching the $Ae^{-\zeta\omega_n t}$ curve. This also enables an analysis of the change in the loss factor during each period, together with the change in the amplitude (and time). It has to be mentioned that the samples are collected at a much higher frequency compared to the vibrations. Even in an extreme situation, where the real extremum is exactly in the middle of the collected samples, the deviation of the determined maximum is not bigger than the sensor resolution.

4.3. Phase Angle Analysis

The developed test stand enables phase angle measurements using a single displacement sensor, as described above in Section 2.2. The phase shift is herein understood as relative displacement phase between points of measurement close to the specimen support (10, 11, or 12) and Point 1 (beam free end). The phase shift is determined using the same measurement results described above (cf. Table 3). The data are additionally normalized to values of 0–360, where 0° denotes synchronized motion, and positive values correspond to the beam end "delay". The curve illustrating the phase shift between points 1 and 10 is adopted for further consideration, as this curve best simulates the above-assumed beam eigenfrequency.

The phase angle can be determined analytically using the following relation [50]:

$$\theta_\zeta(\omega) = tan^{-1}\frac{2\,\zeta r}{1 - r^{2\prime}} \tag{9}$$

where $r = \omega/\omega_n$ is the ratio of forced (ω) and natural (ω_n) frequency.

Based on Equation (9), it is possible to draw a theoretical curve and compare it to the measuring curve (frequency-dependent changes in the phase angle) for known values of the damping ratio and natural frequency. It follows from Equation (9) that a change in natural frequency involves a change in the location of the point where the curve runs through the value of 90 degrees, and a change in damping affects the curve slope or flattening. A rough analysis of the data marked in orange squares in Figure 11 (the phase shift between points 1 and 11) indicates that the natural frequency is close to 45.4 Hz, i.e., $\omega_n = 285.3\ s^{-1}$ (phase angle of 90°). The values were established with a better accuracy using the Microsoft Excel Solver tool. For a frequency of about 45.4 Hz from the range of 44.7–46.1 Hz with the resolution of 0.1 Hz, the approximating function defining the history of frequency-dependent changes in the phase angle was determined. This was achieved by minimizing the functional $\delta_\zeta(\omega)$ of the sum of squares of differences in the measured (θ_{mi}) and theoretically established (θ_{ti}) phase angle values for each of the fifteen frequencies (points) under analysis (frequency closest to the 90° in the middle):

$$\delta_\zeta(\omega) = \frac{1}{n}\sum_{i=1}^{n}(\theta_{ti} - \theta_{mi})^2 \rightarrow min. \quad (10)$$

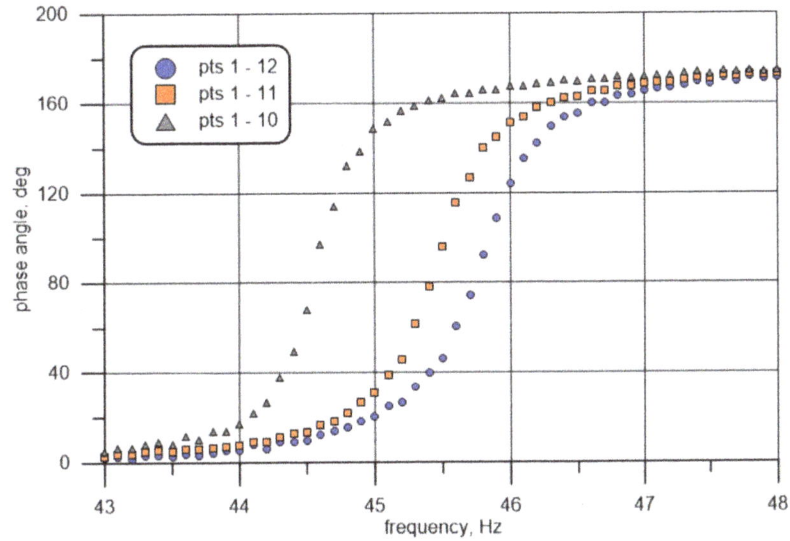

Figure 11. Frequency-dependent phase angle between measuring Point 1 and Point 10—gray, Point 11—orange, and Point 12—blue.

The value obtained in this manner was the minimization criterion for the "solver" function. The initial values for the described procedure algorithm were assumed to be $\eta = 0.07$ (value close to the one obtained from the testing of individually excited, free vibration) and $f_n = 45.4$ Hz. This produced values of $\eta = 0.082$ and $f_n = 45.44$ Hz, which were then adopted as exact resultant values of the determined quantities.

A similar procedure for the phase angle analysis was adopted for the next natural vibration form corresponding to the frequency of about 296 Hz. The theoretical curve was matched based on data from the range of 292–300 Hz. The obtained values of the loss factor and natural frequency were $\eta = 0.057$ and $f_n = 296.3$ Hz, respectively. The matching of the curves, together with the points representing the results of measurements, are presented graphically in Figures 12 and 13. For the used test bench, the uncertainty of the measured phase angle values (for one point) is approximately 0.5° for about 45 Hz and changes proportionally with the frequency. The uncertainty of the measured amplitude (for one point) is approximately 0.001 mm and the uncertainty of the measuring frequency is insignificant.

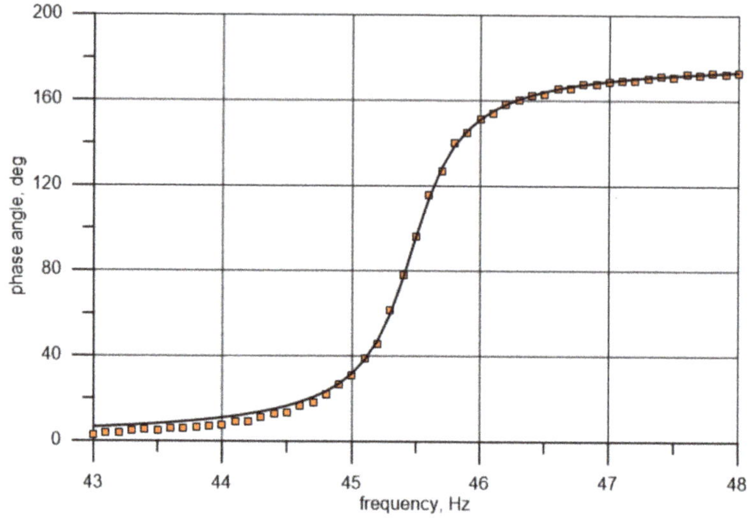

Figure 12. Comparison of the phase angle obtained from measurements and its theoretical distribution after eigenfrequency and damping were matched to natural frequency and damping values (1st bending mode).

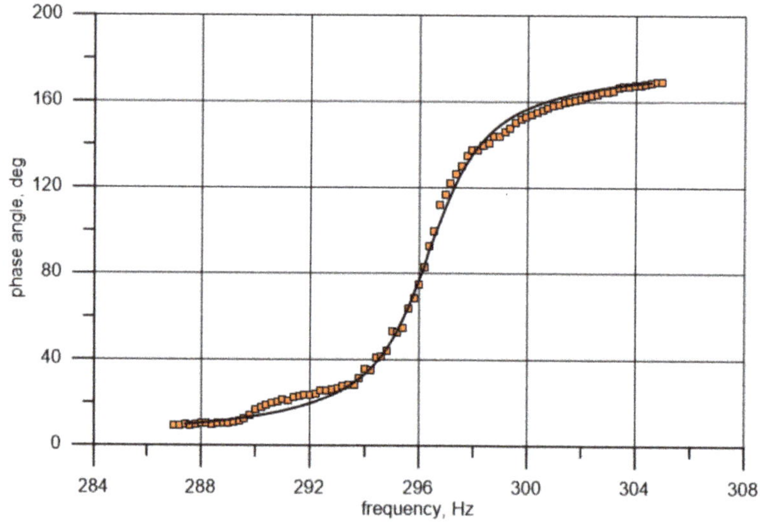

Figure 13. Comparison of the phase angle obtained from measurements and its theoretical distribution after eigenfrequency and damping were matched to natural frequency and damping values (2nd bending mode).

The above procedure was adopted for the other two specimens. Table 4 below presents the results obtained for all three beams (specimens I, II, and III).

Table 4. Vibration data for investigated specimens.

<table>
<tr><th></th><th></th><th>Element</th><th>I
178 × 20 × 6 mm</th><th>II
178 × 20 × 4 mm</th><th>III
98 × 20 × 4 mm</th></tr>
<tr><td></td><td></td><td>calculated frequency I, Hz</td><td>46.616</td><td>31.115</td><td>100.96</td></tr>
<tr><td></td><td></td><td>calculated frequency II, Hz</td><td>290.59</td><td>194.51</td><td>627.61</td></tr>
<tr><td rowspan="12">First bending mode</td><td rowspan="4">Amplification 1</td><td>maximum amplitude, mm</td><td>0.95</td><td>1.56</td><td>0.36</td></tr>
<tr><td>determined frequency (max amplitude), Hz</td><td>45.6</td><td>32.3</td><td>97.1</td></tr>
<tr><td>determined frequency (phase shift), Hz</td><td>45.52</td><td>32.31</td><td>97.06</td></tr>
<tr><td>loss coefficient, -</td><td>0.065</td><td>0.052</td><td>0.06818</td></tr>
<tr><td rowspan="4">Amplification 2</td><td>maximum amplitude, mm</td><td>1.89</td><td>3</td><td>0.87</td></tr>
<tr><td>determined frequency (max amplitude), Hz</td><td>45.5</td><td>32.3</td><td>97.5</td></tr>
<tr><td>determined frequency (phase shift), Hz</td><td>45.44</td><td>32.3</td><td>97.52</td></tr>
<tr><td>loss coefficient, -</td><td>0.082</td><td>0.06</td><td></td></tr>
<tr><td rowspan="4">Amplification 3</td><td>maximum amplitude, mm</td><td>3.71</td><td>5.9 *</td><td></td></tr>
<tr><td>determined frequency (max amplitude), Hz</td><td>45.4</td><td>32.2</td><td></td></tr>
<tr><td>determined frequency (phase shift), Hz</td><td>45.41</td><td>32.29</td><td></td></tr>
<tr><td>loss coefficient, -</td><td>0.083</td><td>0.074</td><td></td></tr>
<tr><td rowspan="12">Second bending mode</td><td rowspan="4">Amplification 1</td><td>maximum amplitude, mm</td><td>0.18</td><td>0.7</td><td>0.36</td></tr>
<tr><td>determined frequency (max amplitude), Hz</td><td>291.2/296.0</td><td>187/193.2</td><td>501.8/511.8</td></tr>
<tr><td>determined frequency (phase shift), Hz</td><td>297.8</td><td>193.04</td><td></td></tr>
<tr><td>loss coefficient, -</td><td>0.053</td><td>0.066</td><td></td></tr>
<tr><td rowspan="4">Amplification 2</td><td>maximum amplitude, mm</td><td>0.85</td><td>1.21</td><td>0.6</td></tr>
<tr><td>determined frequency (max amplitude), Hz</td><td>290.6/296.2</td><td>186.4/193.2</td><td>499.4/512</td></tr>
<tr><td>determined frequency (phase shift), Hz</td><td>296.3</td><td>193.12</td><td></td></tr>
<tr><td>loss coefficient, -</td><td>0.07</td><td>0.068</td><td></td></tr>
<tr><td rowspan="4">Amplification 3</td><td>maximum amplitude, mm</td><td>1.32</td><td>1.76</td><td></td></tr>
<tr><td>determined frequency (max amplitude), Hz</td><td>290.2/295.4</td><td>185.6/192.6</td><td></td></tr>
<tr><td>determined frequency (phase shift), Hz</td><td>295.2</td><td>192.67</td><td></td></tr>
<tr><td>loss coefficient, -</td><td>0.07</td><td>0.075</td><td></td></tr>
</table>

For the first two specimens, which were longer, the data were collected for three different excitement levels. This was achieved by changing the amplification of the channel controlling the vibration exciter, which is expressed quantitatively by the maximum achievable amplitude (the highest amplitude in the entire history at this test point) and described as the "maximum amplitude". In the case of specimen I, the amplitudes were 0.95, 1.89, and 3.71 mm for the first bending mode. For the first vibration form, no matter which criterion was adopted (the maximum amplitude of Point 1 or the ratio between the amplitude at Point 1 and Point 11), the result was the same. The next field for a given data set marked as the "determined frequency (max amplitude)" defines the resonance frequency found by establishing the amplitude maxima (measuring point 1) depending on the frequency. The next two fields below show the results obtained from the phase angle shift between points 1 and 11. In the case of the next bending mode, for a frequency of about 290–300 Hz for specimen 1, the values are presented in the same way as for the first vibration form. Two frequencies are given in the table for each element. The first was established based on the amplitude maximum value at Point 1 and the second was determined from the maximum ratio between the amplitude at Point 1 and in Point 11. For specimen III, due to the beam's higher eigenfrequencies and, consequently, smaller displacement amplitudes that could be achieved, a smaller amount of data was collected. The test stand structure, including the kind of input function and measured quantities (displacement), guarantees a higher resolution and measuring accuracy for low frequencies. In general, a rise in frequency involves a decrease in the accuracy of determination of individual indices in the presented configuration of the test bench.

The result marked with an asterisk (*) defines the specimen end amplitude estimated based on measurements performed at the element's other points (2, 3, 4, …) knowing the beam deflection shape. The result itself is beyond the measuring sensor range (amplitude of 5 mm for a symmetrical configuration).

Based on the data listed in Table 4, it can be concluded that oscillations are damped more intensively for bigger amplitudes, which can be seen in Figure 10. This may be related to another behavioral pattern of the material: Its nonlinearity for high stress and strain values, or to resistance of the air surrounding the specimen as air is pumped along the longer side due to the specimen's motion. At this stage, it is impossible to carry out a quantitative analysis of the impact of individual phenomena. Further numerical, analytical, and experimental investigations are required.

However, using the presented measuring data, it is possible to determine the material's elastic modulus E. Using relation (6) defining the prismatic cantilever beam's natural vibration, it can be stated that the eigenfrequency is proportional to the square root of E.

$$f \sim \sqrt{E} \tag{11}$$

Based on this relation and using the data obtained from numerical calculations, the material stiffness value in the model taken from the material technical data sheet (2852 MPa) can be corrected. The elastic modulus values determined from the measurements are listed in Table 5.

Table 5. Elastic modulus results.

Element	Mode	Frequency, Hz		E, MPa
		Calculated	Measured	
I	1	46.6	45.4	2705
II		31.1	32.3	3073
III		101.0	97.3	2649
I	2	290.6	296.4	2968
II		194.5	192.9	2806
III		627.6	512.0	1898

Differences of about ± 10% between individual specimens can be noticed for the first mode. The situation is similar for the second mode, but not for specimen III, which is characterized by the highest stiffness of all the tested elements. The elastic modulus value is very different in that case. This may be due to frequency-dependent changes in the material elastic properties. It may also be the effect of the significant relative error in amplitude determination (very small amplitude of the input function at the sensor constant error) and higher natural frequency (fewer points in time domain mapping the oscillating motion).

5. Discussion

Based on numerical calculations, it was found that the analysed damping values (determined from the history of changes in the phase angle) have little significant impact on natural vibration values, which is in agreement with theory ($\omega_d = \omega_n\sqrt{1-\zeta^2}$). The square root value for the observed instances of damping exceeds 0.99.

The values obtained from tests where free vibrations were excited for the other elements are listed in Table 6. The presented data illustrate the loss factor for amplitude values higher than 0.5 mm. Several tests were performed for each element. Their number is given in the table and the loss factor determined in each test never differed by more than 0.01 from the element average.

Table 6. Loss factor determined experimentally.

Specimen	I	II	III
Loss factor	0.085	0.067	0.087
Number of tests	2	3	8

The results presented in Table 6 suggest the diversity of intense values of the material stiffness with the specimen thickness. This may be due to stresses and material defects arising during the printing process and related to the manufacturing technology and cooling process. The phenomenon requires further analysis. It may also be an effect of the interaction and relative motion of the material layers.

Certain phenomena characterizing the developed and applied methodology, particularly the measuring and research methodology, were observed during the testing. Firstly, temperature has a substantial impact on the observed indications, and the resonance frequency values are especially affected. The temperature during the testing was maintained at the level of 23 ± 1 °C, but additional tests showed considerable variability of the resonance frequency (of the order of a few/several percent) in the temperature range of 18–32 °C.

The presented vibration testing can be used to determine the elastic modulus of an element made with AM technology. Apart from the base material, the elastic modulus value depends on the infill structure, the orientation of added layers, the infill percentage, and the infill quality. The most important factors are the quality of the layer-to-layer adherence/joint cohesion and possible geometrical discontinuities arising during the printing process. The nozzle temperature is probably also essential, but this requires further analyses. When determining the elastic modulus by means of the method presented herein, care should be taken to appropriately select the tested element's geometrical dimensions to meet the characteristics of the applied measuring apparatus. First and foremost, an opportunity should be created to ensure an adequately high amplitude to minimize the relative measuring error. Another important factor is the ratio between the specimen resonance frequency and the sampling frequency. The smaller the ratio, the more precise the vibration simulation. As a result, the stiffness will be established with a much better accuracy.

The tested element heating due to the work of friction forces inside the material may also have been significant in the testing described herein. To minimize the problem, it was made sure that, during each measurement (especially in the range of resonance frequencies), no temperature gradient higher than 1 °C was created. It was done by thermovision infrared camera. The susceptibility to this effect is strongly dependent on the tested element's geometry, amplitudes, input function forces, and specimen cooling.

Another phenomenon observed during the measurements worth noting is the impact of the test run-up time, i.e., the period from the moment the exciter is activated to the start of data acquisition. For the time of half a second, the values gathered and presented in the manner described above (cf. Figure 11) are shown in Figure 14 (specimen II). A local maximum and minimum can clearly be seen close to 33.2 and 33.8 Hz, respectively. Other characteristic deformations of the curves can also be noticed in the chart. They re-appeared when the tests were repeated and they are visible for all three basic points under analysis (10, 11, and 12). The phenomenon became weaker with an increase in the run-up time. This leads to the probable cause of the curve deformation mentioned above. It may be due to acoustic beats caused by the wave excited by the exciter's continuous motion (quasi-steady state) and the excited wave frequency when motion starts (transient-state condition). All the results presented above were obtained for a sufficiently long run-up time. According to the damping characteristic described above (roughly, a constant part of energy is lost in every period of motion), the time depends on the analysed frequency.

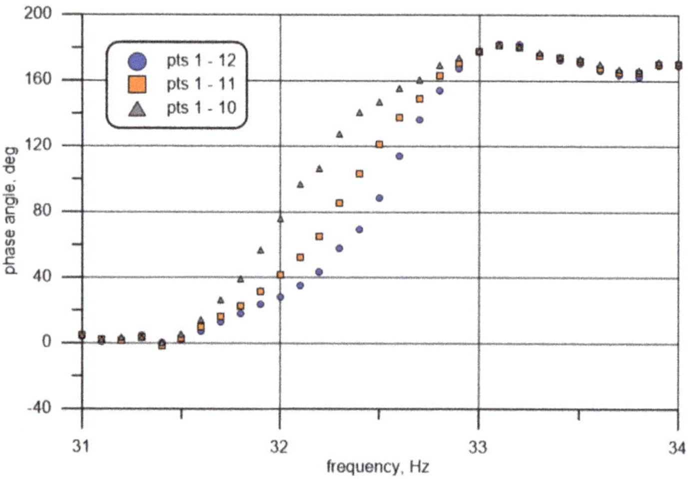

Figure 14. Frequency-dependent phase angle between measuring Point 1 and Point 10 for too short a run-up time: Point 10—gray; Point 11—orange; and Point 12—blue.

When performing numerical calculations and comparing their results to experiments, attention should be focused on the adopted model assumptions, which are usually a simplification. This particularly concerns the support conditions, which may differ from real conditions in many ways. One example is a support which is not stiff enough. Apart from its translational motion, slight rotation also occurs.

6. Summary and Conclusions

This paper presents a novel procedure for vibration measurements using a single measuring sensor. An optical displacement sensor with the measuring range of 10 mm is introduced. Such an approach limits the measuring apparatus costs and enables phase angle measurements for very small elements, where, for reasons related to the element dimensions, the standard use of two sensors is impossible. Also shown are methods of interpretation and usefulness of the data collected in this way. The data make it possible to determine selected mechanical parameters, such as the density, dynamic modulus of linear stiffness, shear modulus, material internal damping, etc. The proposed method enables measurements of damping and the resonance frequency in cases where the history of the frequency-dependent input function force is unknown.

An analysis of selected dynamic properties of elements made with 3D printing technology was conducted. To achieve a comprehensive assessment of the obtained values and to estimate measuring errors, it is necessary to repeat tests for a bigger number of specimens and perform a statistical analysis. It should be emphasized that the presented results depend on a large number of variables, e.g., the printing temperature, measurement ambient temperature, course of the printing process, etc. It seems that, if resonance frequencies are selected appropriately, the proposed method makes it possible to carry out detailed and reliable tests. The key issue is recording the displacement amplitude during the testing as accurately as possible, which depends on the measuring sensor's quality and the sampling frequency. The presented method can also be used with velocity and acceleration sensors. With appropriately selected vibration frequencies, this should improve the testing accuracy. The testing presented in the paper made it possible to record displacement in the range of 100–1600 measurements per period, i.e., with a phase accuracy of about 3.5° to about 0.2°. Lower sampling frequencies will worsen the accuracy, e.g., due to "missing out" the displacement extremum, to the poorer matching of selected approximating functions or to the aliasing. A reduction in the sampling frequency decreases the resolution and phase angle measurement accuracy for the determined waveform.

This paper presents a method of phase angle measurement using a single sensor enabling the determination of selected mechanical properties, in particular, the resonance frequency and damping in cases where functions of the quantitative description of the vibration exciter (force and base excitation) are not constant and their history is unknown. This makes it possible, for example, to use the system of the sound card, amplifier, and oscillator with different characteristics, which are also modified due to the dynamic impact of the fixed specimen.

It was found from the testing results that, in the analysed cases, the model of the loss of a constant part of the oscillating system energy described damping fairly well. No dramatic variations were observed with a change in the vibration frequency in the range of 30–300 Hz. Other aspects worth noting are the support stiffness and the stiffness of the entire exciter system. The support impact was substantially eliminated by an adequate selection of the phase measurement points. Nevertheless, the support stiffness may have affected the results to a certain extent. The results indicate that the metal fixture (the support) displacement was slight, but even small up-and-down motion non-synchronized in phase may, in the case of the described geometry, have a significant effect on the specimen behavior. This problem needs to be analysed further. The selected resonance occurrence criterion (minimization of the amplitude on the specimen-support interface) minimizes the error resulting therefrom.

Author Contributions: Conceptualization, R.B.; methodology, R.B. and G.N.; software, R.B.; validation, R.B. and G.N.; formal analysis, G.N.; investigation, R.B.; resources, R.B. and G.N.; data curation, R.B.; writing—original draft preparation, R.B. and G.N.; writing—review and editing, R.B. and G.N.; visualization, R.B.; supervision, G.N.; project administration, R.B. and G.N.; funding acquisition, R.B. and G.N. All authors have read and agreed to the published version of the manuscript.

Funding: This work was supported by the Silesian University of Technology, Grant No. 08/050/BKM18/0173.

Institutional Review Board Statement: Not applicable.

Informed Consent Statement: Not Applicable.

Data Availability Statement: The data presented in this study are available on request from the corresponding authors.

Conflicts of Interest: The authors declare no known conflicts of interest. The funders had no role in the design of the study; in the collection, analyses, or interpretation of data; in the writing of the manuscript; or in the decision to publish the results.

References

1. Tofail, S.A.M.; Koumoulos, E.P.; Bandyopadhyay, A.; Bose, S.; O'Donoghue, L.; Charitidis, C. Additive manufacturing: Scientific and technological challenges, market uptake and opportunities. *Mater. Today* **2018**, *21*. [CrossRef]
2. Quan, Z.; Wu, A.; Keefe, M.; Qin, X.; Yu, J.; Suhr, J.; Byun, J.; Kim, B.; Chou, T.-W. Additive manufacturing of multidirectional preforms for composites: Opportunities and challenges. *Mater. Today* **2015**, *18*. [CrossRef]

3. Gibson, M.A.; Mykulowycz, N.M.; Shim, J.; Fontana, R.; Schmitt, P.; Roberts, A.; Ketkaew, J.; Shao, L.; Chen, W.; Bordeenithikasem, P.; et al. 3D printing metals like thermoplastics: Fused filament fabrication of metallic glasses. *Mater. Today* **2018**, *21*. [CrossRef]
4. Lu, B.; Li, D.; Tian, X. Development trends in additive manufacturing and 3D printing. *Engineering* **2015**, *1*, 85–89. [CrossRef]
5. Canessa, E.; Fonda, C.; Zennaro, M. Low-cost 3D printing for science, education & sustainable development. In Proceedings of the 2013 ICTP Science Dissemination Unit, Trieste, Italy, 6–8 May 2013; ISBN 92-95003-48-9.
6. Shakerin, S.; Hadadzadeh, A.; Amirkhiz, B.S.; Shamsdini, S.; Li, J.; Mohammadi, M. Additive manufacturing of maraging steel-H13 bimetals using laser powder bed fusion technique. *Addit. Manuf.* **2019**, *29*. [CrossRef]
7. Zindani, D.; Kumar, K. An insight into additive manufacturing of fiber reinforced polymer composite. *Int. J. Lightweight Mater. Manuf.* **2019**, *2*. [CrossRef]
8. Yu, T.; Zhang, Z.; Liu, Q.; Kuliiev, R.; Orlovskaya, N.; Wu, D. Extrusion-based additive manufacturing of yttria-partially-stabilized zirconia ceramics. *Ceram. Int.* **2019**. [CrossRef]
9. Haleem, A.; Javaid, M. 3D printed medical parts with different materials using additive manufacturing. *Clin. Epidemiol. Glob. Health* **2019**. [CrossRef]
10. Urhal, P.; Weightman, A.; Diver, C.; Bartolo, P. Robot assisted additive manufacturing: A review. *Robot. Comput. Integr. Manuf.* **2019**, *59*. [CrossRef]
11. Li, G.; Zhao, J.; Wu, W.; Jiang, J.; Wang, B.; Jiang, H.; Fuh, J.Y.H. Effect of ultrasonic vibration on mechanical properties of 3D printing non-crystalline and semi-crystalline polymers. *Materials* **2018**, *11*, 826. [CrossRef] [PubMed]
12. Qahtani, M.; Wu, F.; Misra, M.; Gregori, S.; Mielewski, D.F.; Mohanty, A.K. Experimental design of sustainable 3D-printed poly(lactic acid)/biobased poly(butylene succinate) blends via fused deposition modeling. *ACS Sustain. Chem. Eng.* **2019**, *7*. [CrossRef]
13. Das, P.; Sahu, S.K. Experimental and numerical study on free vibration of cracked woven fiber glass/epoxy composite beam. *Mater. Today Proc.* **2020**, *33*, 5505–5510. [CrossRef]
14. Gori, Y.; Kumar, P.; Patil, P.P. FEA simulation of vibrating cantilever plate with transverse surface crack. *Mater. Today Proc.* **2017**, *4*. [CrossRef]
15. Gomez-Gras, G.; Jerez-Mesa, R.; Travieso-Rodriguez, J.A.; Lluma-Fuentes, J. Fatigue performance of fused filament fabrication PLA specimens. *Mater. Des.* **2018**, *140*, 278–285. [CrossRef]
16. Chacon, J.M.; Caminero, M.A.; García-Plaza, E.; Nunez, P.J. Additive manufacturing of PLA structures using fused deposition modelling: Effect of process parameters on mechanical properties and their optimal selection. *Mater. Des.* **2017**, *124*, 143–157. [CrossRef]
17. Ezeh, O.H.; Susmel, L. On the fatigue strength of 3D-printed polylactide (PLA). *Procedia Struct. Integr.* **2018**, *9*, 29–36. [CrossRef]
18. Shi, X.; Li, Q.; Zheng, A. Effects of heat treatment on the damping of EVM/PLA blends modified with polyols. *Polym. Test.* **2014**, *35*, 87–91. [CrossRef]
19. An, T.; Selvaraj, R.; Hong, S.; Kim, N. Creep behavior of ABS polymer in temperature-humidity conditions. *ASM Int. J. Mater. Eng. Perform.* **2017**, *26*. [CrossRef]
20. Zhang, D.; Li, Z.; Qin, S.; Han, S. Optimization of vibration characteristics of fused deposition modeling color 3D printer based on modal and power spectrum method. *Appl. Sci.* **2019**, *9*, 4154. [CrossRef]
21. Gietl, J.; Vignola, J.; Sterling, J.; Ryan, T. Characterization of damping properties in 3D printed structures. *J. Phys. Conf. Ser.* **2018**, *1149*, 012002. [CrossRef]
22. Parenti, P.; Cataldo, S.; Grigis, A.; Covelli, M.; Annoni, M. Implementation of hybrid additive manufacturing based on extrusion of feedstock and milling. *Procedia Manuf.* **2019**, *34*. [CrossRef]
23. Lee, C.Y.; Liu, C.Y. The influence of forced-air cooling on a 3D printed PLA part manufactured by fused filament fabrication. *Addit. Manuf.* **2019**, *25*. [CrossRef]
24. Aggogeri, F.; Borboni, A.; Merlo, A.; Pellegrini, N.; Ricatto, R. Vibration damping analysis of lightweight structures in machine tools. *Materials* **2017**, *10*, 297. [CrossRef]
25. Chaitanya, S.K.; Reddy, K.M.; Ch, S.N.S.H. Vibration properties of 3D printed/rapid prototype parts. *Int. J. Innov. Res. Sci. Eng. Technol.* **2015**, *4*. [CrossRef]
26. Wang, R.; Shang, J.; Li, X.; Luo, Z.; Wu, W. Vibration and damping characteristics of 3D printed Kagome lattice with viscoelastic material filling. *Sci. Rep.* **2018**, *8*, 9604. [CrossRef]
27. Nagesha, B.K.; Dhinakaran, V.; Shree, M.V.; Kumar, K.M.; Chalawadi, D.; Sathish, T. Review on characterization and impacts of the lattice structure in additive manufacturing. *Mater. Today Proc.* **2020**, *21*, 916–919. [CrossRef]
28. Matlack, K.H.; Bauhofer, A.; Krödel, S.; Palermo, A.; Daraio, C. Composite 3D-printed metastructures for low-frequency and broadband vibration absorption. *Proc. Natl. Acad. Sci. USA* **2016**, *113*. [CrossRef] [PubMed]
29. Del Giudice, L.; Vassiliou, M.F. Mechanical properties of 3D printed material with binder jet technology and potential applications of additive manufacturing in seismic testing of structures. *Addit. Manuf.* **2020**, *36*, 101714. [CrossRef]
30. Elmadih, W.; Syam, W.P.; Maskery, I.; Chronopoulos, D.; Leach, R. Mechanical vibration bandgaps in surface-based lattices. *Addit. Manuf.* **2019**, *25*, 421–429, ISSN 2214-8604. [CrossRef]
31. Rouse, J.P.; Simonelli, M.; Hyde, C.J. On the use of small ring testing for the characterisation of elastic and yield material property variation in additively manufactured materials. *Addit. Manuf.* **2020**, *36*, 101589, ISSN 2214-8604. [CrossRef]
32. Magarremova, L.; Vasyliev, B.; Kinzburskiy, V. Novel designs of turbine blades for additive manufacturing. In Proceedings of the ASME Turbo Expo 2016: Turbine Technical Conference and Exposition, GT2016, Seoul, Korea, 13–17 June 2016. [CrossRef]

33. Ledezma-Ramírez, D.F.; Tapia-González, P.E.; Ferguson, N.; Brennan, M.; Tang, B. Recent advances in shock vibration isolation: An overview and future possibilities. *ASME. Appl. Mech. Rev.* **2019**, *71*, 060802. [CrossRef]
34. Tian, X.; Liu, T.; Yang, C.; Wang, Q.; Li, D. Interface and performance of 3D printed continuous carbon fiber reinforced PLA composites. *Compos. Part A Appl. Sci. Manuf.* **2016**, *88*. [CrossRef]
35. Hetnarski, R.B.; West, R.A.; Torok, J.S. Damping of vibrations of layered elastic-viscoelastic beams. *ASME. Appl. Mech. Rev.* **1993**, *46*, S305–S311. [CrossRef]
36. Rashidi, M.; Sharafi, P.; Alembagheri, M.; Bigdeli, A.; Samali, B. Operational modal analysis, testing and modelling of prefabricated steel modules with different LSF composite walls. *Materials* **2021**, *13*, 5816. [CrossRef] [PubMed]
37. Zhao, Z.; Ren, J.; Du, S.; Wang, X.; Wei, Z.; Zhang, Q.; Zhou, Y.; Yang, Z.; Lu, T.J. Bending response of 3D-printed titanium alloy sandwich panels with corrugated channel cores. *Materials* **2021**, *14*, 556. [CrossRef] [PubMed]
38. Shiou, F.J.; Liu, M.X. Development of a novel scattered triangulation laser probe with six linear charge-coupled devices (CCDs). *Opt. Lasers Eng.* **2009**, *47*, 7–18. [CrossRef]
39. Wojciechowski, J.A.; Pisarek, J.; Kityk, A.V. Non-contact 1D vibration analysis in temporal digital speckle photography. *Opt. Lasers Eng.* **2010**, *48*, 320–324. [CrossRef]
40. Lyu, C.; Liu, H.; Alimasi, A.; Liu, Y.; Ge, C.; Yang, J. Monitoring ambient vibration pollution based on visual information perception and neural network analysis. *Opt. Lasers Eng.* **2021**, *137*, 106353. [CrossRef]
41. Psota, P.; Mokrý, P.; Lédl, V.; Stašík, M.; Matoušek, O.; Kredba, J. Absolute and pixel-wise measurements of vibration amplitudes using time-averaged digital holography. *Opt. Lasers Eng.* **2019**, *121*, 236–245. [CrossRef]
42. Fu, Y. Low-frequency vibration measurement by temporal analysis of projected fringe patterns. *Opt. Lasers Eng.* **2010**, *48*, 226–234. [CrossRef]
43. Mishra, B.B.; Kumar, A.; Zaburko, J.; Sadowska-Buraczewska, B.; Barnat-Hunek, D. Dynamic response of angle ply laminates with uncertainties using MARS, ANN-PSO, GPR and ANFIS. *Materials* **2021**, *14*, 395. [CrossRef] [PubMed]
44. Opazo-Vega, A.; Rosales-Garcés, V.; Oyarzo-Vera, C. Non-destructive assessment of the dynamic elasticity modulus of eucalyptus nitens timber boards. *Materials* **2021**, *14*, 269. [CrossRef] [PubMed]
45. Hagedorn, P.; Pabst, U. Some remarks on the identification of damping parameters at the boundary of vibrating systems. *ASME Appl. Mech. Rev.* **1995**, *48*, S107–S110. [CrossRef]
46. Avramov, K.V.; Mikhlin, Y.V. Review of applications of nonlinear normal modes for vibrating mechanical systems. *ASME Appl. Mech. Rev.* **2013**, *65*, 020801. [CrossRef]
47. Campiche, A. Numerical modelling of CFS three-story strap-braced building under shaking-table excitations. *Materials* **2021**, *14*, 118. [CrossRef]
48. Dylag, Z.; Jakubowicz, A.; Orlos, Z. *Wytrzymalosc Materiałów (In Polish: Strength of Materials)*; WNT: Warszawa, Poland, 1997; pp. 272–279, ISBN 83-204-2105-5.
49. *Technical Data Sheet, PLA by Innofil3D BV*; BASF 3D Printing Solutions BV. Available online: https://www.ultrafusefff.com/wp-content/uploads/2016/05/TDS-Innofil3D-PLA-160608.pdf (accessed on 22 February 2021).
50. Inman, D.J. *Engineering Vibrations*, 4th ed.; Pearson: London, UK, 2013, ISBN 978-0132871693.

materials

MDPI

Article

Application of Teager–Kaiser's Instantaneous Frequency for Detection of Delamination in FRP Composite Materials

Adam Gałęzia [1,*] and Anita Orłowska-Gałęzia [2]

[1] The Faculty of Automotive and Construction Machinery Engineering, Warsaw University of Technology, Narbutta 84, 02-524 Warsaw, Poland
[2] Department of Intelligent Technologies, Institute of Fundamental Technological Research, Polish Academy of Sciences, Pawińskiego 5B, 02-106 Warsaw, Poland; aorlow@ippt.pan.pl
* Correspondence: adam.galezia@pw.edu.pl

Abstract: Composite materials are widely used in many engineering applications and fields of technology. One of the main defects, which occur in fiber-reinforced composite materials, is delamination. It manifests itself in the separation of layers of material and the damaged structure once subjected to mechanical loads degrades further. Delamination results in lower stiffness and the decrease of structure's carry load capability. Its early detection is one of the tasks of non-invasive structural health monitoring of layered composite materials. This publication discusses a new method for delamination detection in fiber-reinforced composite materials. The approach is based on analysis of energy signal, calculated with Teager–Kaiser energy operator, and comparison of change of the weighted instantaneous frequency for measurement points located in- and outside of delamination area. First, applicability of the developed method was tested using simple models of vibration signals, reflecting considered phenomena. Next, the authors' weighted instantaneous frequency was applied for detection of deamination using signals obtained from FEM simulated response of the cantilever beam. Finally, the methods effectiveness were tested involving real experimental signals collected by the laser Doppler vibrometer (LVD) sensor measuring vibrations of the delaminated glass-epoxy specimens.

Keywords: delamination; Teager–Kaiser energy operator; instantaneous frequency; fiber-reinforced composite material

Citation: Gałęzia, A.; Orłowska-Gałęzia, A. Application of Teager–Kaiser's Instantaneous Frequency for Detection of Delamination in FRP Composite Materials. *Materials* **2021**, *14*, 1154. https://doi.org/10.3390/ma14051154

Academic Editor: Enrique Casarejos

Received: 23 January 2021
Accepted: 18 February 2021
Published: 1 March 2021

1. Introduction

Fiber-reinforced polymer (FRP) composite materials have found application in a wide range of engineering structures—from parts of vehicles, through structures like pedestrian bridges to wind turbine blades. The main cause of their popularity is very good strength to mass ratio.

Both, during manufacturing as well as use, FRP structures are subjected to a number of degradation processes, such as cracks or delamination. Degradation can result from loading the structure with force over accepted level, e.g., as a result of impulse, or due to violating construction's fatigue strength on cyclic load. According to present state of technology, the origination of a defect does not preclude further use of a structure. However, further operation of an object, in accordance with a given level of safety, implies the need for early detection, identification, and localization of the fault as well as controlling fault development. Detection of an early phase of a fault allows to change operation conditions or to plan maintenance and repairs [1,2].

Delamination is the separation of part of a material of FRP which causes local change of its stiffness. Further loading of a structure, will effect, in the growth of delamination area or, in an extreme case, its rapid development causing destruction or irreversible loss of operation ability of a structure, e.g., like in the case of helicopter rotor blade delamination which took place in Israel in 2009 [3]. Due to the importance of this degradation mechanism, as well as the difficulty to detect and localize delamination, a number of diagnostic methods have already been considered by engineers and still new ones are being developed.

Those methods are based on observations of different physical phenomena such as: ultrasonic waves [4,5], acoustic emission [6], thermographic response of the structure excited with the thermal (IR) radiation or external forces [7–9], or vibrations [10,11]. Despite there still being some challenges for ultrasonic and thermographic methods [12], these are the leading methods used for inspections of high-risk composite components like, e.g., aerospace structures [6]. They allow to fully identify relatively small delamination, but the qualified operator has to set the test and the inspection process may take many hours to complete [13].

Due to economical and practical reasons, the methods based on observation and analysis of vibrations have gained big popularity, mainly in structural health monitoring applications. It is caused by the fact that the vibration signal can be easily measured by relatively cheap equipment [14]. There is a broad range of measurement technologies, starting from classical transducers like accelerometers [14], through optical fibers [14], piezoelectric strain sensors [15], micro-electro-mechanical systems (MEMS) sensors [16,17], up to touchless measurement methods realized, e.g., by laser Doppler vibrometry [18]. In addition, most frequently, the operating conditions cause the structure to vibrate, opening opportunity to possess diagnostic information without adding additional excitation energy.

Literature presents many techniques, which are dedicated to dealing with vibrational response of a structure and aim for detection and/or identification of the structural damages, such as delamination. A great number of methods are based on inverse problem analysis [19,20]. For example, in [21], the authors presented results of the examination of the three different inverse algorithms for predicting the location and size of delamination: direct solution using a graphical method, artificial neural network (ANN), and surrogate-based optimization. Discussed algorithms have been validated using numerical data generated from the finite element (FE) model of delaminated beams, as well as data from experimental modal analysis conducted on carbon-fiber reinforced polymer beam specimens. It was shown that all the presented algorithms accurately predicted the delamination parameters, when the FE model validation data were used. However, when experimental data were used, the ANN algorithm turned out to be sensitive to the measurements errors.

In [22], the authors showed that delamination can be identified, solving the inverse problem with the use of gradient optimization, when the virtual distortion method is used for modeling of the healthy structure. The delamination identification scheme was formulated by the inverse problem also in [23,24]. Inverse analysis methods are usually very effective in delamination identification. They are capable of finding the exact delamination location, with the use of input data collected by low number of sensors. From the other hand, they are very time consuming, as they need demanding learning process, e.g., in the case of ANN use, or a detailed model of the investigated structure.

Another group of methods identify delamination by the use of so-called damage indicators. The most popular damage indicators are based on time domain signals or modal characteristics extracted from the vibrational signal. The example of such an indicator is the probabilistic delamination indicator (PDI) presented in [25]. PDI is developed on the basis of delamination induced relative frequency change curves and its relationship to mode shapes. It was proved, that the method is capable to detect mid-plane as well as off-mid-plane delamination in laminated composite beams. It was also shown that the proposed method shows two symmetric, potential locations of delamination and because of that, can be used for preliminary tests, preceding the use of, e.g., an ultrasonic apparatus. Additionally, in [26], the method based on the computation of the damage indicators was presented. The authors suggested damage indicators based on the analysis of phase representation of signals. In this approach, the analyzed signal is presented on a phase plane as the Poincare map. The identification and localization of failure in the form of delamination, is performed based on the comparison analysis between representation of signals from structure with and without a failure. Some other damage indicators are depicted in [27].

The methods based on the use of damage indicators are significantly simpler and less time consuming then the inverse analysis methods, but there are still areas that need to be improved. Many indicators exhibit sensitivity problems, need a reference state, and do not provide the possibility of detecting false alarms, reducing their reliability.

Intelligent signal processing-based algorithms can be pointed as the next group of delamination identification methods. An interesting example of signal processing-based identification is presented in [28]. The author proposed a two-step delamination detection and evaluation procedure, which consists of modal shapes extraction and an advanced signal processing algorithm, based on 2D wavelet transform (WT) with B-spline wavelets of fractional order. Another novel testing procedure of that type, based on the feature extraction capability of multi-level wavelet-based processing, is presented in [29]. Intelligent signal processing-based methods are using advanced mathematical tools. They frequently operate as multi step algorithms and require preliminary selection of signal processing parameters.

As the delamination is a very serious damage and can result in catastrophic failure, there is a very high concern in the scientific community in finding the effective methods allowing for its identification. The papers mentioned above represent only a small fraction of the published literature resources in the field, showing the diversity of scientific tools adopted for delamination identification purposes. In the more broad discussion on the vibration-based delamination assessment methods, their potential and classification are a job for a self-contained review article rather than for the introductory part of the research paper; a general conclusion coming from at least few review papers [20,30,31] can be presented as a good state of the art summary. Conclusions can be drawn that all presented techniques have their own advantages and disadvantages and there is no general technique that allows to identify all kinds of delamination in all kinds of structures. In addition, taking into account that most of the methods presented in literature are examined on simple elements like beams or plates, it seems that in the case of more composed objects, the reliable SHM system should use at least a few identification methods. It legitimizes the search for new, complementary methods of delamination identification.

In the context of the above, the proposed method should not be considered as a stand-alone solution, being a perfect remedy, but as one of possible assisting methods in the vibration-based SHM system. Nonlinear dynamics of breathing deformation, observed in the finite element model of the vibrating specimen as well as in experimental tests, points to the need of selecting proper signal processing and analysis methods allowing to detect this phenomena.

The presented approach, is based on the Teager–Kaiser energy operator (TKEO). The analysis of the time domain signal using the Teager–Kaiser energy operator enables observation of transient disturbances of the signal's instantaneous frequency (IF) [32]. Such disturbances can result from a failure, e.g., crack or delamination of a composite structure. The main assumption underlying the developed method is local, periodic change of stiffness of the tested object due to opening and closing of delamination which will manifest in the change of an instantaneous frequency of vibration. As the change of stiffness in delaminated area occurs within a single period of vibration, the vibration of delaminated beam is characterized by specific half-period fluctuation of instantaneous frequency.

In contrast to many present vibration-based methods, the original TKEO-based delamination detection method does not require knowledge of dynamic behavior of the structure in reference, i.e., healthy state, nor referring to the numerical model. In the considered case of breathing delamination, comparison of instantaneous frequency of vibration signals recorded in measurement points located within and outside delamination allows to create a useful failure indicator.

The paper is organized in the following order: Section 2 discusses disturbance of the vibration signal resulting from opening and closing of delamination. Section 3 presents a simplified model of a vibration signal, in which variation of instantaneous frequency within each half-period of vibration is assumed to result from phenomena of breathing delamination. Section 4 is devoted to discussion on the applicability of selected signal processing methods for detection of considered half-period disturbance of instantaneous frequency. This section also gives some details on the Teager–Kaiser energy operator and introduces the original Teager–Kaiser weighted instantaneous frequency f_w. The analysis described in this section was performed on the signal obtained using the model described in the previous section. Those works were preliminary activities before the FEM model was built and experimental tests were performed. Section 5 presents results of tests performed with the FE model of fiber-reinforced composite specimens. In contrast to the numerical model from Section 3, the FE model allowed to obtain more realistic behavior regarding amplitude and frequency variations. Laboratory tests of specimen with artificially introduced delamination and their results are presented in Section 6. Section 7 concludes performed research as well as discuss advantages and drawbacks of the Teager–Kaiser weighted instantaneous frequency f_w.

2. Breathing Delamination Phenomena

Present-day publications [33,34] as well as numerical models discussed in Section 4 and results of experimental tests presented in Section 5, indicate that in specific mode shapes of delaminated structure, relative motion of the delaminated layer can be observed. This phenomenon is known as opening and closing of delamination, or breathing delamination (Figure 1) and it causes fluctuations of stiffness [35] in the delaminated structure. As a result of the variation of stiffness, the frequency of vibrations undergoes instantaneous changes. It can be assumed that delamination reveals itself in a specific signal pattern with uneven half-periods. During the opening of delamination (Figure 1a), the structure has lower stiffness, which is reflected in lower eigenfrequency. When delamination is closed (Figure 1c), the stiffness of a specimen is close to the stiffness of an undamaged structure. As a result, eigenfrequency is higher compared to the situation when delamination is opened. During excited vibrations, e.g., in first mode, for a single period of excitation, part of the structure's motion takes place for opened delamination, while the other part takes place for closed delamination. However, the complete period of the structure's motion is consistent with excitation frequency.

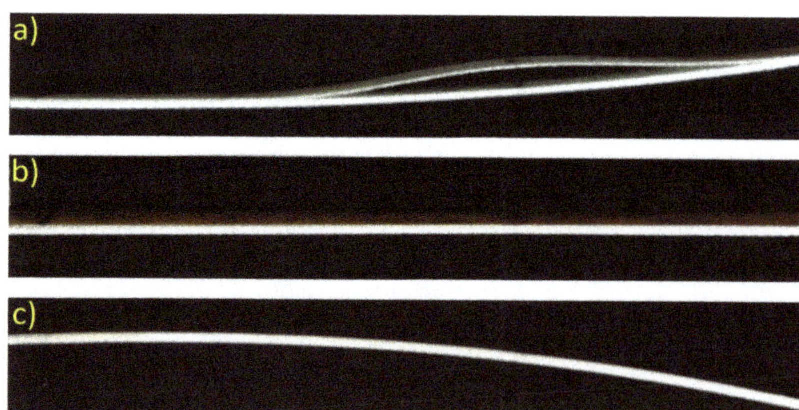

Figure 1. Breathing delamination of the beam structure during vibration: (**a**) opening of delamination when the beam tip goes up; (**b**), beam in equilibrium; (**c**) closing of delamination when the beam tip goes down.

3. Simplified Model of Vibration Signal

To reflect nonlinear dynamic behavior of the structure with breathing delamination and to perform preliminary analysis, the dedicated numerical model of a vibration signal was created. The model must be considered simplified because it is not reflecting any particular breathing delamination of any specific structure. Additionally, it assumes that change of instantaneous frequency takes place in a contiguous way without rapid transitions.

For the preliminary tests, it was assumed that half-periods of the created signal will differ by 2 Hz and instantaneous frequency will continuously vary from 9 to 11 Hz and back. In the first approach, indicated as Model 1, the numerical model was created by combining half-periods of two single harmonic signals with different frequencies using the cubic spline interpolation function $S(t)$ [36], smoothly connecting sine half-periods. Equation (1) describes single period T of the modeled signal:

$$x(t)_{Model1} = \begin{cases} A\cdot sin(2\pi\cdot f_1\cdot t) & for\ t\ \epsilon\ (0.1T, 0.45T\ ) \\ A\cdot sin(2\pi\cdot f_2\cdot t) & for\ t\ \epsilon\ (0.625T, 0.925T) \\ S(t)\ for\ t\ \epsilon\ \langle 0.925T, 0.1T\rangle\ and\ t\ \epsilon\ \langle 0.45T, 0.625T\rangle \end{cases}. \quad (1)$$

To build the numerical signal, the upper half-period of 9 Hz waveform and the bottom half-period of 11 Hz waveform were joined without interval. Due to different curvature of half-periods, the connection point was smoothed with spline interpolation. Due to spline smoothing, the zero-crossing point is shifted from the original zero-crossing of 9 and 11 Hz waveforms. In the $x(t)_{Model1}$ signal, all connection points of half-periods were smoothed, while the beginning and the end of the signal were modified to keep the same curvature of half-periods.

The Model 2 ($x(t)_{Model2}$) is created by combining two harmonic signals with appropriately defined amplitudes A_1 and A_2, according to the following analytical Equation (2):

$$x(t)_{Model2} = A_1 \sin(2\pi\cdot f\cdot t) + A_2 \sin(2\pi\cdot 2\cdot f\cdot t). \quad (2)$$

The selection of amplitude ratio A_1/A_2 as well as value of frequency f allowed to correctly model the non-symmetry of the signal. Figure 2 presents the comparison of the single period for both models with the assumption of representation of the signal with 9 and 11 Hz half-periods.

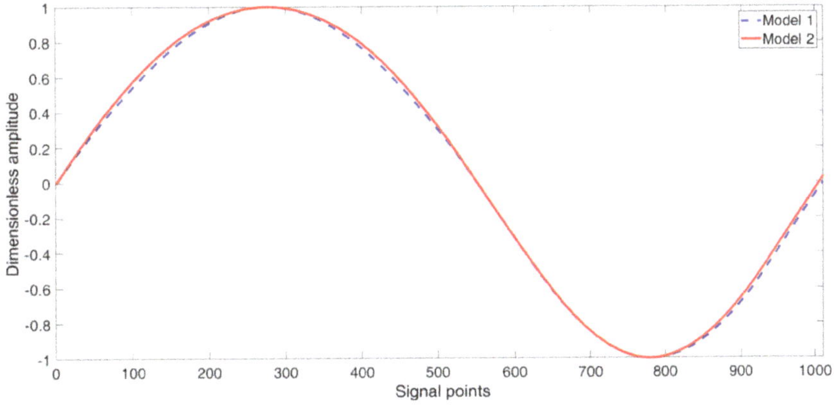

Figure 2. Comparison of waveforms from models.

Main differences between the models are related to spline interpolation smoothing used in Model 1. The smoothing causes modification of part of the curvature of the waveform as well as the shifting of the beginning of the consecutive full period of the signal. Because the signal created by the use of the Model 2 does not suffer from those drawbacks, it was used in further analysis. Next, the section refers only to use of the $x(t)_{Model2}$ signal and it will be denoted as $x(t)$.

4. Detection of Frequency Fluctuation

The section discusses application of three different tools for detection of instantaneous change of the signal frequency. The signal $x(t)$ (Equation (2)) was analyzed using: spectral analysis, Hilbert transform-based demodulation and the Teager–Kaiser energy operator-based demodulation. The mathematical principia of the presented methods as well as results of the conducted analysis were presented in the next subsections.

4.1. Spectral Analysis

Spectral analysis, which is based on the Fourier transform mathematical theorem, is one of most popular tools used in signal analysis [37], e.g., in machines' condition monitoring. The method allows for decomposition of a time series into harmonic oscillations and represents its amplitude and phase spectrum. The results of the numerical spectral analysis are strongly dependent on the duration of the investigated signal. It is commonly used in measurement practice, which in case of multi-harmonic signals, at least few fundamental periods are analyzed, in order to correctly grasp all components of interest. The result of spectral analysis, conducted on the signal $x(t)$, is presented in Figure 3. The analyzed signal consisted of 30 complete periods.

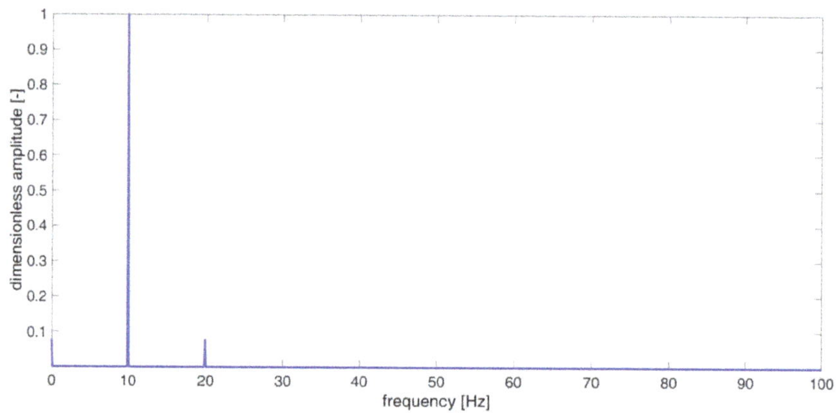

Figure 3. Spectrum of signal $x(t)$.

The spectrum has 3 harmonic components: DC component, main component with frequency 9.95 Hz, and its second harmonic—19.9 Hz. The frequency of the main component results from the period's duration of the analyzed signal, which is composed of two different half-periods. Investigating structure, for which it is not possible to refer to the healthy state, the frequency observed in spectrum can be interpreted as modal frequency of the undamaged structure. The spectrum reveals no information regarding deviation of frequency.

4.2. Hilbert Transform Demodulation Analysis

The demodulation based on Hilbert transform is well established in the engineering community and found application in many tasks such as, for example, bearing diagnostics [2]. It allows for identification of modulation phenomena and, particularly, for estimation of instantaneous frequency [38].

The complex sum of the analyzed signal $s(t)$ and its Hilbert transform $\hat{s}(t)$ creates the analytical signal $\tilde{s}(t)$ defined as Equation (3):

$$\tilde{s}(t) = s(t) + j\hat{s}(t). \tag{3}$$

The analytical signal allows to estimate envelope (Equation (4)) and instantaneous frequency (Equation (5)) of the analyzed signal:

$$|\tilde{s}(t)| = \sqrt{s^2(t) + \hat{s}^2(t)}, \tag{4}$$

$$\phi(t) = arg\{\tilde{s}(t)\} = arctg\frac{\hat{s}(t)}{s(t)}. \tag{5}$$

In the considered case of the nonlinear signal with half-period frequency variation, the Hilbert transform demodulation does not allow to properly detect changes occurring in the signal. According to this approach, the investigated signal has weak amplitude and frequency modulations. The spectrum of the obtained instantaneous frequency signal reveals existence of the 9.95 Hz component as well as the contribution of its higher harmonics.

The modeled discrepancy of frequency is not correctly represented in the instantaneous frequency (Figure 4). The Hilbert demodulation indicates that instantaneous frequency has small variations around mean value equal to 9.95 Hz. The extreme values of the obtained IF occur for instants of transition of half-periods. Hilbert transform demodulation results do not point to the frequencies related to half-periods and the same does not identify well enough the breathing delamination symptoms.

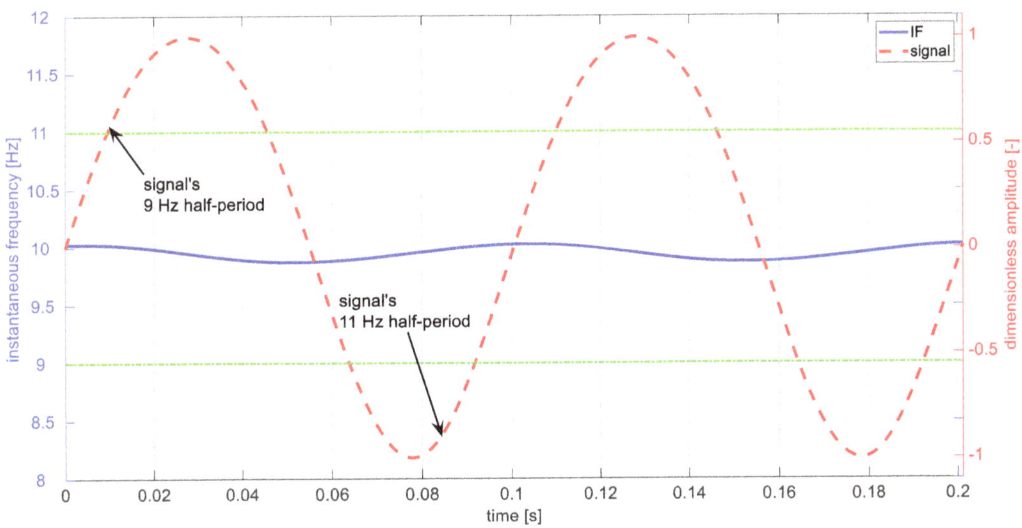

Figure 4. Analyzed signal (red), and time-domain fluctuation of Hilbert demodulation instantaneous frequency (IF) (blue). Green horizontal lines indicate values of the IF expected for the considered case.

4.3. Signal Processing Using the Teager–Kaiser Energy Operator

The Teager–Kaiser energy operator is a differential operator and was presented in [39] while its properties were discussed in [40–43]. Many successful applications of TKEO and related measures in condition monitoring of gears and bearings can be found in publications [44–46]. One of the features of TKEO is that it calculates the signal's energy point-by-point. As a result, TKEO is more sensitive to transient changes occurring in signal. On one hand, it is sensitive to noise, but on other hand, it allows for easier, comparing to solutions based on, e.g., Hilbert transform, detection of instantaneous changes in the signal's amplitude or frequency, e.g., such as transient changes of IF caused by breathing delamination.

Teager–Kaiser energy $E_{TK}(t)$ is the time-domain signal obtained by the operator $\Psi(s(t))$ (Equation (6)) acting on the analyzed signal $s(t)$:

$$\Psi(s(t)) = \dot{s}^2(t) - s(t)\ddot{s}(t). \tag{6}$$

In [41], the authors presented a compact and easy to implement form of the TKEO for discrete signals (Equation (7)):

$$\Psi_d(s_n) = {s_n}^2 - s_{n-1}s_{n+1}. \tag{7}$$

For a certain class of AM-FM signals, described as (Equation (8)):

$$s(t) = A(t)\cos(\omega(t)t), \tag{8}$$

it can be shown that (Equation (9)) [47]:

$$E_{TK}(t) = \Psi(s(t)) \approx A(t)^2\omega(t)^2. \tag{9}$$

Extending this concept, the TKEO (Equation (6)) can be applied for determination of the envelope (Equation (10)) and instantaneous frequency (Equation (11)):

$$A^2(t) = \frac{\Psi(s(t))^2}{\Psi(\dot{s}(t))}, \tag{10}$$

$$\omega^2(t) = \frac{\Psi(\dot{s}(t))}{\Psi(s(t))}. \tag{11}$$

The demodulation procedure based on Equations (10) and (11) is known as the energy separation algorithm (ESA) and was presented in [43,48]. Discussion on properties of the algorithms, both for continuous (CESA) and discrete (DESA) signals is presented in [47].

It is important to emphasize here, that because in the general case, the base formula (9) allows for determination of approximate value of Teager–Kaiser energy, Equations (10) and (11) are accurate for harmonic signals only. In the case of signals with AM and FM modulations, both for CESA and DESA, the additional high frequency components arise in related Teager–Kaiser energies. In consequence, some additional conditions have to be imposed on the signal to minimize discrepancy, such as [49]: the analyzed signal is changing slowly, in relation to sampling frequency and the modulation bandwidth and value are significantly smaller from the carrier frequency.

Several DESA algorithms exist [48] and although all known TKEO-based demodulation algorithms were tested for the analyzed signal (Equation (2)), the DESA-2 algorithm was chosen as giving the best results with lowest errors. According to DESA-2 [48], instantaneous frequency is defined by Equation (12) and envelope by Equation (13):

$$\omega_n \approx \frac{1}{2} arccos\left[1 - \frac{\Psi_d(x_{n+1} - x_{n-1})}{2\Psi_d(x_n)}\right], \tag{12}$$

$$A_n \approx \frac{2\Psi_d(x_n)}{\sqrt{\Psi_d(x_{n+1} - x_{n-1})}}. \tag{13}$$

One must be aware of the limitations of this approach. The vibration signal coming from breathing delamination is not an AM-FM class signal and does not fulfill the conditions allowing for proper demodulation. As a result, it is not possible to fully correctly estimate changes occurring in the signal. The value of the instantaneous frequency is overestimated. Additionally, as [50] discusses, mutual influence of modulation phenomena can arise. In case of the considered signals, in order to improve obtained estimation results, the Teager–Kaiser weighted instantaneous frequency f_w is analyzed (Equation (14)):

$$f_w = \frac{\omega_n \cdot A_n}{2\pi}. \tag{14}$$

Figure 5 presents part of the modeled signal x(t) and its Teager–Kaiser weighted instantaneous frequency f_w. The periodic changes of weighted instantaneous frequency f_w (blue) are correlated with signals half-periods (red). However, due to imperfection of the used signal's model $x(t)$ (Equation (2)), resulting from superposition of harmonic components, the transition between half-periods is not smooth, causing inflections seen in f_w. The mentioned above limitations of the DESA-2 algorithm cause the method to give acceptable quantitative estimates of weighted instantaneous frequency—the lower (8.26 Hz) as well as higher frequency (11.37 Hz) differ from the modeled parameters (9 and 11 Hz) with an acceptable margin. The presented IF estimation method gives good qualitative results, indicating the existence of breathing delamination symptoms. In contrast to the Hilbert transform approach, the extreme values of Teager–Kaiser weighted instantaneous frequency are concurrent in time with half-periods of the signal. Despite the divergence of the values, the method can find application in practice because it reveals existence of phenomena of interest.

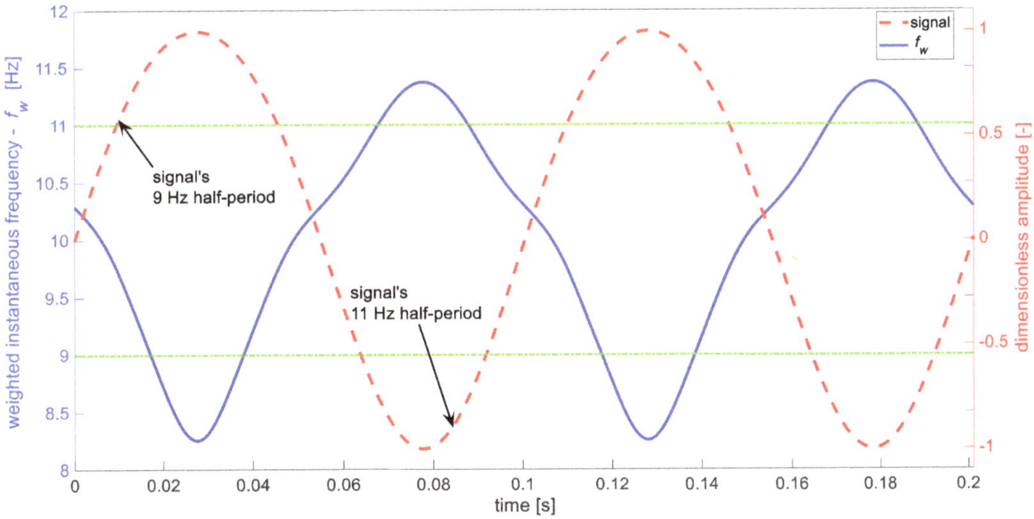

Figure 5. Analyzed signal (red) and calculated weighted instantaneous frequency (blue). Green horizontal lines indicate values of the IF expected for the considered case.

Comparison of analyzed methods clearly show, that only the one based on TKEO demodulation allowed for estimation of modeled changes of IF and revealed phenomena on interest. The spectrum analysis as well as Hilbert demodulation did not make it possible to identify variation of instantaneous frequency.

4.4. Estimation of Instantaneous Frequency of Decaying Signal

To define applicability of the described Teager–Kaiser weighted instantaneous frequency (Equation (14)), a number of test cases were investigated. Both steady-state and decaying signals (Equation (15)) were analyzed:

$$x_{decaying}(t) = A_d{}^{-t} \cdot x(t), \tag{15}$$

where: $x(t)$—the signal $x(t)_{Model2}$, A_d—amplitude decaying factor, t—time.

The reason for performing such tests was the higher ease of FE modeling of the structure with breathing delamination producing decaying signals, which were later used as input data for analysis of the method effectiveness in the case of the cantilever beam with delamination (Section 5.4). However, during laboratory measurements of structure, described in Section 6, vibrations with steady amplitude were excited due to technical ease of the experiments. The presented research task ensued to validate the TKEO method for different modeled signals.

Signals with amplitude decaying factor from 1 (no decaying) to 0.1 (fastest decaying) were analyzed to identify whether decaying increases estimation error. For each considered case, the investigated signal was limited to the time range from the beginning till the moment when the signal's absolute envelope reached the value of 0.2. Statistical parameters for each case, including the signal's duration, are presented in Table 1. The obtained results differ from the assumed values—11 and 9 Hz, but it is possible to observe the deviation of instantaneous frequency related to the half-periods duration.

Table 1. Statistical parameters of instantaneous frequency of the decaying signal.

Decaying Factor A_d	Signal Duration [s]	Mean Frequency [Hz]	Maximum Frequency [Hz]	Minimum Frequency [Hz]	Standard Deviation of Frequency
1	10	10.0162	11.3782	8.2789	0.9693
0.9	10	10.0155	11.6189	7.6486	0.9706
0.8	4.8994	10.0141	11.4111	7.2562	0.9704
0.7	2.3489	10.0032	11.4144	7.2461	0.9753
0.6	1.3455	9.9901	11.4354	7.2396	0.9805
0.5	0.8517	9.9698	11.3830	7.2315	0.9833
0.4	0.5732	9.9718	11.3523	7.2258	0.9824
0.3	0.4007	10.0263	11.3871	7.2232	0.9978
0.2	0.2855	10.0047	11.5053	7.2376	1.0157
0.1	0.2013	10.0841	11.7125	7.2946	1.0396

The quality of the instantaneous frequency estimation was strongly related to the time length of the signal taken to the analysis. For the signals with low values of A_d, the calculated weighted instantaneous frequencies f_w, had higher estimation errors. From performed tests, a conclusion can be made that although the estimation of instantaneous values of frequency becomes less correct with the increase of the decaying factor, still the phenomenon of frequency variation can be observed. Additionally, it is worth to recall that even for signals with short duration, the DESA-2 based approach allowed to perform useful analysis.

Figure 6 presents part of the decaying signal (Equation (15)) and change of its weighted instantaneous frequency f_w. Similar to the case of signal with steady amplitude, one can observe periodic changes of estimated instantaneous frequency corresponding to the signal's half-periods.

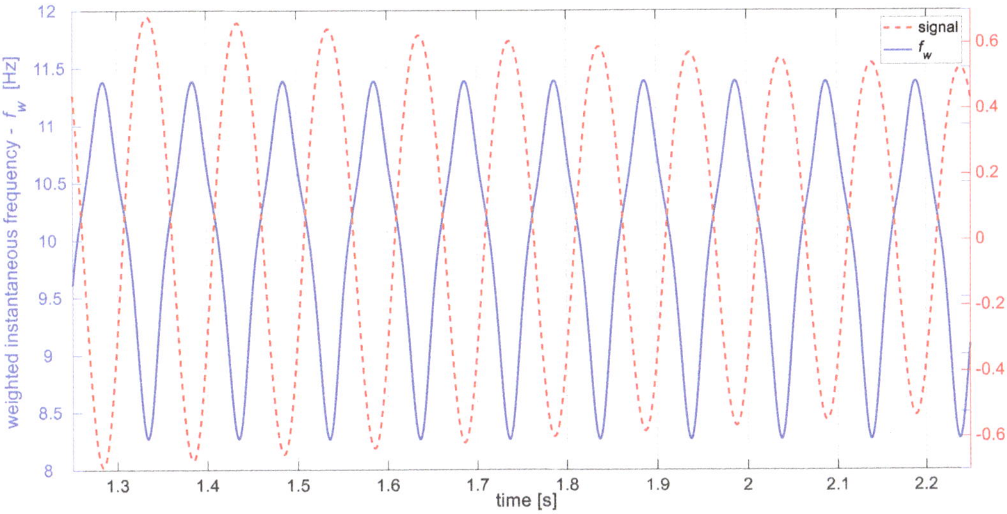

Figure 6. Part of the decaying signal (red) $A_d = 0.8$ and estimated weighted instantaneous frequency (blue).

5. Estimation of Instantaneous Frequency of Signal from FEM Simulation

5.1. General Description of Modeling of the Delaminated Beam

The finite element modeling for simulation of a delaminated beam behavior was performed with the use of the commercial finite element software ANSYS. The delaminated beam was composed of four blocks, as it is presented in Figure 7a. The longer inner edge of the block number 3 was modeled by the sinusoidal function. That allowed to reflect realistic crack conditions [51]. The adjoining edges of the blocks, numbered consecutively as 1, 3, 4 and 2, 3, 4, were connected by the constraints imposed on nodal displacements *ux* (horizontal axis) and *uy* (vertical axis). Besides, the CF boundary conditions scheme was applied to the whole delaminated beam, meaning that the beam was fixed in one end. The standard surface-to-surface contact interaction introduced between two faces of the delamination, allowed neither penetration nor separation among the sublaminate structures.

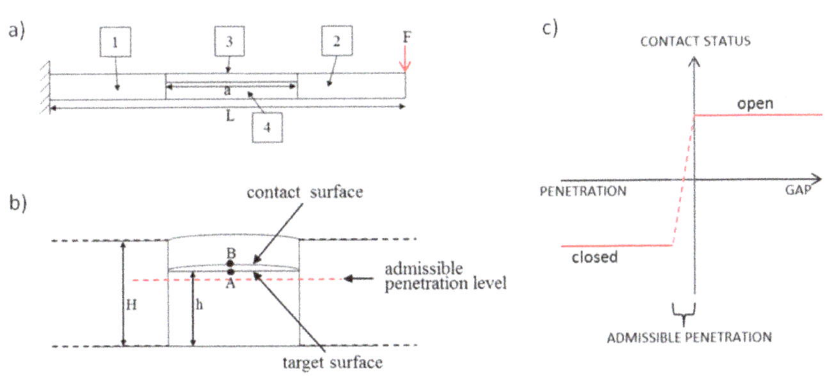

Figure 7. Scheme of the beam parts connections (**a**), zoom of part of the beam with delamination (**b**), contact status definition with penalty-based method (**c**).

Four-nodal plane elements with a plane strain condition defined in the width direction were used. This is a rather rarely used practice in analysis of beams upon out of plane loading, but this approach was verified by the comparison of the eigenvalues obtained from the discussed model with the eigenvalues of full 3D model. There were no meaningful frequency differences for the first three modes for considered geometry. The use of four-nodal plane elements allows to reduce time cost of computation in comparison to use of 3D brick elements. The additional asset is, that those elements allow to create more realistic conditions at the delamination tips than standard beam elements.

5.2. General Description of the Contact Modeling

As delamination splits part of the beam into two distinct segments, which are connected in the delamination tips and can come into interaction during the vibrations, the additional boundary conditions, called contact conditions [52–54], have to be fulfilled for each time step of numeric simulation. In ANSYS software, those conditions are defined via the special type of contact elements, appropriate for specific types of contact problems. The type used in the presented simulation, was 2D surface-to-surface contact element. The pair-based contact definition was chosen for this purpose, which means that one of the delamination boundaries was established as the so-called target surface, while the other one as contact surface. Both surfaces are specified in Figure 7. Since the part situated above the delamination (for the purpose of publication called upper sublaminate) is thinner and more flexible than the opposite part, the elements which represent contact surface (dedicated rather for deformable body) were overlaid on the exterior of upper surface. Similarly, the elements which represent target surface and are applicable for rigid as well as flexible bodies, were overlaid on the exterior of lower surface (called lower sublaminate), which is thicker and less flexible.

To enforce maintaining the constraints in contact modeling, the penalty method was used with contact status defined as it is shown in Figure 7c. In this method, the contact force is the function of the penetration distance and some degree of penetration is involved. The computation algorithm used in simulation identified separation and penetration of layers based on the distance between contact surfaces of delamination. For separation, the distance between the layers is positive and as a result, the contact force, used in equations of motion, equals zero. For penetration, the distance has negative value and as a result, the contact force (Equation (16)) is added to equations of motion. The normal contact force F_n is defined in the penalty method by the use of so-called normal contact stiffness k_n and user-defined admissible penetration x_p:

$$F_n = k_n \cdot x_p. \tag{16}$$

The value of contact stiffness k_n was as small as possible, but also high enough to prevent penetration. Friction was neglected.

The number of equilibrium equations in this method, for most of the cases, is much smaller compared to other methods and the time of the computation can be shortened.

5.3. Case Study for 40% Delamintion

For the purpose of demonstrating the breathing delamination effect, transient analysis were performed with the initial condition imposed as the displacement field. Those conditions recreated deformation of the beam, caused due to application of static, perpendicular force in the beam tip. Delamination of the length a equal to 40% of the total beam length L, with the thickness-wise location equal 75% of the beam's height H, localized in the middle of the beam, was analyzed. The beam material was assumed as anisotropic and its properties are presented in Table 2.

Table 2. Mechanical properties of the beam.

E_{11} [GPa]	E_{22} [GPa]	G_{12} [GPa]	ν_{12}	ρ [kg/m ^{3}]
17.4	17.4	1.45	0.3	2075

Figure 8a shows time evolution of the vertical displacements in points A and B, for the discussed case. As it is shown in Figure 7b, points A and B are localized in the middle of the delamination area, but point A belongs to the lower sublaminate, while point B belongs to the upper sublaminate. When the beam goes downwards, delamination is closed and displacements of the points A and B are practically the same, while when the beam goes upwards, delamination surfaces remain detached from each other and point B reaches a higher amplitude than point A. Figure 8b allows to compare the difference of the vertical displacements in points A and B with the tip displacements.

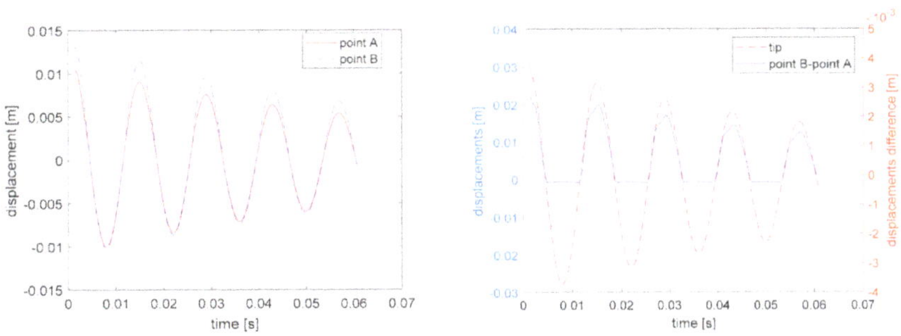

(a) (b)

Figure 8. Displacements evolution in points A and B for the case of free vibrations (**a**), the difference of displacement of point A and B and normalized tip displacements (**b**).

5.4. Analysis of FEM Signals

The numerical FE model described in Section 5.2, was used to simulate the vibrations of the beam with different delamination lengths. For each case, delamination was located in the middle of the beam's length and at 75% of the beam's height. The ratios of delamination size with regard to specimen length x_a/L, ranging from 0.2 to 0.6, were examined. For each case, the displacements time signal from two points: the one located in the middle of the beam and the second located in its tip, were collected and analyzed.

To obtain the band-limited signal, low-pass filtration was applied in order to remove influence of higher frequencies. As it can be seen in Figure 9 for the case of $x_a/L = 0.4$, periodic variations of Teager–Kaiser weighted instantaneous frequency occur due to the opening and closing of delamination. The same phenomena are observed for all investigated cases. Table 3 presents calculated statistical parameters for estimated f_w in relation to x_a/L ratio for the node located in the delamination area (point "delam") and the node located at the free end of the specimen (point "tip"). For big delamination, i.e., $x_a/L = 0.5$ and $x_a/L = 0.6$, the difference between lowest and highest f_w identified in the signal exceeds 5 Hz.

It can be observed that the deviation of f_w for given x_a/L ratio is significantly smaller in the case of "tip" node comparing to the "delam" node located over the delamination. For big delamination, the frequency deviation can also be observed in the point located at the free end. This is related to the influence of large delamination on dynamical behavior of the whole structure. Figure 10 shows comparison of standard deviation of weighted instantaneous frequency f_w for the point located in the delamination area (red, dashed) and the free end of the FE model (blue), revealing significant difference between points located on the same structure. As shown, the point in the delamination area demonstrates higher difference of half-period frequencies then the point located outside this area, which gives the prospect not only for the detection of delamination existence but also identification of its size.

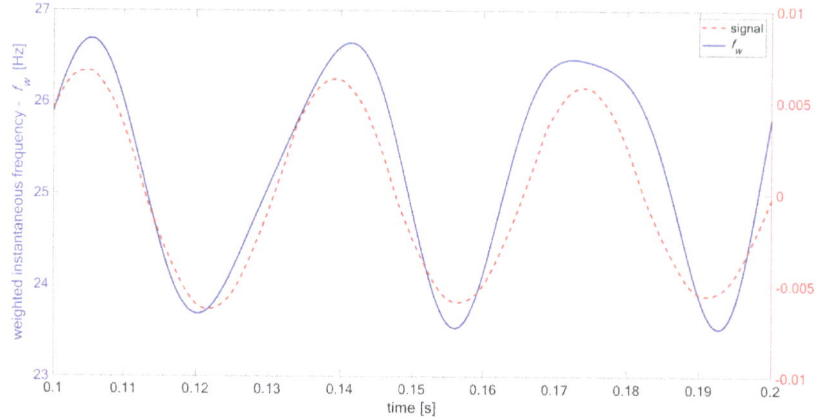

Figure 9. Part of the decaying signal (**red**) from the FEM simulation and its Teager–Kaiser weighted instantaneous frequency (**blue**), case $x_a/L = 0.4$.

Table 3. Influence of the delamination size on weighted instantaneous frequency of the signals from delamination area (point "delam") and from the free end (point "tip").

x_a/L	Point	Teager–Kaiser Weighted Instantaneous Frequency f_w [Hz]			
		Mean	**Maximum**	**Minimum**	**Standard Deviation**
0.2	delam	25.915	26.328	25.469	0.204
	tip	25.993	26.293	25.743	0.153
0.3	delam	25.624	26.304	24.968	0.310
	tip	25.993	26.293	25.743	0.153
0.4	delam	25.147	26.691	23.525	1.065
	tip	24.995	25.451	24.663	0.192
0.5	delam	24.739	27.641	21.970	1.563
	tip	24.313	25.158	23.817	0.326
0.6	delam	20.658	24.784	16.526	2.651
	tip	22.949	24.041	22.258	0.401

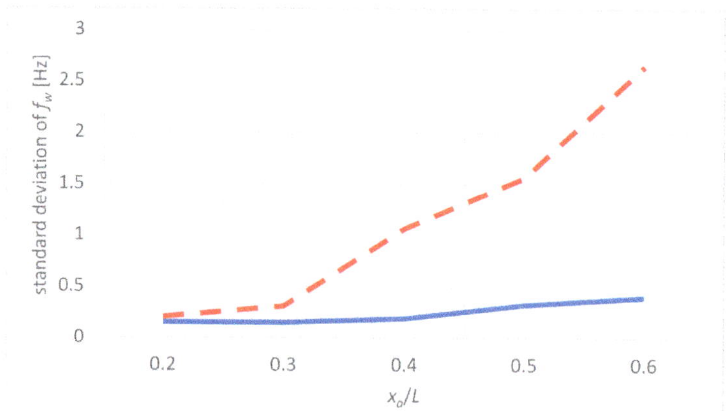

Figure 10. Standard deviation of f_w for the delamination center (**red, dashed**) and free-end (**blue**).

For validation purposes, the signals were also analyzed with the Hilbert transform demodulation approach. In contrast to the TKEO-based method, which allowed to identify variation of f_w between successive half-periods for all tested signals, the Hilbert transform demodulation approach revealed 2.5 Hz variation of the IF only for the biggest delamination ($x_a/L = 0.6$). For other tested cases, the IF was not diagnostically informative. For this reason, results presented in the consecutive part of the publication are limited to the ones obtained with the estimated Teager–Kaiser weighted instantaneous frequency approach.

6. Experimental Validation of Teager–Kaiser Weighted Instantaneous Frequency

6.1. The Specimens

For preparation of the tests specimens, an 8th layered composite plate was manufactured from bidirectional E-glass fabric (Saerbeck, Germany) of basis weight 163 g/m^2 and epoxy resin ARALDITE LY3505 (Woodlands, TX, USA) with hardener XB 3403 (Woodlands, TX, USA), by the use of vacuum-bag method. During the hand lay-up of the fabric, three 0.0001 m thick Teflon inserts of different dimensions were placed between the 6th and 7th layer of the composite. Four individual test specimens of nominal dimension 0.18 m × 0.02 m × 0.0014 m were cut from the laminate after resin solidification and the inserts were removed after cutting out specimens. The inserts allowed to create a partially separated layer with controlled length and edges. The created delamination span through the whole width of the beams. Throughout vibration tests, specimens were clamped along one of the tips with 0.03 m distance and active length was equal to 0.15 m. The center of delamination was located at 0.075 m from the free end, i.e., in the center of the active part of the specimen. During the experiment, four specimens were tested: one without delamination ($x_a/L = 0$, further indicated as 00) and three with different ratios of delamination length x_a to active length L, respectively: $x_a/L = 0.2$ (indicated as 20), $x_a/L = 0.4$ (indicated as 40), and $x_a/L = 0.6$ (further indicated as 60). The photo of specimens is presented in Figure 11, while the specimens' geometry, layer arrangement and the photo of the experimental stand during the samples preparation are shown in Figure 12.

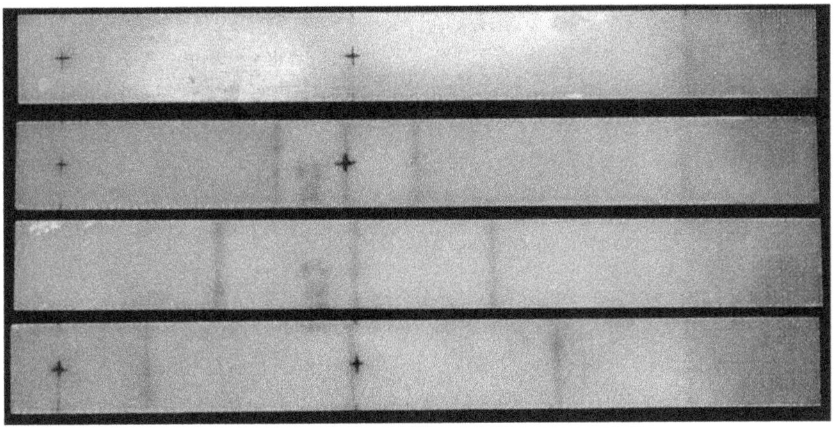

Figure 11. The specimens, from top to bottom respectively: 00, 20, 40, 60.

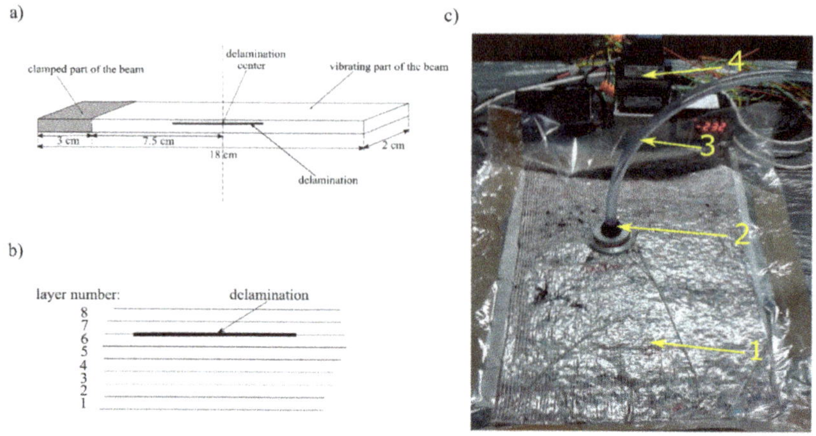

Figure 12. Specimens' geometry and delamination placement (**a**), general scheme of layer arrangement for test sample (**b**), experimental stand during samples preparation (**c**), where: 1—form with constituent materials in vacuum-bag; 2—vacuum connector; 3—vacuum hose; 4—data acquisition system controlling process parameters.

6.2. The Experimental Setup

The schematic diagram is presented in Figure 13. The tested specimen was placed in a mounting clamp attached to the armature of the modal shaker ModalShop 2100E11. The modal shaker and the mounting clamp excited vertical vibrations which caused bending modes of specimens. The specimens had the CF boundary conditions—one end of the tested beam was fixed and the other was vibrating freely. During all experiments, bolts of the mounting clamp were screwed with force of 7 Nm using torque wrench. The Bruel&Kjaer 4507-B-004 accelerometer sensor was mounted on top of the mounting clamp for reference measurement of excitation signal. Vibrations of the specimen were recorded using Polytec PSV-400-3D laser Doppler vibrometer, set in "single measurement head" configuration. The LDV sensor measured bending vibrations in direction perpendicular to beams surface, in two measurement points: one located in the center of delamination area (indicated M1) and second located at the free end of the specimen (indicated M2). To record signals from both measurement points, the scanning head was shifted between measurement locations using special frame, allowing for repetitive setting of the head. The shaker excited the specimen to vibrate according to control signal generated by the LDV's inner generator.

6.3. The Measurements

The measurements were conducted in two steps. In the first one, all specimens were excited with pseudo-random noise signal and modal analysis was performed in the range (0, 2000) Hz. This band covered the first five bending modes of the specimens. The results are presented in Table 4. The modal analysis reveals differences in modal frequencies between specimen 00 and specimens with delamination. However, the monotonic trend is observed only for third and fifth modal frequency. Direct application of the modal frequencies for structural health monitoring requires that modal frequencies of an undamaged structure are known in advance, which allows to identify the difference resulting from failure [55]. This method does not allow for the localization of a failure. To identify the localization of delamination, methods such as mode shape curvature could be used [56].

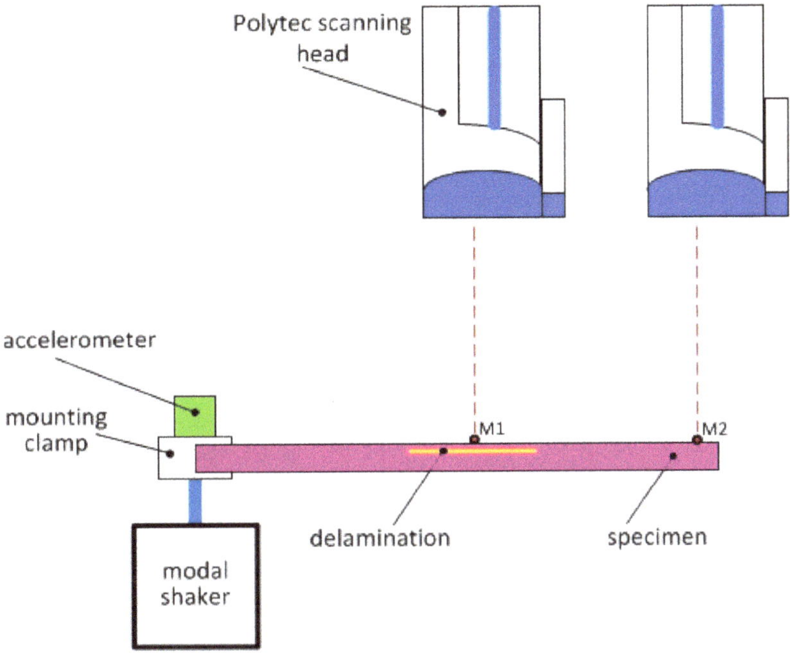

Figure 13. Experimental setup—scheme with measurement points M1 and M2.

Table 4. First modal frequencies for analyzed specimens.

Specimen (x_a/L Ratio)	Bending Modal Frequency [Hz]					Weight [g]	Average Width [m]
	1st	2nd	3rd	4th	5th		
00—without delamination (0)	29.07	184.72	522.41	1014.36	1694.77	8.9	0.00143
20—with delamination (0.2)	30.07	184.73	499.66	1036.46	1534.63	8.8	0.00153
40—with delamination (0.4)	28.43	192.13	429.18	914.43	1451.81	9	0.00159
60—with delamination (0.6)	28.45	185.33	403.95	780.89	1320.15	9	0.00149

It must be pointed out, that in the case of the first two modes, the differences between modal frequencies of the individual specimens can be related to its weight and width variation, as well as influence of mounting errors. First modal frequencies might not be an objective criterion for technical state assessment.

In the second step, the specimens were excited with the harmonic signal with the frequency equal to the identified first modal frequency. During the test of a single specimen, the Polytec vibrometer recorded signals from measurement points: M1 (center of specimen) and M2 (free end of specimen—out of delaminated area). For each measurement point, the vibrometer recorded the simultaneously time domain vibration signal of specimen and reference vibration of mounting clamp. The sampling frequency was equal to 12,800 Hz. Duration of the recorded waveforms was 5.12 s.

Because fluctuation of instantaneous frequency, discussed in the Sections 3 and 5, manifests strongly in displacement signal, therefore, velocity signals, recorded by the laser vibrometer, were integrated using numerical procedure from the Matlab environment [36]. Although the excitation signal was harmonic, the mechanical system of the shaker, as well as the environment, produced a measurement noise. The presence of noise in analyzed signals decreases quality of results obtained using the Teager–Kaiser energy operator. To reduce the influence of the noise and increase the signal-to-noise ratio, each measurement signal was time synchronously averaged (TSA). This procedure used zero-crossing of rising edge of the reference signal. Averaging caused reduction of measurement signals to realizations with duration of 10 full periods of the reference signal. For such averaged signals, f_w was calculated and further analyzed.

6.4. Experimental Results

TSA displacement signals from measurement points M1 and M2 of the successive specimens as well as reference signal, measured by the sensor located on mounting clamp, were processed with the use of the DESA-2 algorithm to obtain f_w.

The weighted instantaneous frequency of reference signal was oscillating close to the assumed frequency of the excitation signal as it is presented in Figure 14. The small fluctuations might result from work conditions of the used modal shaker, which is rather dedicated to heavy structures.

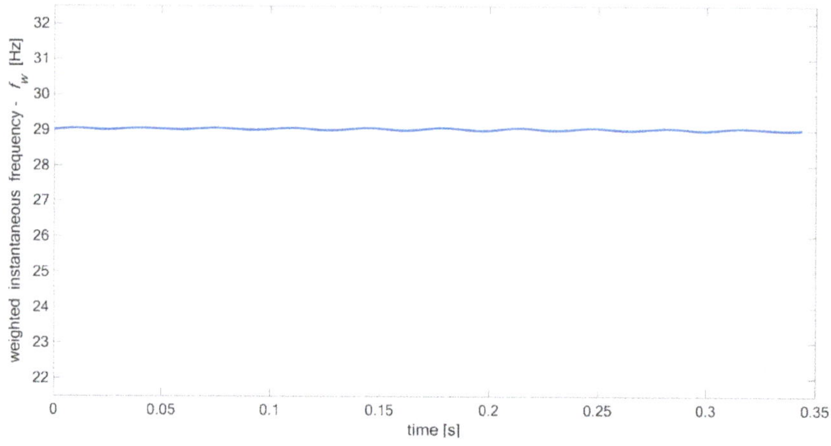

Figure 14. Weighted instantaneous frequency f_w of reference signal (specimen 00).

Figures 15–18 present time fluctuations of Teager–Kaiser weighted instantaneous frequency f_w related to M1 and M2 measurement points for samples 00, 20, 40, and 60 respectively. The consistent scale was used for the sake of comparison. Each figure presents deviation of f_w for both measured signals. For the sample without delamination, the deviation of the weighted instantaneous frequency for both points is similar and is close to the deviation of f_w of the reference signal. It was observed that with the increase of delamination size, the fluctuations of the weighted instantaneous frequency from point M2 increase, but still remain smaller than the fluctuations measured in the point M1, which is located directly above the delamination.

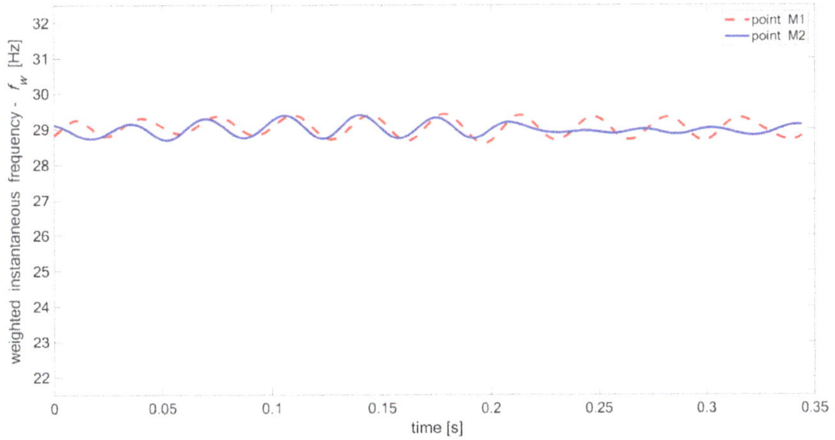

Figure 15. Teager–Kaiser weighted instantaneous frequency for sample 00.

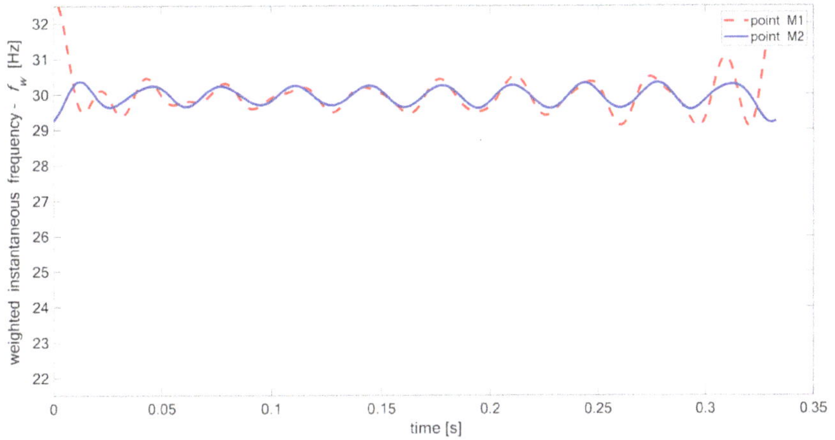

Figure 16. Teager–Kaiser weighted instantaneous frequency for sample 20.

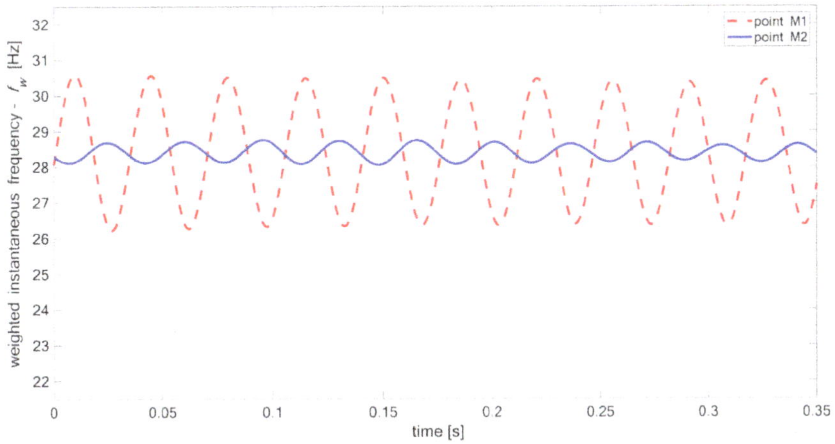

Figure 17. Teager–Kaiser weighted instantaneous frequency for sample 40.

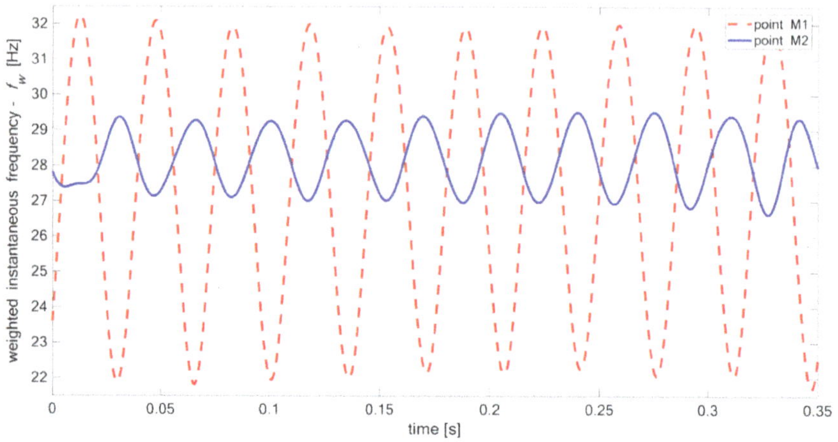

Figure 18. Teager–Kaiser weighted instantaneous frequency for sample 60.

For comparison of the presented signals and evaluation of delamination, statistical parameters can be used such as: standard deviation (Equation (17)), coefficient of variation (Equation (18)), or values span (Equation (19)) of the weighted instantaneous frequency f_w:

$$S = \sqrt{\frac{1}{N-1} \sum_{i=1}^{N} |x_i - \overline{x}|^2}, \tag{17}$$

where x_i—successive samples of signal, N—number of samples of signal, $\overline{x}$—mean value of signal $\overline{x} = \frac{1}{N} \cdot \sum_{i=1}^{N} x_i$.

$$V = \frac{S}{\overline{x}}, \tag{18}$$

where $\overline{x}$—mean value of signal, S—standard deviation (17).

$$D = \max(x) - \min(x). \tag{19}$$

Values of those parameters computed for all the samples are presented in Table 5.

Table 5. Characteristic parameters of waveforms of instantaneous frequency for measurement points.

Sample	Meas. Point	$\overline{x}$ [Hz]	S [Hz]	V	D	$\frac{S_{M1}}{S_{M2}}$	$\frac{V_{M1}}{V_{M2}}$
00	M1 M2	29.0347 28.9857	0.2268 0.1688	0.0078 0.058	0.824 0.707	1.3436	0.1345
20	M1 M2	29.9945 29.9481	0.5557 0.2504	0.0185 0.0084	3.508 1.169	2.2192	2.2024
40	M1 M2	28.4434 28.4013	1.4609 0.2064	0.0514 0.0073	4.366 0.704	7.0780	7.0411
60	M1 M2	27.2422 28.1925	3.5267 0.8441	0.1295 0.0299	10.532 2.886	4.1781	4.3311

For the performed measurements, the most informative parameters were standard deviation S and coefficient of variation V. Change of standard deviation between samples with increasing size of delamination is presented in Figure 19.

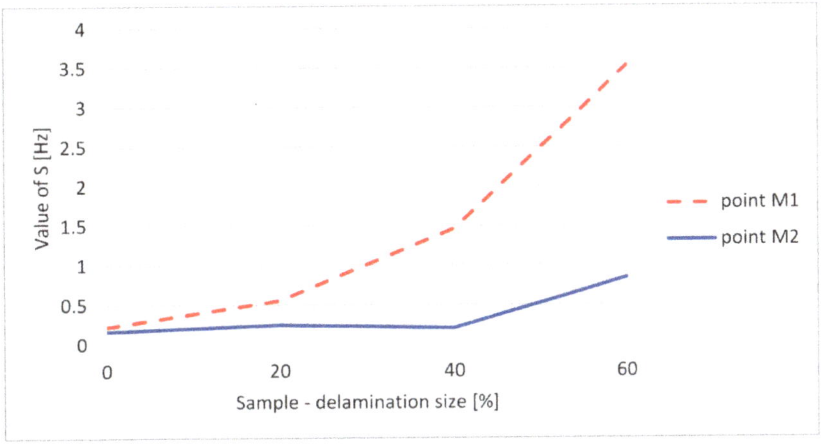

Figure 19. Standard deviation of vibration data for different delamination size.

The experimental work confirmed periodic changes of weighted instantaneous frequency f_w caused by opening and closing of breathing delamination. Based on the measures like standard deviation, it is possible to build a failure indicator.

7. Conclusions

The paper discussed the application of the original Teager–Kaiser weighted instantaneous frequency f_w dedicated for analysis of vibrational response of the beams with breathing delamination. In the framework of the presented research work:

- numerical models were created, reflecting breathing delamination phenomena in vibration signal;
- the conducted experimental tests confirmed half-period variation of instantaneous frequency resulting from breathing delamination;
- the weighted instantaneous frequency f_w, a new approach for estimation of instantaneous frequency, has been developed and its effectiveness was tested;
- a failure indicator and a diagnostic method were proposed for detection of breathing delamination type failure.

The main findings of the conducted research can be summarized as:

- the use of the Teager–Kaiser operator for the construction of the new weighted instantaneous frequency parameter gives much better prospects for delamination detection than in the case of use of instantaneous frequency obtained using demodulation with the Hilbert transform;
- standard deviation of f_w can be a good indicator of delamination existence;
- the method is efficient for the analysis of the real-world vibrational signals, when proper signal processing is applied.

The performed research allowed to identify several limitations of the method, e.g., that the presence of noise in the signal causes that the calculation of f_w requires additional signal processing steps such as averaging. Another drawback of f_w is that it is based on the DESA-2 algorithm dedicated for demodulation of AM-FM signals. Despite the fact that the analyzed signal does not belong to AM-FM class, the method gives qualitative information on failure existence, although obtained results are overestimated. Another limitation of the potential SHM method, based on analysis of the f_w, is that excitation must allow delamination to open and close. In the considered case frequency, the first mode was used. It might be needed to refer to the FEM model of a structure to identify suitable excitation frequency.

The main benefit of f_w is that, as long as the signal has high sampling frequency, short signals can be used for analysis because duration does not influence the result, as is in the case of integration-based methods. Based on the comparison of fluctuations of weighted instantaneous frequency for measurement points located in- and outside of the delamination area, one can deduce regarding delamination. Finally, it is important to stress that the main advantage of the proposed assisting damage detection approach is that it does not require analysis of the mode shapes of the structure, or any foreknowledge of the undamaged structure. This is for sure the key element that makes the method linked to the nature of the damage as such rather than to the structure.

The proposed approach can be useful for monitoring of simple beam-like structures e.g., composite bridge decks. However, the structure's boundary conditions must allow for delamination breathing. This occurs for at least some of the boundary condition schemes, such as CF (clamped-free) or CC (clamped-clamped). As it was shown, in the case of a simple cantilever beam, the first mode was sufficient for delamination detection. In the case of more complicated structures, the proper excitation must be implemented to force breathing delamination phenomena to occur in all suspected components of the structure. In such a situation, multimodal testing can be necessary.

The experimental research presented in the paper was limited to the case of unidirectional arrangement of reinforcing layers in the specimens. In order to verify the method's effectiveness for other stacking sequences, future research work will cover testing of specimens with various orientation of layers and the ratio of the stiffness of the separated part (upper sublaminate in Figure 7b) to the stiffness of the undamaged specimen. The mutual stiffness ratio will affect the amplitude of fluctuations of f_w identifying the method's applicability range.

The definition of the applicability scope of the method to other types of failures and development of the algorithm, allowing for identification of size of delamination, is an open issue and will be the subject of future work.

Author Contributions: The individual contributions of the authors: conceptualization, A.G.; methodology, A.G. and A.O.-G.; software, A.G.; validation, A.G. and A.O.-G.; formal analysis, A.G.; investigation, A.G. and A.O.-G.; resources, A.O.-G.; writing—original draft preparation, A.G. and A.O.-G.; writing—review and editing, A.G. and A.O.-G.; visualization, A.G.; supervision, A.G.; project administration, A.G.; funding acquisition, A.G. All authors have read and agreed to the published version of the manuscript.

Funding: This research was supported by the Polish National Agency for Academic Exchange grant number PPN/BEK/2018/1/00468/U/00001 in framework of the Bekker programme.

Institutional Review Board Statement: Not applicable.

Informed Consent Statement: Not applicable.

Data Availability Statement: The data presented in this study are available on request from the corresponding author.

Conflicts of Interest: The authors declare no conflict of interest.

References

1. Radkowski, S.; Jasiński, M. Use of condition monitoring in the proactive maintenance strategy. In *Proceedings of the 8th World Congress on Engineering Asset Management, Hong Kong, China, 30 October–1 November 2013*; Springer International Publishing: Cham, Switzerland, 2015; pp. 601–610.
2. Randall, R.B.; Antoni, J. Rolling element bearing diagnostics—A tutorial. *Mech. Syst. Signal Process.* **2011**, *25*, 485–520. [CrossRef]
3. Christensen, K. With the Robinson R44's safety issues, the FAA has taken a more hands-off approach. *Los Angeles Times* **2018**. Available online: https://www.latimes.com/la-me-ln-robinson-heli2-20181118-story.html (accessed on 18 September 2020).
4. Liu, Z.; Yu, H.; He, C.; Wu, B. Delamination damage detection of laminated composite beams using air-coupled ultrasonic transducers. *Sci. China Phys. Mech. Astron.* **2013**, *56*, 1269–1279. [CrossRef]
5. Zaiß, M.; Jank, M.H.; Netzelmann, U.; Waschkies, T.; Rabe, U.; Herrmann, H.G.; Thompson, M.; Lanza, G. Use of Thermography and Ultrasound for the Quality Control of SMC Lightweight Material Reinforced by Carbon Fiber Tapes. *Procedia CIRP* **2017**, 33–38. [CrossRef]

6. Dahmene, F.; Yaacoubi, S.; Mountassir, M. EL Acoustic Emission of Composites Structures: Story, Success, and Challenges. *Phys. Procedia.* **2015**, *70*, 599–603. [CrossRef]
7. Derusova, D.A.; Vavilov, V.P.; Pawar, S.S. Evaluation of equivalent defect heat generation in carbon epoxy composite under powerful ultrasonic stimulation by using infrared thermography. In Proceedings of the International Scientific Conference on "Radiation-Thermal Effects and Processes in Inorganic Materials", Tomsk, Russia, 3–8 November 2014. [CrossRef]
8. Uhl, T.; Szwedo, M.; Bednarz, J. Application of active thermography for SHM of mechanical structures. In Proceedings of the 4th European Workshop on Structural Health Monitoring, Cracow, Poland, 2–4 July 2008; DEStech Publications, Inc.: Lancaster, PA, USA; pp. 149–156.
9. Katunin, A.; Wachla, D. Analysis of defect detectability in polymeric composites using self-heating based vibrothermography. *Compos. Struct.* **2018**, *201*, 760–765. [CrossRef]
10. Zhou, W.; Li, H.; Wu, Z. Damage Detection in the Composite Fuel Tank by Vibration Measurement Approach. In Proceedings of the 14th World Conference on Earthquake Engineering, Beijing, China, 12–17 October 2008.
11. Kessler, S.S.; Spearing, S.M.; Atalla, M.J.; Cesnik, C.E.S.; Soutis, C. Damage detection in composite materials using frequency response methods. *Compos. Part B Eng.* **2002**, *33*, 87–95. [CrossRef]
12. Garnier, C.; Pastor, M.-L.; Eymab, F.; Lorrain, B. The detection of aeronautical defects in situ on composite structures using Non Destructive Testing. *Compos. Struct.* **2011**, *93*, 1328–1336. [CrossRef]
13. Towsyfyan, H.; Biguri, A.; Boardman, R.; Blumensath, T. Successes and challenges in non-destructive testing of aircraft composite structures. *Chinese J. Aeronaut.* **2019**. [CrossRef]
14. Das, S.; Saha, P. A review of some advanced sensors used for health diagnosis of civil engineering structures. *Measurement* **2018**, *129*, 68–90. [CrossRef]
15. Tuloup, C.; Harizi, W.; Aboura, Z.; Meyer, Y.; Khellil, K.; Lachat, R. On the use of in-situ piezoelectric sensors for the manufacturing and structural health monitoring of polymer-matrix composites: A literature review. *Compos. Struct.* **2019**, *215*, 127–149. [CrossRef]
16. Forke, R.; Scheibner, D.; Dötzel, W.; Mehner, J. Measurement unit for tunable low frequency vibration detection with MEMS force coupled oscillators. *Sensors Actuators A.* **2009**, *156*, 59–65. [CrossRef]
17. Wang, S.; Wei, X.; Zhao, Y.; Jiang, Z.; Shen, Y. A MEMS resonant accelerometer for low-frequency vibration detection. *Sensors Actuators A Phys.* **2018**, *283*, 151–158. [CrossRef]
18. Kudela, P.; Wandowski, T.; Malinowski, P.; Ostachowicz, W. Application of scanning laser Doppler vibrometry for delamination detection in composite structures. *Opt. Lasers Eng.* **2017**, *99*, 46–57. [CrossRef]
19. Hossain, M.S.; Chao, O.Z.; Ismail, Z.; Noroozi, S.; Khoo, S.Y. Artificial neural networks for vibration based inverse parametric identifications: A review. *Appl. Soft Comput.* **2017**, *52*, 203–219. [CrossRef]
20. Gomes, G.F.; Mendéz, Y.A.D.; Alexandrino, P.D.S.L.; da Cunha, S.S., Jr.; Ancelotti, A.C., Jr. The use of intelligent computational tools for damage detection and identification with an emphasis on composites—A review. *Compos. Struct.* **2018**, *196*, 44–54. [CrossRef]
21. Zhang, Z.; Shankar, K.; Ray, T.; Morozov, E.V.; Tahtali, M. Vibration-based inverse algorithms for detection of delamination in composites. *Compos. Struct.* **2013**, *102*, 226–236. [CrossRef]
22. Orłowska, A.; Kołakowski, P.; Holnicki-Szulc, J. Detecting delamination zones in composites by embedded electrical grid and thermographic methods. *Smart Mater. Struct.* **2011**. [CrossRef]
23. Rezaiee-Pajand, M.; Shahabian, F.; Tavakoli, F.H. Delamination detection in laminated composite beams using hybrid elements. *Compos. Struct.* **2012**, *94*, 2777–2792. [CrossRef]
24. Ihesiulor, O.K.; Shankar, K.; Zhang, Z.; Ray, T. Delamination detection with error and noise polluted natural frequencies using computational intelligence concepts. *Compos. Part B Eng.* **2014**, *56*, 906–925. [CrossRef]
25. Sha, G.; Cao, M.; Radzieński, M.; Ostachowicz, W. Delamination-induced relative natural frequency change curve and its use for delamination localization in laminated composite beams. *Compos. Struct.* **2019**, *230*, 111501. [CrossRef]
26. Manoach, E.; Warminski, J.; Kloda, L.; Teter, A. Numerical and experimental studies on vibration based methods for detection of damage in composite beams. *Compos. Struct.* **2017**, *170*, 26–39. [CrossRef]
27. Yang, C.; Oyadiji, S.O. Damage detection using modal frequency curve and squared residual wavelet coefficients-based damage indicator. *Mech. Syst. Signal Process.* **2017**, *83*, 385–405. [CrossRef]
28. Katunin, A. Vibration-based spatial damage identification in honeycomb-core sandwich composite structures using wavelet analysis. *Compos. Struct.* **2014**, *118*, 385–391. [CrossRef]
29. Chiariottia, P.; Martarelli, M.; Revel, G.M. Delamination detection by Multi-Level Wavelet Processing of Continuous Scanning Laser Doppler Vibrometry data. *Opt. Lasers Eng.* **2017**, *99*, 66–79. [CrossRef]
30. He, M.; Zhang, Z.; Ramakrishnan, K.R. Delamination Identification for FRP Composites with Emphasis on Frequency-Based Vibration Monitoring-A Review. *Struct. Durab. Heal. Monit.* **2018**, *12*, 213–256. [CrossRef]
31. Fan, W.; Qiao, P. Vibration-based Damage Identification Methods: A Review and Comparative Study. *Struct. Heal. Monit.* **2010**, *10*, 83–111. [CrossRef]
32. Gałęzia, A.; Gumiński, R.; Jasiński, M.; Mączak, J. Application of energy operators for detection of failures in gearboxes. *Mech. Res. Commun.* **2017**, *82*, 3–8. [CrossRef]
33. Luo, H.; Hanagud, S. Dynamics of delaminated beams. *Int. J. Solids Struct.* **2000**, *37*, 1501–1519. [CrossRef]

34. Singh, A.K.; Chen, B.Y.; Tan, V.B.C.; Tay, T.E.; Lee, H.P. Finite element modeling of nonlinear acoustics/ultrasonics for the detection of closed delaminations in composites. *Ultrasonics* **2017**, *74*, 89–98. [CrossRef] [PubMed]
35. Shen, M.-H.H.; Grady, J.E. Free vibrations of delaminated beams. *AIAA J.* **1992**, *30*, 1361–1370. [CrossRef]
36. Matlab R2020a. Available online: https://nl.mathworks.com/help/matlab/ref/spline.html (accessed on 2 May 2020).
37. Brandt, A. *Noise and Vibration Analysis: Signal Analysis and Experimental Procedures*; John Wiley & Sons: Hoboken, NJ, USA, 2011. [CrossRef]
38. Feldman, M. *Hilbert Transform Applications in Mechanical Vibration*; John Wiley & Sons: Hoboken, NJ, USA, 2011. [CrossRef]
39. Kaiser, J.F. On a simple algorithm to calculate the 'energy' of a signal. In Proceedings of the International Conference on Acoustics Speech, and Signal Processing, Albuquerque, NM, USA, 3–6 April 1990; pp. 381–384.
40. Maragos, P.; Potamianos, A. Higher order differential energy operators. *IEEE Signal Process. Lett.* **1995**, *2*, 152–154. [CrossRef]
41. Kaiser, J.F. On Teager's Energy Algorithm and its generalization to continuous signals—A simple algorithm to calculate the 'energy' of a signal. In Proceedings of the 4th IEEE Digital Signal Processing Workshop, New Paltz, NY, USA, 16–19 September 1990.
42. Kvedalen, E. Signal Processing Using the Teager Energy Operator and Other Nonlinear Operators. Masters Thesis, University of Oslo, Oslo, Norway, May 2003. Available online: https://citeseerx.ist.psu.edu/viewdoc/download;jsessionid=7A993F17F2C3E6F0ED246214466F856F?doi=10.1.1.216.4519&rep=rep1&type=pdf (accessed on 21 January 2001).
43. Maragos, P.; Kaiser, J.F.; Quatieri, T. On amplitude and frequency demodulation using energy operators. *IEEE Trans. Signal Process.* **1993**, *41*, 1532–1550. [CrossRef]
44. Liang, M.; Soltani, B. An energy operator approach to joint application of amplitude and frequency-demodulation for bearing fault detection. *Mech. Syst. Signal Process.* **2010**, *24*, 1473–1494. [CrossRef]
45. Antoniadou, I.; Howard, T.P.; Dwyer-Joyce, R.S.; Marshall, M.B.; Naumann, J.; Dervilis, N.; Worden, K. Envelope analysis using the Teager-Kaiser Energy operator for condition monitoring of a wind turbine bearing. *Appl. Mech. Mater.* **2014**, *564*, 170–175. [CrossRef]
46. Gałęzia, A.; Barczewski, R.; Jakubek, B. Possibilities of Faults Detection of Rolling Bearings Using Energetic Descriptors of Vibrations Signals. In *Condition Monitoring of Machinery in Non-Stationary Operations*; Timofiejczuk, A., Chaari, F., Zimroz, R., Bartelmus, W., Eds.; Springer: Berlin/Heidelberg, Germany, 2018; pp. 329–337. [CrossRef]
47. Maragos, P.; Kaiser, J.F.; Quatieri, T.F. Energy separation in signal modulations with application to speech analysis. *IEEE Trans. Signal Process.* **1993**, *41*, 3024–3051. [CrossRef]
48. Maragos, P.; Kaiser, J.F.; Quatieri, T.F. On separating amplitude from frequency modulations using energy operators. In Proceedings of the ICASSP-92: 1992 IEEE International Conference on Acoustics, Speech, and Signal Processing, San Francisco, CA, USA, 23–26 March 1992. [CrossRef]
49. Salzenstein, F.; Boudraa, A.-O.; Cexus, J.-C. Generalized higher-order nonlinear energy operators. *J. Opt. Soc. Am. A.* **2007**, *24*, 3717–3727. [CrossRef]
50. Potamianos, A.; Maragos, P. Speach formant frequency and bandwidth tracking using multiband energy demodulation. *J. Acoust. Soc. Am.* **1996**, *99*, 3795–3806. [CrossRef]
51. Goyal, K.; Jorge, V.R. Micro-Scale Crack Propagation Using the eXtended Finite Element Method. In Proceedings of the 57th AIAA/ASCE/AHS/ASC Structures, Structural Dynamics, and Materials Conference, San Diego, CA, USA, 4–8 January 2016; pp. 1–24. [CrossRef]
52. Johnson, K.L. *Contact Mechanics*; Cambridge University Press: Cambridge, UK, 1987.
53. Wriggers, P.; Laursen, T. *Computational Contact Mechanics*; Springer: Berlin/Heidelberg, Germany, 2006.
54. Ansys 17.1, Help. Available online: https://www.ansys.com/-/media/ansys/corporate/files/pdf/solutions/it-professionals/platform-support/platform-support-by-application-171.pdf?la=en&hash=E6566109CA6A9B691A45F5D9C19E4BE1F5985BAF (accessed on 1 March 2021).
55. Cawley, P.; Adams, R.D. The location of defects in structures from measurements of natural frequencies. *J. Strain Anal. Eng. Des.* **1979**, *14*, 49–57. [CrossRef]
56. Pandey, A.K.; Biswas, M.; Samman, M.M. Damage detection from changes in curvature mode shapes. *J. Sound Vib.* **1991**, *145*, 321–332. [CrossRef]

materials

MDPI

Article

Extraction of Magnetic Field Features to Determine the Degree of Material Strain

Przemysław Szulim * and Szymon Gontarz

Institute of Vehicles and Construction Machinery, Warsaw University of Technology, 02-524 Warsaw, Poland; szymon.gontarz@pw.edu.pl
* Correspondence: przemyslaw.szulim@pw.edu.pl

Abstract: Currently, to realize the reliable operation and proper exploitation of complex machines and structures, information regarding the material condition must be obtained. This information should ideally be acquired in a noninvasive manner. In addition, contemporary rapid technological development is conducive to the research and advancement of new methods, including magnetic methods. This publication describes the methods that can enable the extraction of information from the magnetic field, which is valuable for determining the material effort state and performing technical diagnostics. The issue of using the magnetic field to assess the technical condition of structures is a promising trend in technical diagnostics. Moreover, new ways to process the magnetic field information are being identified to connect the observed surface changes in the magnetic field with the significant diagnostic symptoms. This work provides an extensive introduction to the theoretical basis and diagnostic techniques based on measurements of the magnetic field obtained in close proximity to the structure of interest. The key limitations of the method and associated possibilities are highlighted. The model considerations were taken into account to provide a mathematical description of the extraction process and possible interpretations of the acquired signals. According to the received guidelines, the plan and implementation of two experiments are described along with the obtained results, which demonstrated the possibility of identifying valuable information that can be used to determine the state of the material stress and perform diagnostics of steel structures.

Keywords: technical diagnostics; magnetic field; material effort; steel structures; nondestructive diagnostics

Citation: Szulim, P.; Gontarz, S. Extraction of Magnetic Field Features to Determine the Degree of Material Strain. *Materials* **2021**, *14*, 1576. https://doi.org/10.3390/ma14061576

Academic Editor: Salvatore Grasso

Received: 2 February 2021
Accepted: 18 March 2021
Published: 23 March 2021

1. Introduction

Many materials and machines that can present a real risk of a catastrophe due to fatigue wear, exceeding stress limits, or the appearance of plastic deformation have magnetic properties that can affect the local magnetic field, making it possible to increase the variety of nondestructive techniques, which are necessary for the modern management of technical facility operations [1–3]. For this reason, along with the technological development of magnetic sensors and measuring instruments, magnetic methods have recently become very attractive. Additionally, bearing in mind the consequences of unforeseen architectural failures and failures of technical facilities, it is necessary to advance science toward the technology of detecting early stages of damage, which is possible thanks to the use of magnetic methods [4–6].

In general, magnetic techniques are divided into active and passive techniques. Currently, the group of active magnetic methods is quite large and widely used. The main methods representing this family include the magnetic noise method (Barkhausen method), eddy current method, powder technique, and a method based on magnetic flux leakage. These types of tests require specialized, usually complex, measuring equipment, and the obtained results are sometimes ambiguous. However, these methods enable the detection of the most dangerous defects: flat surface defects and narrow gap defects. Although they play a very important role in technical diagnostics, the methods are adapted to detect existing defects. In parallel with the development of active diagnostic methods, the development of a group of passive diagnostic methods can be observed, which have all the advantages of active methods but, at the same time, do not require the use of artificial sources of magnetic fields, which is associated with the use of complex and expensive devices. The methods that can detect damage in the nucleation phase are the following methods: magnetoacoustic emission (EMA), metal magnetic memory (MMM) method, and other passive techniques [7–13], which only use the existence of the natural magnetic field of the Earth. These methods are becoming increasingly popular and are the subject of intensive research.

However, due to the variety and complexity of magnetomechanical effects, it is difficult to identify the quantitative measures that allow us to precisely define the diagnostic thresholds for specific magnetic signals. Recently, the most widely used approach is to acquire the magnetic signal by scanning the spatial distribution of the magnetic field of the sample [14] (e.g., MPM). This method uses the phenomenon of memorizing the effects of cyclical and boundary loads on the structure of the material and analyzes the residual magnetic field [15]. Such an assumed phenomena model of an object is very simplified because it does not even take into account changes in magnetic properties, e.g., permeability of the material, with a change in the stress state of the material [16]. Moreover, the current state of the art shows that new models are needed to match the magnetomechanical phenomena [17]. The relationships existing in the field of plastic deformation seem to be particularly problematic because traditional magnetic–stress coupling models are based on the magnetoelastic theory, which cannot explain magnetic phenomena in the plastic deformation range [18]. The results obtained by scientists indicate new possibilities of describing and modeling phenomena, which better describe the observed reality. Along with the new description of physical interactions, new measurement and analysis methods are necessary that will use the emerging possibilities of obtaining information regarding the magnetic signal [19], both in terms of the elastic and plastic deformations. In both these cases, changes in the mechanical properties of the construction material entail changes in the magnetic properties, mainly the magnetization and permeability of the material [20]. The influence of both parameters is observed by the measuring equipment in the magnetic field around the structure but without the possibility of separating them. The possibility of a separate analysis of both factors could significantly contribute to improving the quality of the estimation of the technical condition of the structure.

Regarding this issue, the next chapter indicates potential physical phenomena that are related to the magnetic signal, which may contain useful diagnostic information for the objects under consideration. Next, the article discusses the issue of selecting a sensor with the appropriate sensitivity and the problem of how to observe the magnetic signal, i.e., whether it is to be a measurement at a point, on a certain observation plane, or whether a spatial measurement is needed. These considerations lead to the idea of fusing information from multiple additional sensors, which may prove pivotal in the proposed approach. The next part of the article describes the key stage from the point of view of diagnostics, i.e., finding a magnetic diagnostic parameter that will be sensitive to degradation changes taking place in a technical facility [21,22]. Chapter 3 presents the mathematical basis of the method of information separation from the magnetic field to extract the information that is most closely related to the technical condition of the material. The following chapters describe two experiments and an exemplary interpretation of the results. The first experiment involved flat samples made of 3H13 stainless steel in various stress states. The second involved an analysis of changes in the distribution of the magnetic field on a sample that was carried through successive states of stress until breaking. The proposed methodology, including the process from planning the experiment and selecting the measurement path, through the measurement, to the extraction of diagnostic information, cannot be clearly assigned to either the passive or active magnetic technique. However, it has been shown that the optimal performance of the entire proposed process may lead to the acquisition of qualitatively and quantitatively appropriate magnetic information. By appropriately transforming this information, it was possible to identify specific magnetic, electromagnetic, or magnetomechanical phenomena. The analysis of these phenomena related to the working conditions and the condition of the technical object led to finding an appropriate diagnostic measure that, in accordance with the nature of magnetic methods, described the early stages of the occurring damage.

2. Theoretical Background

To utilize the new diagnostics information pertaining to the functional properties of technical objects, we recommend the analytical estimation of the cross-effects, especially those pertaining to the mechanics and magnetics [21,23–27]. Certain researchers indicated that as the stress condition changes in a material with magnetic properties, the material exhibits a magnetic state, which can be shown by the change in the material magnetization intensity. The magnetization intensity can be attributed to the magnetomechanical phenomena, which might be affected by the static or dynamic loads applied to the material specimen and can be appropriately recorded and analyzed if required.

Furthermore, this impact varies depending on the "quantity and quality" of the material that the specific object is made from. Certain bodies can generate a magnetic field around them. Other objects, when placed in an external magnetic field, lead to a distinct change in this field. To model this aspect, the magnetic properties of a ferromagnetic object can be described as a set of magnetic dipoles ($\mathbf{m}_i$) that are bound together with microcurrents. According to this concept, the magnetic features of a body of a set of dipoles can be characterized by the magnetization vector **M**, which can be defined as a resultant magnetic moment per unit of volume:

$$\mathbf{M} = \frac{\sum \mathbf{m}_i}{\Delta V}. \tag{1}$$

To render the equation transparent and more usable from the diagnostics viewpoint, the following expression can be formulated on the macroscopic scale:

$$m = f(\lambda, \omega, E, v, H, T). \tag{2}$$

λ and ω are the components of magnetostriction ($\lambda = \Delta l/l$ and $\omega = \Delta V/V$ for linear and volumetric measures, respectively), and E and v denote Young's modulus and Poisson's ratio, respectively.

When a permanent layout of the magnetic moments occurs, the specimen exhibits external magnetic features (poles). This behavior is typical for permanent magnetic materials, including ferromagnets. In general, ferromagnetic materials include iron, cobalt, nickel, and certain transitional metals pertaining to iron and rare earth elements. When performing diagnostics on engineering structures that are built mainly from steel or cast steel, it is preferable to operate with ferromagnetic soft materials, which usually lose nearly all their magnetization once the magnetizing field is removed and exhibit only residual magnetism, which is considerably smaller than the maximum magnetism. For such materials, the resulting value **M** changes substantially due to an external magnetic field or other physical influences on the object in the presence of such an external field [16].

The ferromagnetic behavior can be explained with reference to Weiss' theory of domains, which indicates that each ferromagnetic material consists of magnetic domains wherein the atoms are arranged in 3D crystal lattices and demonstrate magnetic moments (nonvanishing magnetization vectors). In practical magnetization, the phenomena related to changes in the magnetization state **M**, are highly complex; however, these phenomena can be approximated through a first magnetization curve, as shown in Figure 1.

First, the evolution of the domain structure is reorganized by domain wall displacement. Next, the moment rotation attempts to align the moments of individual areas of spontaneous magnetization parallel to the external field. Finally, the paraprocess occurs, in which after the technical saturation point, as many domains as possible follow the direction of the outer field. A further increase in the magnetic induction occurs with an increase in the **H** field intensity.

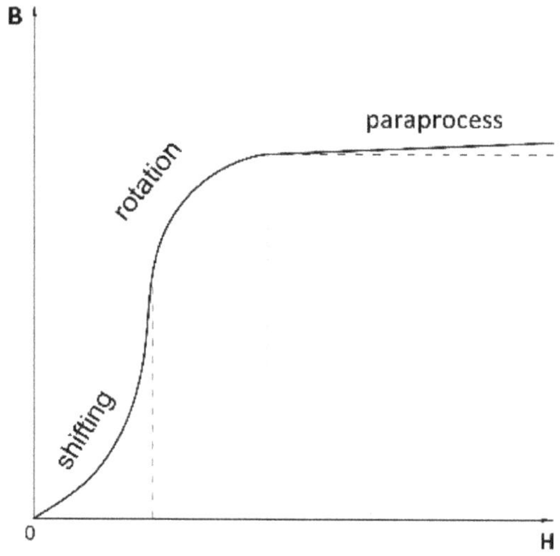

Figure 1. First magnetization curve of ferromagnetic magnetization.

The described process refers to an idealized case and is a classic description that is appropriate for the response of a material to the excitation in the form of a change in the magnetic field intensity. However, the actual process is highly complicated because the movements of the walls and rotary movements may overlap, and the nature of the movements is strongly dependent on the structural features of the ferromagnetic material. In this scenario, the specific border points of the domain changes may be ambiguous. In addition, in the presence of other physical influences (see Equation (2)) on the object placed within even a weak magnetic field, the magnetization involving the rotary movement of the domains cannot be excluded, as indicated by the irreversible and nonlinear effects [16].

The described phenomena and behavior of the magnetic material are analogous in the case of magnetization involving magnetomechanical effects, as the mechanical interactions may modify the primary magnetization curve.

According to Equation (2), which presents the magnetization model, and the results of the experiments, a change in the material structure (degradation), stress distribution, material temperature, or magnetic field intensity can influence the magnetization intensity distribution layout $\mathbf{M}(x,y,z)$ and magnetization intensity distribution of the magnetic induction $\mathbf{B}(x,y,z)$ in the vicinity of the object being tested. Therefore, the reading of the magnetometer should be understood as an association between the magnetization **M**, the intensity of the external magnetic field **H**, and the magnetic induction **B**, as expressed in the constitutive law equation:

$$\mathbf{B} = \mu\mathbf{H} = \mu_0(\mathbf{H} + \mathbf{M}), \tag{3}$$

where μ is relative permeability and μ_0 is permeability of free space. The relationship between the magnetization **M**, the intensity of the external magnetic field **H**, and the magnetic induction is a scientifically proven theorem to support nondestructive magnetic testing and monitoring of the structures using active and passive methods.

In general, one can identify three methods of interaction that produce magnetization changes that occur in accordance with the previously described domain processes:

- Stress-affected change via deflection (deflection change in the elastic strain or plastic strain range);
- Change caused by thermal effects (heating the material to a temperature in excess of the Curie point);
- Electromagnetic change (by the action of an external magnetic field).

Electromagnetic magnetization is specific to active methods, while temperature-affected magnetization corresponds to temperatures close to the Curie point. In contrast, in the case of passive methods, the key issue pertains to the stress-affected magnetization.

The stress-affected magnetization procedure in a weak magnetic field (at the level of the natural magnetic field on the Earth's surface) is essentially different from the technical magnetization procedure conducted under a strong (>>50 A/m) permanent or variable magnetic field, as is currently implemented in non-destructive testing (NDT) inspections. The comprehension of only these variances could provide adequate bases to realize quantitative and qualitative interpretation of the measurement results and enable the fully controlled employment of the observer of the state to diagnose technical objects. In this regard, it is desirable to simplify the numerical model of stress magnetization in a weak magnetic field, which is highly complex compared to that of technical magnetization. The associated division is based on the need to separate both the reversible effects (the Villari effect and its derivatives) and irreversible effects (ΔE effect, energy dissipation via Barkhausen noise, change in the magnetization observed after unloading the material, austenite–martensite phase transformation), as indicated in Equations (4)–(6). This division is consistent with the experimentally confirmed mechanism of reversible moving domains and the corresponding irreversible rotation. This division can be expressed in the analytical form as follows:

$$M(\sigma, t) = M_{dir}(\sigma, t) + M_{irr}(\sigma = 0, t) \tag{4}$$

$$M_{dir}(\sigma, t) \rightarrow \mu_0 \left(\frac{\partial M}{\partial \sigma}\right)_H = \left(\frac{\partial B}{\partial \sigma}\right)_H = \left(\frac{\partial \lambda}{\partial H}\right)_\sigma \tag{5}$$

$$M_{irr}(\sigma = 0, t) \rightarrow M = \chi H + v H^2 \tag{6}$$

where σ is stress, t is temperature, v is Rayileigh coefficient, λ is magnetostriction and χ is magnetic susceptibility. The adopted division highlights a certain analogy. The influence of stresses on the magnetic induction produced by an object in a constant external magnetic field can be equated with the change in the magnetostriction under the influence of an external magnetic field at a constant level of stress. However, for the area in which the magnetoplastic phenomenon occurs, the magnetization increases in proportion to the number of cycles of mechanical excitation. With regard to the proposed model, it is advisable to introduce the notion of the "eigenmagnetic field" [16] as a global signal of the magnetic field recorded by the magnetometer. The "eigenmagnetic field" describes the behavior of a magnetoelastically activated material in a weak magnetic field surrounding the material. The concept is broader than that of the diffuse magnetic field because it considers the complex reversible and irreversible effects. This concept allows for the description of a greater number of physical phenomena that more accurately define the state of stress and strain of the material. This description is valuable; however, to obtain this detailed information, the observation method must be extended to include all components of the magnetic field (in the case of self-magnetic flux leakage (SMFL), two components are sufficient) and use appropriate techniques that enable the separation of the effects of interest. The starting point for this concept may be a model description of the reversible/irreversible phenomena, which can be expressed as

$$\mathbf{B_P} = \mathbf{S} \cdot (\mathbf{R} \cdot \mathbf{B_E} + \mathbf{B_M}), \tag{7}$$

where the magnetic induction $\mathbf{B_P}$ measured in the vicinity of the sample is described. The measurement value includes the component of the terrestrial magnetic induction $\mathbf{B_E}$, and component $\mathbf{B_M}$ can be attributed to the permanent magnetization of the sample. The diagonal matrix **S** defines the effect of the closeness of the magnetic material with relative permeability $\mu_r \geq 1$. Matrix **R** represents the transformations of the local terrestrial field components that are bound in an inertial system to a local magnetic field when observed in a relative system coupled with the tested sample.

A significant limitation to the practical use of this type of model is the complexity of the material structure, access to a limited number of measurements, and the presence of external disturbances. In this regard, the proper recording of the magnetic signals, processing, analysis, and interpretation can enable the enhancement of the capabilities of magnetic diagnostic methods. Consequently, methods that can more accurately estimate the degree of material strain or material degradation are being developed by extracting the information that is not normally available in the measurements. The two proposed methods are discussed in the next section.

3. Measurement Model

The idea of separating the measured values is based on the formulation of a measurement model that considers the components of the magnetic field in various reference systems. The first model assumes that the measurement sensor or the sensor matrix is attached to the surface of the tested specimen and that the system is rotated. For this case, we establish two coordinate systems: a local system that is associated with the sensor and sample and a fixed frame of reference. In the case of the latter system, the specimen remains stationary with the sensors, while a slowly changing magnetic flux of low amplitude passes through the sample, causing a change in the field distribution on the sample's surface.

3.1. Measurement Model with Specimen Rotation

The value of the magnetic field in the vicinity of the tested object may consist of many components. The key components pertain to the Earth's magnetic field and the field resulting from other sources located near the sensor, including residual magnetization (which, among other factors, can be attributed to the stress acting on the tested object). Moreover, the change in the magnetic permeability that occurs as a result of the stresses in the ferromagnetic material leads to local disturbances in the field around the specimen. The magnetic induction $\mathbf{B_b}$ at a specific point in space, as measured by the sensor, can be represented as the product of the real induction $\mathbf{B_n}$ (in the reference frame) and the rotation matrix $\mathbf{R_n^b}$, thereby allowing for the representation of any vector from the sensor coordinate system to the reference coordinate system:

$$\mathbf{B_b} = \mathbf{R_n^b}\mathbf{B_n}. \tag{8}$$

However, the measurement $\mathbf{y^b}$ obtained by the sensor is not equal to the $\mathbf{B_b}$ field due to the presence of disturbances. All the magnetic field disturbances around the sensor are assumed to be constant in time and space. In the nomenclature related to navigation based on the magnetic field, for which calibration algorithms of magnetometers have been developed [28], the concept of magnetically soft and magnetically hard effects is introduced. Hard magnetic effects manifest as the sources of magnetization in the material of the tested sample. These components contribute to the external magnetic field of the Earth. Moreover, considering the methodology adopted in the experiment, these components are related to the coordinate system of the sensor and are designated as the local component of the $\mathbf{B_l^b}$ field. Soft magnetic effects are related to the magnetic susceptibility of the material near the sensor. These effects manifest as components of the additional field induced by the presence of an external field and contribute to the total field recorded by the sensor. However, these components depend on the direction of the external field and can change not only the size of the measured field but also its direction. This effect is modeled using the matrix **S** (3×3). Consequently, the matrix **S**'s coefficients are related to the magnetic permeability of the material. For specimens with no magnetic susceptibility (μ_r = 1), this matrix reduces to the identity matrix. In the case of samples made of ferromagnetic materials, the matrix coefficients take different values depending on the local changes in the magnetic permeability of the sample, as well as on the shape and dimensions of the sample, as observed in the experiments. Notably, this matrix can be reduced to a diagonal matrix to focus only on the components with the most significant physical interpretation for the given research. The effects of the magnetic field induction are generally nonlinear; however, assuming a weak forcing field (at the level of the Earth's magnetic field), a linear relationship **B**(**H**) can be assumed:

$$\mathbf{y^b} = \mathbf{S}\mathbf{B_b} + \mathbf{B_l^b}, \tag{9}$$

where:

$$\mathbf{y^b} = \begin{bmatrix} y_x^b \\ y_y^b \\ y_z^b \end{bmatrix}, \mathbf{S} = \begin{bmatrix} S_x & 0 & 0 \\ 0 & S_y & 0 \\ 0 & 0 & S_z \end{bmatrix}, \mathbf{B_l^b} = \begin{bmatrix} B_{lx}^b \\ B_{ly}^b \\ B_{lz}^b \end{bmatrix}.$$

Under the assumption that the magnitude of the local magnetic field is constant (the magnetic field around the sample is homogeneous), the abovementioned equations can be transformed to a form that can be used to estimate the unknown components of the diagonal matrix **S** and vector $\mathbf{B_l^b}$:

$$\left|\mathbf{S}^{-1}\left(\mathbf{y^b} - \mathbf{B_l^b}\right)\right| = \left|\mathbf{R_n^b}\mathbf{B_n}\right| = \text{const.} \tag{10}$$

This matrix equation can be simplified to a single nonlinear scalar equation:

$$\frac{y_x^{b2}}{S_x^2} + \frac{y_y^{b2}}{S_y^2} + \frac{y_z^{b2}}{S_z^2} - \frac{2y_x^b B_{lx}^b}{S_x^2} - \frac{2y_y^b B_{ly}^b}{S_y^2} - \frac{2y_z^b B_{lz}^b}{S_z^2} + \frac{B_{lx}^{b2}}{S_x^2} + \frac{B_{ly}^{b2}}{S_y^2} + \frac{B_{lz}^{b2}}{S_z^2} = \left|\mathbf{R_n^b}\mathbf{B_n}\right|^2 = c. \tag{11}$$

This equation can be converted to a form that clarifies the known components and unknown factors a_1–a_7:

$$y_x^{b2}a_1 + y_y^{b2}a_2 + y_z^{b2}a_3 - y_x^b a_4 - y_y^b a_5 - y_z^b a_6 + a_7 = c, \tag{12}$$

where:

$$a_1 = \frac{1}{S_x^2},\ a_2 = \frac{1}{S_y^2},\ a_3 = \frac{1}{S_z^2},\ a_4 = \frac{2B_{lx}^b}{S_x^2},\ a_5 = \frac{2B_{ly}^b}{S_y^2},\ a_6 = \frac{2B_{lz}^b}{S_z^2},\ a_7 = \frac{B_{lx}^{b2}}{S_x^2} + \frac{B_{ly}^{b2}}{S_y^2} + \frac{B_{lz}^{b2}}{S_z^2}.$$

For each experiment, many (n) measurements of y_x^b, y_y^b, and y_z^b were obtained to build a matrix equation based on the above equation.

$$\begin{bmatrix} \mathrm{y_{x_1}^b}^2 & \mathrm{y_{y_1}^b}^2 & \mathrm{y_{z_1}^b}^2 & -\mathrm{y_{x_1}^b} & -\mathrm{y_{y_1}^b} & -\mathrm{y_{z_1}^b} & 1 \\ & & & \vdots & & & \\ \mathrm{y_{x_n}^b}^2 & \mathrm{y_{y_n}^b}^2 & \mathrm{y_{z_n}^b}^2 & -\mathrm{y_{x_n}^b} & -\mathrm{y_{y_n}^b} & -\mathrm{y_{z_n}^b} & 1 \end{bmatrix} \begin{bmatrix} a_1 \\ a_2 \\ a_3 \\ a_4 \\ a_5 \\ a_6 \\ a_7 \end{bmatrix} = [c] \tag{13}$$

The search for the coefficients of **S** and $\mathbf{B_l^b}$ was performed in two stages. First, certain unknown coefficients were determined using the linear regression method, and these results were considered to be the starting point to identify the remaining coefficients using Newton's method.

3.2. Measurement Model with Excitation

The experiment with sample rotation can only be performed when the tested object is small. To enable the application of the proposed method to large objects, a reverse situation was engineered by creating an artificial magnetic field with the magnitude of the

Earth's field rotating in any direction. This approach enabled the continuous observation of changes in the magnetic properties of the sample material during the stress experiment, which helped to further clarify the phenomena occurring in the material. In the first approach, to reduce the complexity of the stand, the sample was simulated with an artificial field created by the yoke, thereby limiting the range of changes in the magnetic field to one plane. Sinusoidal changes in the excitation were assumed. Due to the assumed low value of the generated flux and current control in the excitation circuit, this assumption could be considered reasonable. In such cases, the local value of the magnetic field on the sample surface can be described through a simple relationship:

$$B_{Mi} = S_i \sin(\omega t + \varphi_i) + B_{Ci}, \tag{14}$$

where i indicates the x-, y-, or z-direction relative to the sample, and ω is the frequency of the current excitation

The coefficient S_i is related to the magnetic permeability of the sample and describes the degree of scattering of the excitation magnetic field on the sample's surface. When $S_i = 0$, the entire forcing flux is confined to the sample volume and does not leak. The φ_i coefficient is related to the presence of the phase shift between the excitation and registered field, while the B_{Ci} component represents the total influence of the sample's own magnetic induction and the surrounding field. In this scenario, the fields cannot be separated

Both models were designed to extract parameters that could effectively describe the target characteristics of the material for diagnostic purposes. These models were employed in various types of experiments, as described in the subsequent section.

4. Test Stand Description

A detailed description of the test stand and the conditions under which the experiment was conducted for the first variant can be found in the literature [8]. However, in the present study, we primarily focused on the experiment pertaining to the novel variant, which involved the model with a certain stimulation.

4.1. Stand for the Tests in the Nonstationary Variant (Measurement with Sample Rotation)

The common feature of both experiments was the use of a customized measurement device (MagMouse). The device, as shown in Figure 2, consisted of a matrix of three-axis magnetoresistive sensors, an optical displacement sensor (not used in the experiments), and a microcontroller to realize communication with an external data recording application. Honeywell HMC5983 magnetic sensors (Honeywell, Charlotte, NC, USA) were used to construct the device. The HMC sensor is a digital sensor that converts analog field information from a triad of anisotropic magnetoresistance (AMR) sensors into the digital form. It is possible to program the sampling frequency of the internal analog/digital (A/D) converters and control the gain of the analog signal, thereby enabling variations in the sensitivity and measurement range of the sensor. During the tests, the measurement acquisition frequency and range were set to 75 Hz and 0.8 mT, respectively. In the first variant, the device was rigidly connected to the test sample and rotated in various configurations while recording the field values.

To enable coverage along the entire length of the sample, the experiment was performed several times by shifting the matrix of the sensors in relation to the tested sample in specific intervals after each experiment. The measurement points on the sample (magnetic field measurement points subject to further evaluation) are illustrated in Figure 3. During the research, data from only one row of matrix sensors were used.

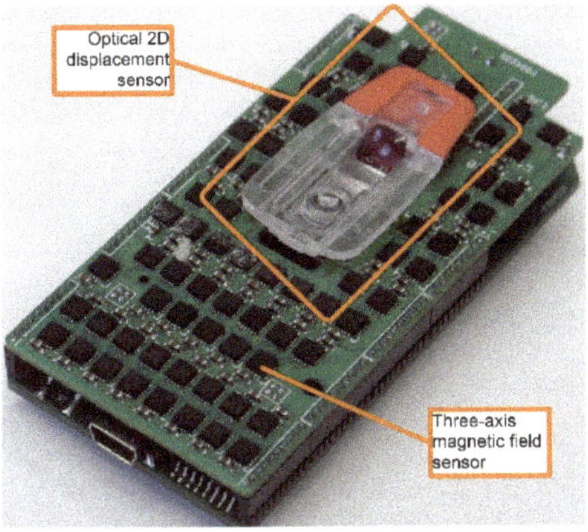

Figure 2. Image of the MagMouse measurement device [8].

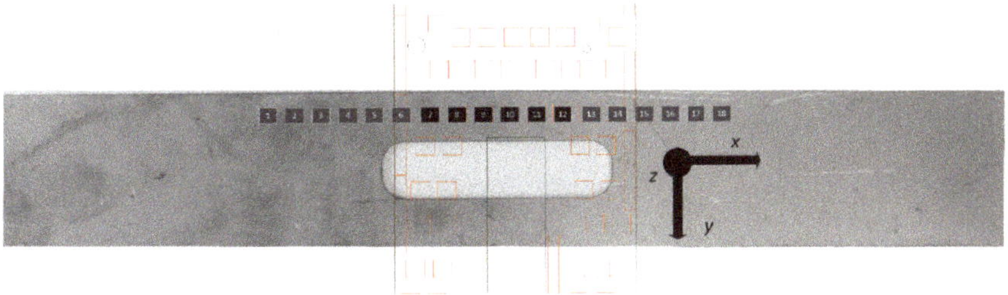

Figure 3. Arrangement of the magnetic field sensors (measurement points) relative to the tested sample.

The experimental stage is presented in Figure 4.

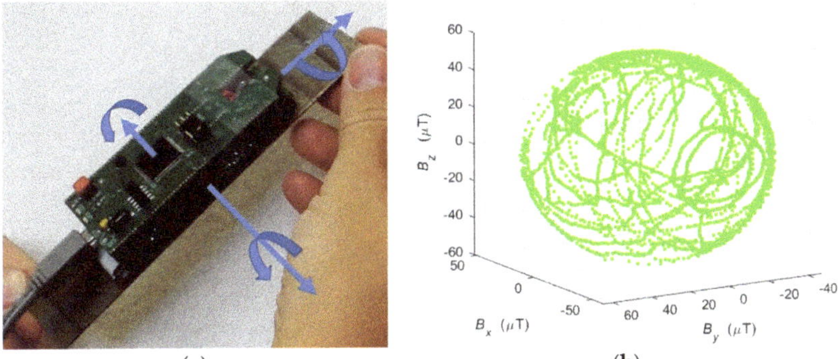

(**a**) (**b**)

Figure 4. Sample rotation procedure with the sensor matrix (**a**) sample result of the recorded magnetic field creating a characteristic sphere for one selected sensor from the matrix (**b**).

After the series of rotations was performed, a magnetic field data set was obtained for each sensor in the matrix. The visualization of the registered field for one measurement point is presented in Figure 4b. A characteristic surface resembling a sphere was created. Next, the set of parameters was extracted according to the model from the first variant. In each experiment, several thousand samples of the magnetic field induction vector were recorded for each measurement point. After estimating the six searched parameters, the effect of filtering out information from noise was obtained. The influence of the sensor measurement noise was significantly reduced. Many series of tests were conducted for specimens in various states of degradation: elastic deformation, plastic deformation, and rupture. The sample specimens used in both variants were made of 3H13/(1.4028 according to EN 10027-2 [29]) steel. The results of the experiment are presented in the subsequent section.

4.2. Stand for the Tests in a Stationary Variant (with a Stationary Sample)

To enhance the knowledge of the processes occurring in the sample during the degradation of the technical condition, and to simultaneously extract the various parameters from the magnetic field signal, the previous variant of the experiment was adopted. The test sample was placed in a testing machine. A yoke forcing a controlled magnetic flux and an array of sensors to measure the field distribution on the surface were placed on the sample. The specimen model, along with the arrangement of the sensors of the magnetic field of the matrix relative to the tested sample, is shown in Figure 5.

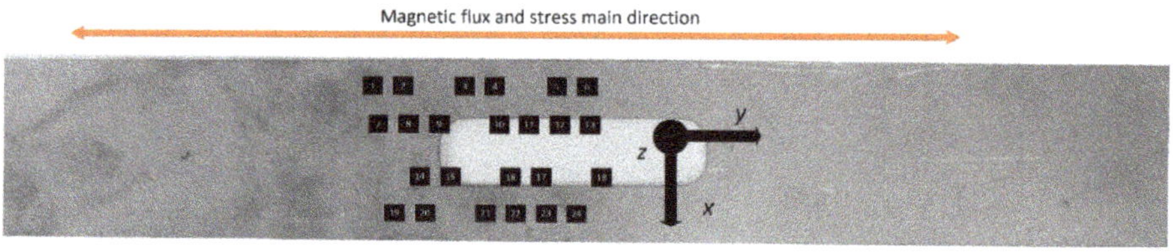

Figure 5. Arrangement of the magnetic field sensors of the matrix relative to the tested sample.

The excitation level was experimentally selected to ensure that the range of field changes was similar to that in the experiment in the first variant. A controlled current source was used to set the current in the yoke. The registration process and signal generation were managed using an application written in the LabView environment. The scheme of the test stand is illustrated in Figure 6.

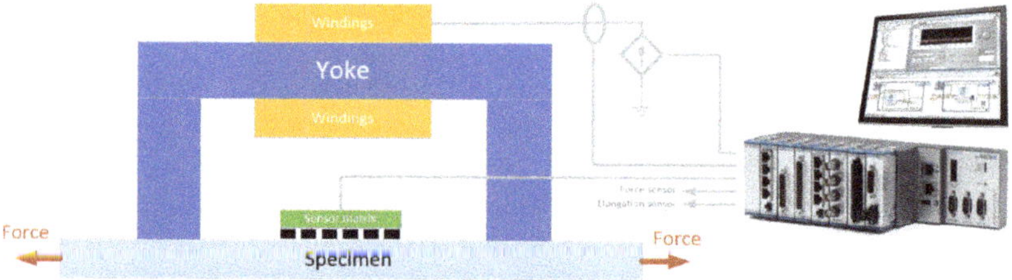

Figure 6. Diagram of the test stand for the second variant of the experiment.

Signals, such as those of the yoke current, tensile force, sample elongation, and magnetic field, were recorded. The experiment was performed on samples made of 4H13 steel (1.4034 according to EN 10027-2 [29]). The experiment involved several steps. At each step, a specific value of the excitation force was applied and excitation based on the magnetic flux was generated. Subsequently, the measurement signals were recorded for four periods of the excitation signal. The frequency of the forcing signal was set to be a constant value of 0.05 Hz. The magnetic field sampling frequency was limited by the capabilities of the magnetic field sensors and set to 75 Hz. The yoke, sample, and matrix during the experiment on the testing machine are shown in Figure 7.

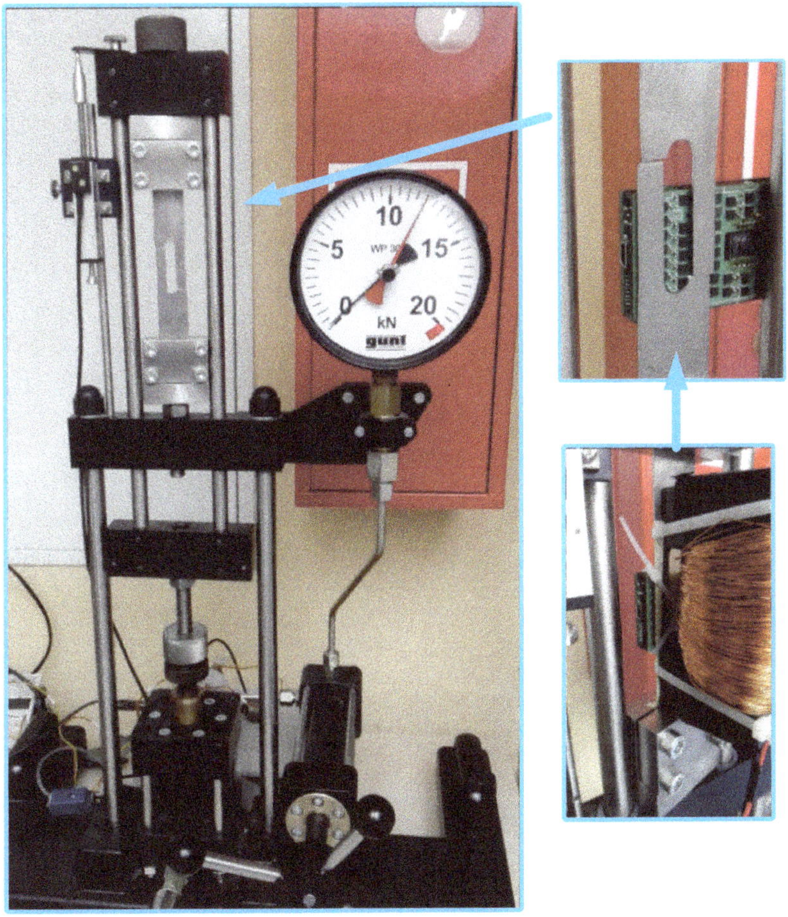

Figure 7. Mutual positions of the yoke, sample, and sensor matrix during the experiment.

5. Test Stand Investigation

The results of the tests for both variants of the experiment are presented in this section. The results were analyzed in accordance with the assumed models (Section 3). Consequently, we could extract the parameters describing the behavior of the magnetic field, which changed depending on the material effort state.

5.1. Experiment in the Nonstationary Variant (Measurement with Sample Rotation)

Figures 8 and 9 present the results obtained for the experiment conducted on three specimens with different degradation states. The results pertain to the separated components of the sample's magnetic field related to the magnetization of the ferromagnetic material.

Figure 8 illustrates the various quantitative and qualitative changes in the magnetic field, which varied with the degradation of the steel material. In the case of parameter **B**, the largest changes occurred in the case of the damaged sample. However, under elastic and plastic deformation, the situation may not be clear. In this scenario, the **S** parameter could provide additional information to recognize the strain states (Figure 9).

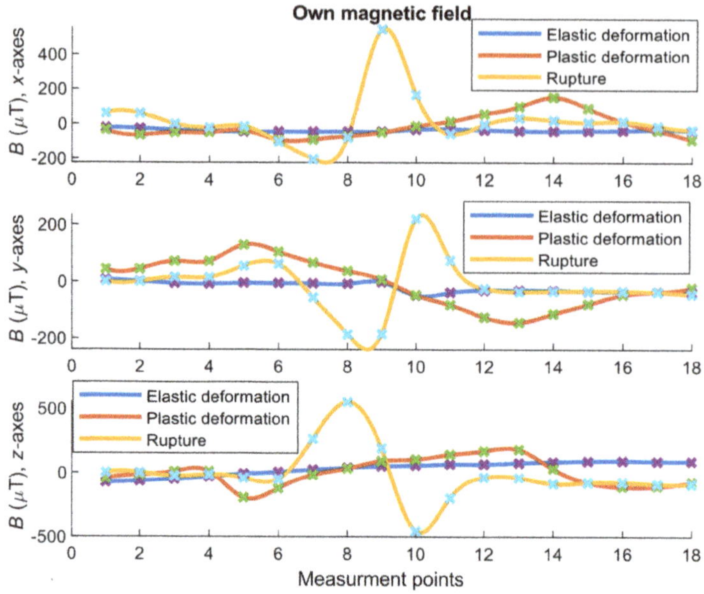

Figure 8. Reversible and irreversible effects associated with different states of the sample strain at successive points on the specimen's surface (measurement points). The x, y, and z charts correspond to measurements from the x-, y-, and z-axes of magnetic sensors, respectively. The x-axis direction was tangential to the main axis of the sample, and the z-direction was normal to the sample surface.

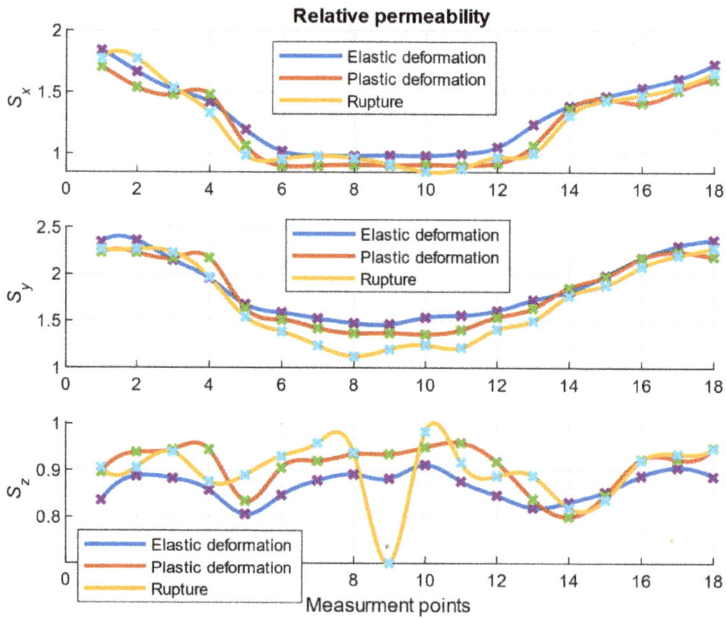

Figure 9. Summary of the changes in the parameter related to the magnetic permeability of the specimen for each axis and different states of stress at successive points on the specimen's surface (measurement points).

Figure 9 shows the influence of the parameter **S** for different effort states of the material. The graph shapes for the *x*- and *y*-axes indicate the changes in the shape of the sample, the width of which was between measurement points 6 and 12. The changes pertaining to the *z*-axis were smaller and may have been related to the constant thickness of the sample. The estimated parameter thus clarified the sensitivity to the shape of the sample. A gradual quantitative change in **S** with the degradation level of the specimen (in all directions) was observed. This observation can help to identify the instant at which the material became plasticized. In raw measurements of the magnetic field, many phenomena affect the result such that degradation estimation is challenging. Using the proposed separation technique, the degradation could be distinguished. Notably, on the axis perpendicular to the sample surface (i.e., the *z*-axis), a larger change occurred in the parameter value for the broken specimen near the material discontinuity. This result was consistent with subsequent tests that demonstrated the significant influence of the changes in the permeability of the material for the range of plastic strains.

Although the described variant (with sample rotation) is not practical in the real world, it proved that even small changes in the structure of the material when interacting with the Earth's magnetic field yielded a characteristic signal in the form of a magnetic field near the sample. This aspect could be identified using the proposed separation for the parameters **B** and **S**. The analyzed samples differed significantly in terms of the state of deformation; therefore, the magnetic effects related to the effort of the material were significant and easy to register. When less force was applied to the sample and the measurement cannot be repeated, a new approach is required to observe similar diagnostic symptoms in the magnetic field. Such an approach may also require highly sensitive apparatus and special processing. The consideration of real-world issues based on the spatial measurement, including additional information regarding the location, can be useful from a diagnostic viewpoint.

5.2. Experiment in the Stationary Variant (with a Stationary Sample)

Considering the noteworthy results obtained in the first experiment, the authors tested samples pertaining to various states of stress. The strength characteristics for this experiment are described in Figure 10, which clarifies the range of elastic and plastic deformations. Additionally, the characteristic points are marked and were further analyzed. Due to the high dynamics of the changes in various parameters for different states of material stress, we compared these changes as a function of the subsequent measurements and not force.

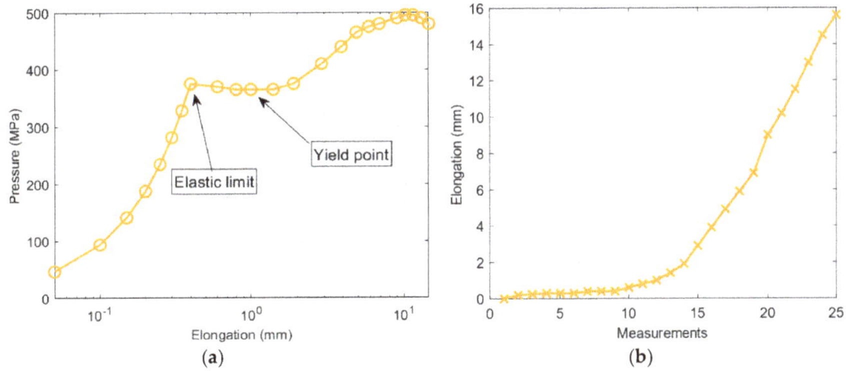

Figure 10. Strength characteristics for the tested specimen (**a**) and changes in the elongation for subsequent measurements (**b**).

The artificial field forced by the yoke flux was selected to match the measurement range of the sensors and to ensure that the work was in the Rayleigh region for which the nonlinearities of the **B**(**H**) characteristic are not clear. Figure 11a shows the **B**(I) graphs for a sample sensor and Figure 11b shows the corresponding time graphs of the field recorded in a single measurement (for a specific material intensity state).

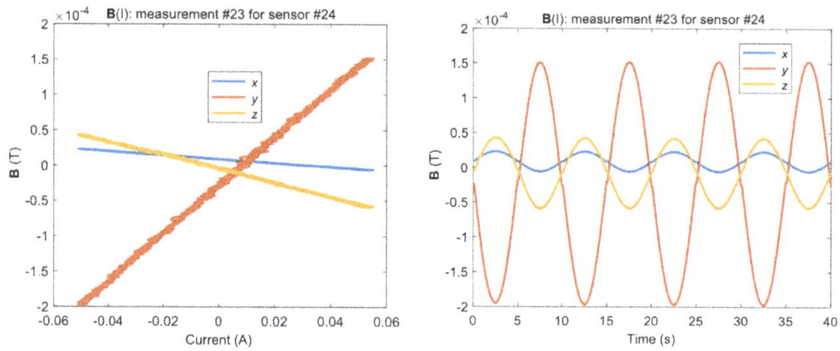

(**a**) (**b**)

Figure 11. Influence of the alternating excitation current on the measurement of the magnetic field for the selected sensor and measurement (**a**) and the corresponding time plot of the magnetic field (**b**).

Figure 12 shows the waveforms of the **S** parameter estimation for the three directions. The highest values pertained to the *y*-direction, as expected, because this direction was the main direction of the magnetic flux of the yoke. In areas 13–15 of the measurement, a characteristic jump occurred near the yield point of the strength characteristic (Figure 10a), i.e., in the range of plastic deformation. Moreover, the *y*-direction changes in this parameter were sustained in the subsequent steps of the experiment; however, for the *x*- and *z*-directions, the changes occurred only in the region close to the point. The estimation of the **S** parameter in the *x*- and *z*-directions highlighted the occurrence of linear changes in the value to promote further plasticization of the material. The *z*-direction appeared to clearly reflect the relationship between the plasticization stage and the change in the material permeability. However, this aspect must be further examined in terms of the influence of the direction of the exciting magnetic field on the estimation of the **S** parameter. Figure 13 shows the estimation of the magnetic field B_{Ci} recorded by sensor 24. As shown in Figure 5, this sensor was located in the middle of the right arm of the specimen. The range of elastic loads corresponded to measurements 1–8 (according to Figure 10a). The measurement at point 8 was assumed to be the elastic limit. In this area, the most significant changes in the magnetic field vector occurred. The maximum magnitude of the vector corresponded to half the elastic range (approximately 170 MPa) of the specimen's material. Moreover, the maximum changes in the magnetic field occurred in the *x*-direction (perpendicular to the strain) and the *z*-direction (normal to the specimen surface).

The measurements presented in Figures 12 and 13 indicate the changes in the estimated parameters pertaining to the characteristic point of the samples. To clarify the spatial distributions of the parameters, the maps for one row of sensors (19–24) are presented. The selected row corresponded to the region of the specimen arm that was subjected to the highest deformation. Figure 14 shows the maps representing the spatial distribution of the **S** parameter for successive measurements.

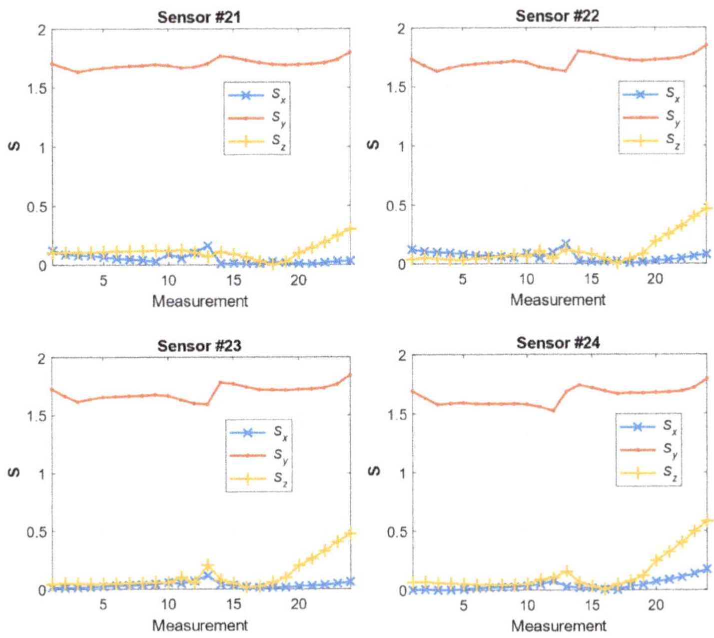

Figure 12. Estimates of the **S** coefficients for successive stress states for sensors 21–24.

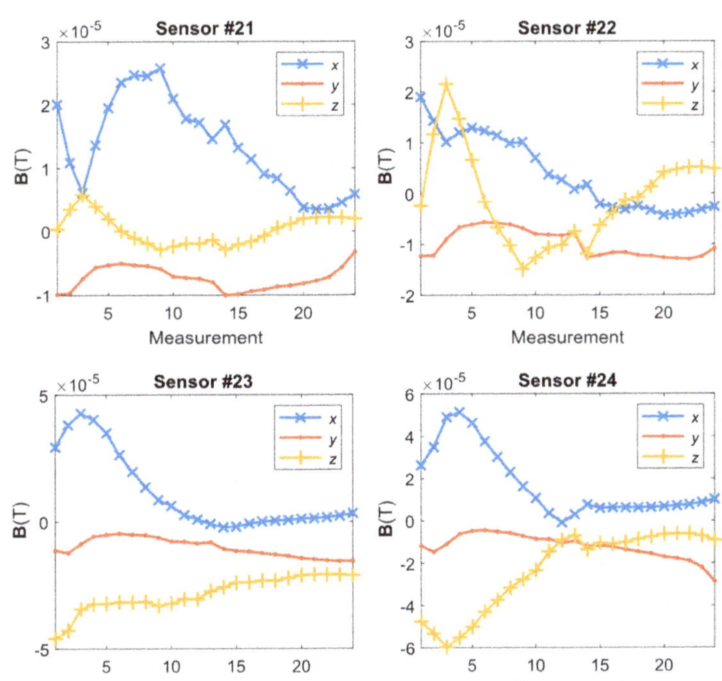

Figure 13. The x-, y-, and z-direction magnetic induction for sensors 21–24 for the specimen discontinuity area.

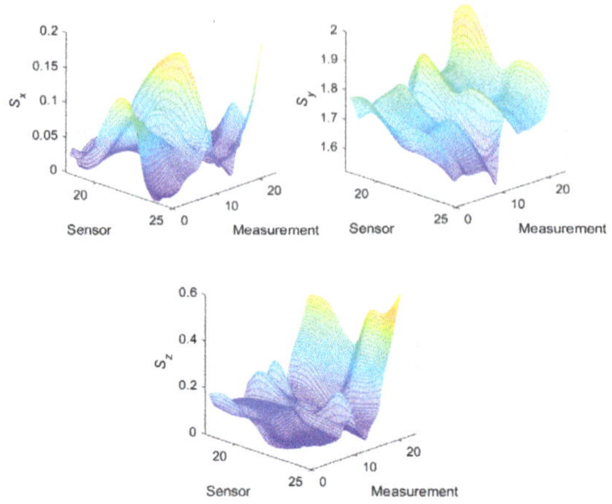

Figure 14. Distribution of the **S** parameter in three directions along the arm for successive states of the material stress.

For the single waveform that is shown in Figure 12 and the remaining measurements on the specimen arm, a characteristic flat fragment of the graph for the elastic deformations was observed. Moreover, beyond the yield point, in the vicinity of measurements 13–15, a characteristic ridge appeared, and a significant increase in the field dispersion occurred, especially in the normal (z) direction and the direction perpendicular to the stress (x).

Figure 15 shows the maps representing the spatial distribution of the magnetic flux for successive measurements. Similar to Figure 13, the greatest spatial changes in the field distribution occurred for the elastic region (measurements 1–9), especially in the case of the x-direction. These changes were also visible in the plastic area, although the changes were smaller. The polarization of the magnetic pole around the fracture area increased, even in the area of plastic deformation, which was most clearly visible in the measurements pertaining to the y-axis. However, the values of the field change for the y-axis were more than twice as small as those for the z-component (normal) of the magnetic field. The sample broke in areas 20 and 21 of the sensor, in which the field intensity was approximately zero.

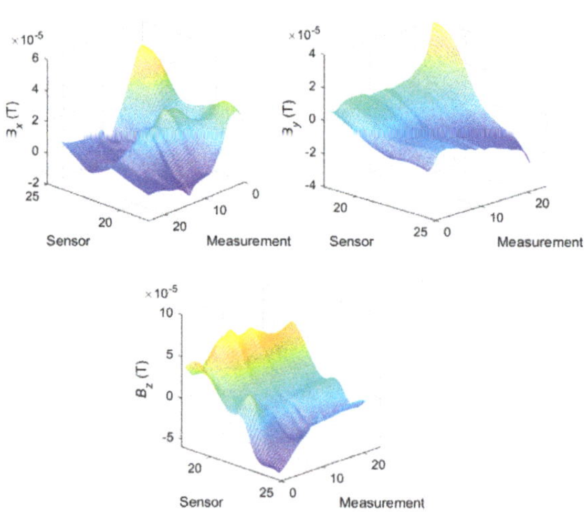

Figure 15. Distribution of the **B** magnetic induction for the broken arm of the specimen for successive states of the material's stress.

The spatial analysis of the degradation of steel elements is highly complex. However, the experiments highlighted several aspects to be considered to clarify the characteristic behaviors, such as an increase in the stress, plasticization, and material rupture.

6. Summary

The tests conducted on steel samples showed that an area of stress concentration caused local magnetization of the material, in accordance with the theoretical assumptions. This phenomenon can be used to identify the stress state of a material in the elastic region. However, in the plastic state, the spatial fluctuations of the field were considerably smaller and exhibited different characteristics. These results could be obtained using the proprietary model approach and the conduction of an experiment appropriate for this model. However, the proposed technique involved certain practical limitations due to the required movement of the examined object in relation to the external magnetic field. Therefore, a second experiment was performed, in which stimulation through a modeled external weak magnetic field was incorporated while the sample remained in place. The results for the simplified experiment indicated that this scenario did not disrupt the effects observed in the first experiment. However, in the field of plastic deformation, changes in the field dispersion parameter occurred, which were considerably smaller in the elastic region. Using the proposed method of information extraction, two pieces of information from the eigenmagnetic field signal of an object could be isolated, which reflected the different effects occurring in different states of the material strain. The information obtained from the magnetic field without separation corresponded to a considerably challenging interpretation of the obtained image of the magnetic field distribution. Both effects related to the sample's inherent magnetization and change in the magnetic permeability affected the resultant distribution of the magnetic field (eigenmagnetic field), and consequently, the measurement of the recorded magnetic field. The boundary of the passage through the two characteristic areas of material strain was very difficult to detect. In addition, the analysis of the individual components of the magnetic field in the described tests enabled the realization of a more accurate inference regarding the direction of stress (the preferred direction of magnetization). Both types of estimated parameters, specifically, the sample's inherent magnetic field and changes in the permeability, indicated a strong correlation between the directions of the recorded field and force. Therefore, the extension of the experiment from the variant with stationary sample II while controlling the spinning of the forcing field in three planes facilitated the understanding of the relationships between the estimated parameters and state of stress in the sample material.

This presented research did not include cyclic loading of the material. Although such an experiment would be interesting, it is possible to conclude that our magnetic parameters will still be usable and show characteristic behaviors. According to Figures 6 and 7 in [16], in the elastic strain area, some irreversible effects could appear. The proposed diagnostic magnetic parameters should agree with this effect in a quantitative manner that is proportional to the share of reversible and irreversible effects. In the same situation, one can predict the plastic range [30], where the quantitative changes will only influence the proposed parameters in the range of values considered in the present paper. On the other hand, the benefit of extracting magnetic field features should be emphasized in the availability of diagnostic information in the face of low-energy changes or disturbances. Nevertheless, it would be worth doing special research on this topic. Finally, we concluded that the observation of both parameters can be realized to enable a more reliable assessment of the stress state of the material. During the analyses, a characteristic jump of the **S** parameter was observed at Yield point (Figure 10a) of the material strength curve, which was related to the change in the permeability of the sample. This behavior is interesting and the resulting phenomena will be examined in future work.

Author Contributions: Conceptualization, P.S. and S.G.; methodology, S.G.; software, P.S.; validation, P.S. and S.G.; formal analysis, S.G.; investigation, P.S.; resources, S.G.; data curation, P.S.; writing—original draft preparation, P.S. and S.G.; writing—review and editing, S.G.; visualization, P.S. All authors have read and agreed to the published version of the manuscript.

Funding: The project was financed by the National Center for Research and Development under the "Security and Defense" program (project no. DOB-BIO9/04/02/2018).

Institutional Review Board Statement: Not applicable.

Informed Consent Statement: Not applicable.

Data Availability Statement: Data available on request. The data presented in this study are available on request from the corresponding author.

Conflicts of Interest: The authors declare no conflict of interest.

References

1. Stanislaw, R.; Krzysztof, S. Use of vibroacoustic signals for diagnosis of prestressed structures. *Eksploat. i Niezawodn.* **2012**, *14*, 84–91.
2. Schabowicz, K. Non-Destructive Testing of Materials in Civil Engineering. *Materials* **2019**, *12*, 3237. [CrossRef] [PubMed]
3. Gałęzia, A.; Gontarz, S.; Jasiński, M.; Mączak, J.; Radkowski, S.; Seńko, J. Distributed System for Monitoring of the Large Scale Infrastructure Structures Based on Analysis of Changes of its Static and Dynamic Properties. *Key Eng. Mater.* **2012**, *518*, 106–118. [CrossRef]
4. Jackiewicz, D.; Kachniarz, M.; Bieńkowski, A. Investigation of the Magnetoelastic Villari Effect in Steel Truss. *Adv. Hum. Factors Bus. Manag. Train. Educ.* **2016**, *519*, 63–70.
5. Kachniarz, M.; Salach, J. Characterization of magnetoelastic properties of Ni–Zn ferrite in wide range of magnetizing fields for stress sensing applications. *Measurement* **2021**, *168*, 108301. [CrossRef]
6. Szumiata, T.; Hibner, K.; Dziewiecki, K.; Mazur, Z.; Gockiewicz, A.; Gzik-Szumiata, M.; Górka, B.; Witoś, M. Stress Monitoring in Steel Elements via Detection of AC Magnetic Permeability Changes. *Acta Phys. Pol. A* **2018**, *133*, 719–721. [CrossRef]
7. Koson-Schab, A.; Szpytko, J. Investigation of the Impact of Load on the Magnetic Field Strength of the Crane by the Magnetic Metal Memory Technique. *Materials* **2020**, *13*, 5559. [CrossRef]
8. Gontarz, S.; Szulim, P.; Lei, Y. Identification of magnetomechanical phenomena in a degradation process of loaded steel elements. *J. Magn. Magn. Mater.* **2018**, *467*, 29–36. [CrossRef]
9. Gontarz, S.; Mączak, J.; Szulim, P. *Online Monitoring of Steel Constructions Using Passive Methods*; Springer: Berlin, Germany, 2015; Volume 19.
10. Szulim, P.; Gontarz, S. Using the surrounding magnetic field in diagnosis of the bldc motor. *J. Electr. Eng.* **2015**, *66*, 193–198.
11. Witoś, M. High Sensitive Methods for Health Monitoring of Compressor Blades and Fatigue Detection. *Sci. World J.* **2013**, *2013*, 218460. [CrossRef]
12. Kohut, P.; Holak, K.; Uhl, T.; Mączak, J.; Szulim, P. Application of Vision Based Damage Detection for Real Civil Engineering Structure. *Key Eng. Mater.* **2013**, *588*, 22–32. [CrossRef]
13. Rucki, M.; Gockiewicz, A.; Szumiata, T. Evaluation of the residual magnetic field measurement system for early identification of railway defects. *Metrol. Meas. Syst.* **2019**, *26*, 687–696. [CrossRef]
14. Dubov, A.A. Detection of Metallurgical and Production Defects in Engineering Components Using Metal Magnetic Memory. *Metallurgist* **2015**, *59*, 164–167. [CrossRef]
15. Wilson, J.W.; Tian, G.Y.; Barrans, S. Residual magnetic field sensing for stress measurement. *Sens. Actuators A* **2007**, *135*, 381–387. [CrossRef]
16. Gontarz, S.; Radkowski, S. Impact of Various Factors on Relationships Between Stress and Eigen Magnetic Field in a Steel Specimen. *IEEE Trans. Magn.* **2011**, *48*, 1143–1154. [CrossRef]
17. Shi, P. Magneto-elastoplastic coupling model of ferromagnetic material with plastic deformation under applied stress and magnetic fields. *J. Magn. Magn. Mater.* **2020**, *512*, 166980. [CrossRef]
18. Wang, Z.D.; Deng, B.; Yao, K. Physical model of plastic deformation on magnetization in ferromagnetic materials. *J. Appl. Phys.* **2011**, *109*, 083928. [CrossRef]
19. Weng, G.; Wang, J.; Liu, Y.; Zhu, X.; Dai, J. Magnetic Stress Sensing System for Nondestructive Stress Testing of Structural Steel and Steel Truss Components Based on Existing Magnetism. *Sensors* **2020**, *20*, 4043. [CrossRef]
20. Papadopoulou, S. Correlating elastic and plastic deformation with magnetic permeability values. *J. Phys.* **2017**, *939*, 012003. [CrossRef]
21. Atherton, D.; Jiles, D. Effects of stress on magnetization. *NDT Int.* **1986**, *19*, 15–19. [CrossRef]
22. Birss, R.R.; Faunce, C.A. Stress-induced magnetization in small magnetic fields. *Le J. Phys. Colloq.* **1971**, *32*, C1-686–C1-688. [CrossRef]
23. Krause, T.W.; Makar, J.; Atherton, D. Investigation of the magnetic field and stress dependence of 180° domain wall motion in pipeline steel using magnetic Barkhausen noise. *J. Magn. Magn. Mater.* **1994**, *137*, 25–34. [CrossRef]

24. Atherton, D.; Rao, T.; De Sa, V.; Schonbachler, M. Thermodynamic correlation tests between magnetostrictive and magnetomechanical effects in 2% Mn pipeline steel. *IEEE Trans. Magn.* **1988**, *24*, 2177–2180. [CrossRef]
25. Craik, D.J.; Wood, M.J. Magnetization changes induced by stress in a constant applied field. *J. Phys. D* **1970**, *3*, 1009–1016. [CrossRef]
26. Jiles, D.C.; Atherton, D.L. Theory of ferromagnetic hysteresis (invited). *J. Appl. Phys.* **1984**, *55*, 2115–2120. [CrossRef]
27. Jiles, D.C.; Atherton, D.L. Theory of the magnetisation process in ferromagnets and its application to the magnetomechanical effect. *J. Phys. D* **1984**, *17*, 1265–1281. [CrossRef]
28. Szulim, P.; Gontarz, S.; Maczak, J. Calibration of magnetic field sensors used for diagnosis of steel construction. *J. Electr. Eng.* **2015**, *66*, 203–207.
29. EN 10027-2. *Designation Systems for Steels—Part 2: Numerical System*; CEN: Brussels, Belgium, 2015.
30. Gontarz, S.; Gumiński, R. New approach to the evaluation of the effort state of steel based on magneto-mechanical effects. *Mech. Res. Commun.* **2017**, *82*, 14–20. [CrossRef]

materials

MDPI

Article

Assessment of Internal Damage in Sandwich Structures by Post-Processing of Mode Shapes Using Curvelet Transform

Andrzej Katunin [1,*] and Sandris Ručevskis [2]

[1] Department of Fundamentals of Machinery Design, Faculty of Mechanical Engineering, Silesian University of Technology, Konarskiego 18A, 44-100 Gliwice, Poland

[2] Institute of Materials and Structures, Riga Technical University, Kipsalas Iela 6A, LV-1048 Riga, Latvia; sandris.rucevskis@rtu.lv

* Correspondence: andrzej.katunin@polsl.pl; Tel.: +48-323-271-069

Abstract: Identification and quantification of structural damage is one of the crucial aspects of proper maintenance of mechanical and civil structures, which is directly related to their integrity and safety. The paper presents a novel approach for detecting various types of damage in sandwich structures by processing the mode shapes using a hybrid algorithm based on the curvelet transform and the standardized damage index concept. The proposed approach uses the properties of directional selectivity, absence of the boundary effect, typical of such a class of transforms, and excellent filtration capabilities of the curvelet transform as well as the classification hypothesis in the standardized damage index, which allows the exclusion of irrelevant information and emphasizes proper damage location and shape. The proposed hybrid algorithm allowed to successfully identify a subsurface core damage in sandwich structures, such as local lack of a core or its debonding from facings. The performed quantification study aimed to evaluate the correctness of identified damage shape confirmed the validity and accuracy of the proposed algorithm not only for the damage detection and localization but also for the estimation of the size of structural damage.

Keywords: damage identification; damage quantification; sandwich structures; modal analysis; curvelet transform; standardized damage index

Citation: Katunin, A.; Ručevskis, S. Assessment of Internal Damage in Sandwich Structures by Post-Processing of Mode Shapes Using Curvelet Transform. *Materials* **2021**, *14*, 4517. https://doi.org/10.3390/ma14164517

Academic Editor: Alessandro Pirondi

Received: 22 June 2021
Accepted: 10 August 2021
Published: 11 August 2021

1. Introduction

Sandwich composite structures have found wide application in numerous civil and mechanical structures, offering great performance in terms of their strength and stiffness and simultaneously being lightweight, which makes it possible to significantly reduce a mass of construction without losing its mechanical properties [1–4]. In spite of the excellent mechanical performance in terms of a possibility of carrying significant loading of various types, sandwich structures are susceptible to defects occurring during the manufacturing process, such as debonding, defects of a core, buckling defects, etc., as well as operational damage, such as low-velocity impact damage [1,5]. Such defects and damage may significantly decrease the mechanical properties of a sandwich structure, causing a drop of stiffness, which, in turn, could be an initiator of the development of a failure mechanism, causing final failure and disintegration. Therefore, defect and damage identification is one of the crucial tasks during the operational inspection of sandwich structures.

Recently, numerous non-destructive testing (NDT) techniques have been adapted to detect, localize, and identify surface and internal flaws in sandwich structures. These techniques, depending on the industrial conditions, in which a sandwich structure is operated, include visual testing [6], ultrasonic testing [7–9], thermography [10,11], eddy-current testing [12], shearography [7], guided waves-based techniques [1,13,14], electrical impedance [15], terahertz testing [16], etc. as well as by various embedded sensors in smart structures [17]. An overview of defects and damage detectability in sandwich structures can be found e.g., in the review paper of Nsengiyumva et al. [18]. However, the above-mentioned techniques are usually demanding in setup preparation and are complex in operation as well as costly. In spite of this, the vibration-based damage identification techniques with a possibility of high-precision measurements (e.g., using laser Doppler vibrometers (LDVs) or Digital Image Correlation (DIC) measurement technique) and appropriate processing of testing results can provide a good and cost-effective alternative for defect and damage identification in sandwich structures.

Vibration-based damage identification methods are usually based on the processing of mode shapes or modal curvatures acquired from the vibrating structures. This is because, from the parameters, which can be acquired from a modal analysis (i.e., natural frequencies, and corresponding mode shapes and damping factors), the mode shapes and their derivative representations are the most sensitive to structural damage. Moreover, analysis of mode shapes gives a possibility to not only detect damage but determine its spatial location and sometimes also a type and other properties of damage. Another great advantage of this approach is the possibility of damage identification without a baseline, i.e., data from a healthy structure or its model is not necessary for detecting damage, the damage identification procedure is performed solely on a tested (damaged) structure. This is possible due to the presence of local changes in structural properties, which also affect the modal response locally. Consequently, to enhance the detectability of damage, most processing algorithms are applied to mode shapes.

Among the considered processing algorithms, one can distinguish two main groups: the algorithms based on various types of damage indices (DIs) and those based on space-frequency analysis. The first group covers numerous approaches based on energy operators [19–22], curvature mode shape differences [23,24], modal criteria and parameters [5,25–27], and many other. The second group implies a direct application of filtering procedure to mode shapes or modal curvatures using the algorithms primarily based on wavelet transforms (WTs) [28,29] and hybrid algorithms using them [30,31]. As it was shown in numerous previous studies (see e.g., [32,33]), the effectiveness of WTs in damage identification problems is significantly influenced by the properties of a selected wavelet, i.e., its type, number of vanishing moments, compactness, power distribution, etc. This, in turn, has a direct influence on the filtering ability, since WTs, in particular their discrete representations, can be considered as a set of low- and high-pass filters. Unfortunately, the selection of a proper wavelet basis function is not guided by any standard rules, and in many cases, this selection is performed by the trial-and-error method. Additionally, the two-dimensional (2D) wavelets are unable to represent properly the curvilinear damage signatures, since at each level of decomposition they operate on a square grid [34]. This means that in the case of the existence of damage with a curved boundary, it would not be possible to track its shape precisely. Another problem appearing during the application of WTs is the presence of the boundary effect, causing a significant increase of the resulting wavelet coefficients in the vicinity of borders of a tested structure. To overcome these deficiencies, some non-parametric methods were applied for damage identification, including cross-correlation [35], Vigner-Wille distribution [36], and S-transform [37]. However, a problem with properly filtering out the damage signatures from processed mode shapes or curvatures remains open in the area of structural damage identification. Fortunately, the above-mentioned problems with the boundary effect, ineffective filtering, and some other directly related to the classical tensor-product construction of 2D wavelet bases, like poor directional selectivity or non-smooth windowing [38], are resolved by the family of X-let transforms being developed over the last 20 years and currently containing hundreds of application-oriented transforms with enhanced properties with respect to classical WTs.

Like in the case of many other filtering approaches, the origin of X-lets is the image processing, where the filtration ability together with a proper representation of features of a processed image, such as discontinuities, artifacts, smooth edges, etc., are the key properties of the applied tools in numerous applications, primarily in processing of biological and medical images. Further, due to their proven advantageous properties, the X-lets started to be applied in numerous applications in technical sciences, including mechanical and civil engineering applications, and, in particular, quality assurance, NDT, and structural damage identification. Mumtaz et al. [39] used contourlets for tracking cracks from images of aircraft structures. Xu et al. [40] used shearlets for the classification of surface defects of machined elements, similarly to Umamaheswara Raju et al. [41], who used curvelets for an evaluation of a surface quality after machining. The filtering approach based on the curvelet transform (CT) was proposed by Tzanis in [42], where he applied it for fracture detection in ground probing radar data. Anandan and Sabeenian [43] used CT for the detection of fabric defects. There were also a few applications of X-lets in damage identification problems. Bagheri et al. [44,45] used CT for identification of simulated structural damage in plates using their mode shapes, De Marchi et al. [46] used curvelets in their structural health monitoring approach for improvement of a damage detectability in a guided wave-based technique, while the recent studies of Vafaie and Salajegheh [47] proposed an application of wavelets and contourlets for vibration-based damage identification in plates.

The lack of systematic studies on the application of X-let transforms (XTs) in damage identification means that this problem remains open. Moreover, the variety of transforms within the XTs family raises the question of a suitable transform for damage identification. The previous comparative studies on the performance of wavelets vs. X-lets [47,48] clearly show significant improvement of filtering ability and sensitivity to damage of the latter ones, while the comparisons of the performance of X-lets between each other [44,47,48] indicates that curvelets provide the best filtering performance, which makes them suitable for improvement of sensitivity to damage in damage identification problems.

The aim of this study is to analyze damage detectability in sandwich composite plates with simulated damage based on the processing of their mode shapes using CT and standardized damage index (SDI) concept and to highlight its advantages not only in filtering performance, but also in the evaluation of a shape of damage due to enhanced directional selectivity of CT. Thanks to merging the CT-based algorithm with SDI determination in the second step of processing, precise quantification of shapes of the considered damage in sandwich plates is possible.

2. Tested Structures and Damage Identification Algorithm

Sandwich structures are specific composites that have found application in numerous mechanical and civil constructions. Their primary advantage, namely a very high stiffness-to-mass ratio, defines the type of the most critical damage for these structures, like damage of a core or debonding between a core and facings. These damage scenarios were simulated in the tested plates and are described in detail in Section 2.1. In Section 2.2 the testing setup and results of the modal analysis are presented. Then, the fundamentals of the CT and its advantages over other transforms of a similar type are presented in Section 2.3, and its incorporation in the damage identification algorithm with additional post-processing is the subject of Section 2.4.

2.1. Pre-Damaged Sandwich Structures

The tested sandwich structures with the spatial dimensions of 300 × 300 mm and a total thickness of 4.1 mm are composed with a honeycomb core made of Nomex® (DuPont, Wilmington, DE, USA) aramid paper saturated with a phenolic resin with a thickness of 3 mm and two facings made of glass-fiber reinforced polymer (GFRP) composite with a thickness of 0.6 mm each (the decrease of a total thickness is a result of a manufacturing process). The core has a density of 29 kg/m^3 and a diameter of a single cell of 2.5 mm. The GFRP facings were designed as intentionally transparent, which made it possible to see the shape and location of simulated internal damage with a naked eye. More details on constituents used for manufacturing the plates can be found in [28]. According to the performed quasi-static tests [49], the tensile strength for this structure equals 21.45 MPa, while the buckling strength limit in the three-point bending testing mode equals 52.61 MPa.

The sandwich plates were manufactured and supplied by the PPHU Surfpol (Rawa Mazowiecka, Poland) in the vacuum-assisted resin transfer molding manufacturing technique and simulated damage of a core was introduced at the manufacturing stage. Three damaged plates were considered in this study. They consisted of the partial local lack of the core with an irregular boundary placed in the center of the plate (the thickness of a core in this location was reduced twice) {1}, the full local lack of the core with irregular boundary placed in the center of the plate {2}, and the debonding between the core and the upper facing with a square shape and dimensions of 60 × 60 mm placed in the center of the plate {3}. The images of the damaged plates together with their schemes and dimensions are presented in Figure 1.

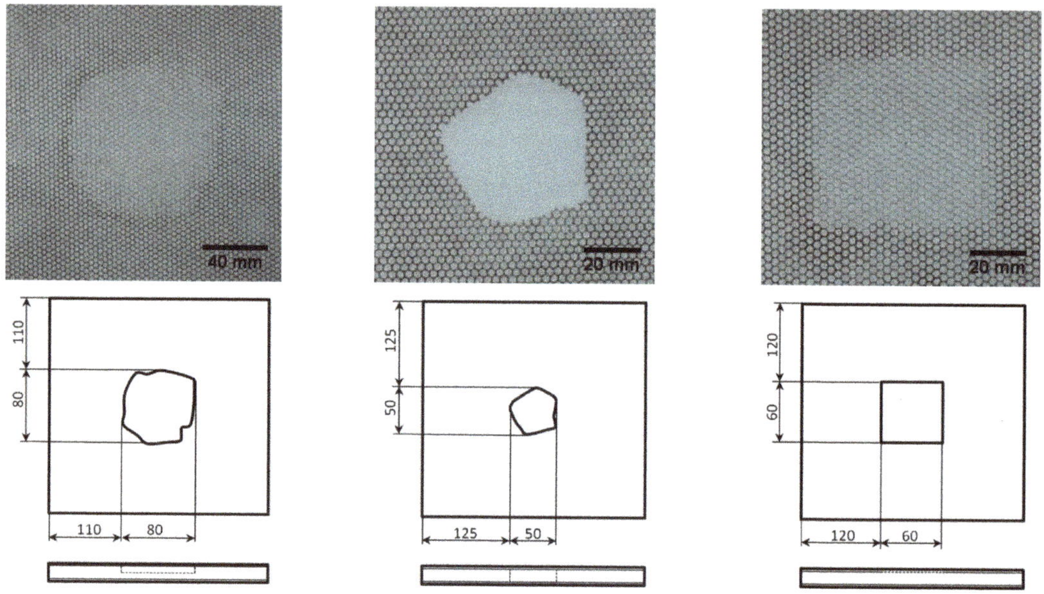

(a) (b) (c)

Figure 1. The images and schemes of the damaged sandwich plates: (**a**) partial local lack of the core {1}, (**b**) full local lack of the core {2}, (**c**) debonding between the core and the upper facing {3} [28,37] with permission from Elsevier.

2.2. Acquisition of Mode Shapes

The modal analysis was performed for the fully clamped specimens described in Section 2.1 using two LDVs. The tested structures were clamped in a steel square frame with 24 bolts on its perimeter using the Jonnesway® (Taipei, Taiwan) T07030N dynamometric wrench to ensure the same clamping conditions and simultaneously do not affect the structural properties due to clamping. The clamping resulted in a reduction of a scanning area to 250 × 250 mm. In scenarios {1} and {3}, the damage sites were located close to the scanned side of the plates. Prior to testing, the upper surfaces of the tested structures were covered with the Helling® (Heidgraben, Germany) anti-glare spray to ensure the proper reflection ability of a laser beam of LDV. Then, in the Polytec® (Waldbronn, Germany) dedicated software, 64 × 64 equidistant measurement points were defined in a scanning area. The frame was mounted on the TIRA® (Schalkau, Germany) TV-51120 electrodynamic shaker, connected with the TIRA® BAA 500 power amplifier, which was used for vibration excitation. The scanning LDV Polytec® (Waldbronn, Germany) PSV-400, connected to the vibrometer controller Polytec® (Waldbronn, Germany) OFV-5000 with the built-in velocity decoder and to a PC, was used for a measurement of a vibration velocity in the defined grid of measurement points, while the point LDV Polytec® (Waldbronn, Germany) PDV-100 was focused on a clamping frame to acquire the reference signal, which made it possible to separate vibrations of a frame from vibrations of a whole system, which, in turn, allowed acquiring vibration of the tested structures. The experimental setup is presented in Figure 2.

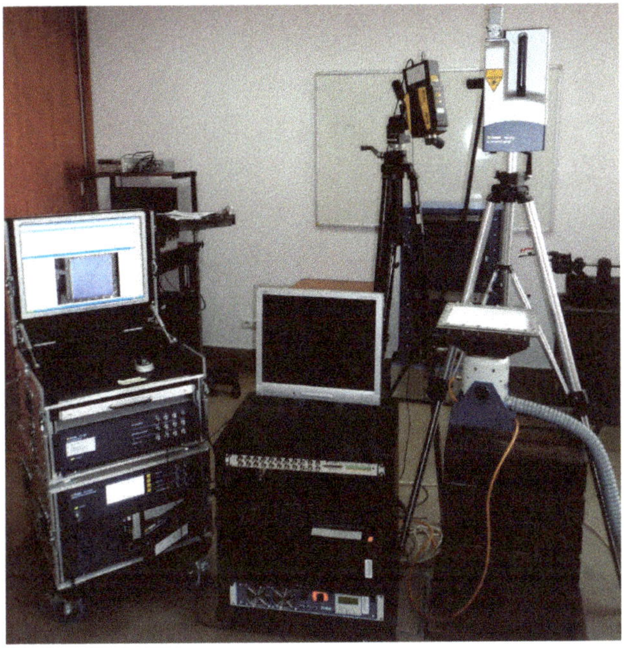

Figure 2. The experimental setup used for the acquisition of mode shapes [28] with permission from Elsevier.

The tested structures were excited with a pseudo-random signal in the frequency band of 0–2000 Hz with a frequency resolution of 1.25 Hz to excite all eigenfequencies in the defined frequency band. To increase the accuracy of the obtained results, five measurements were performed in each defined point and the acquired signals were averaged. Since the magnitude of vibration is directly related to the damage detectability, the following rule was applied: a given mode shape was taken into consideration if its magnitude in the frequency response function (FRF) was at least 20% of the magnitude of the highest observed peak in FRF. This assumption was made due to the presence of measurement noise in signals, and the assumed threshold was determined empirically based on preliminary analysis. After applying this threshold, the mode shapes significantly biased by measurement noise were excluded from consideration, since due to the high level of noise the detectability of damage using these mode shapes was very low. The resulting FRFs of the tested structures are presented in Figure 3, while the eigenfrequencies and the corresponding mode shapes for the tested structures are presented in Table 1 and Figures 4–6, respectively. As it can be observed in Table 1, due to small differences between the determined eigenfrequencies, even detection of structural damage based on analysis of changes in eigenfrequencies can be difficult, which additionally justifies the necessity of effective post-processing of mode shapes.

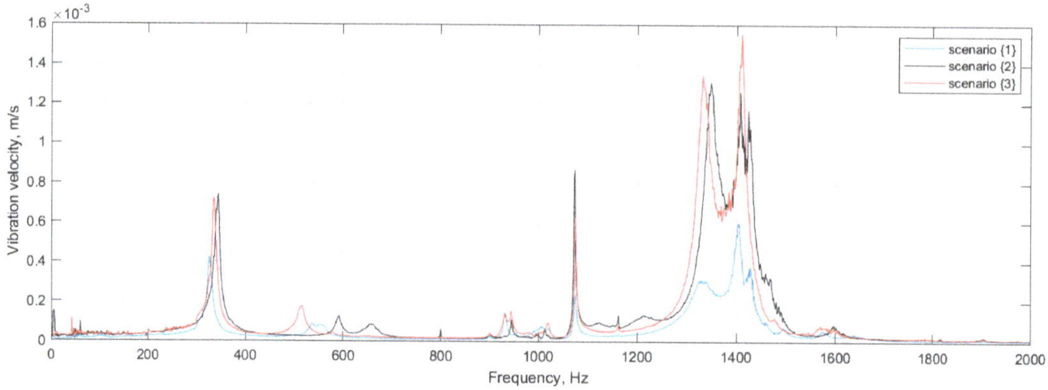

Figure 3. FRFs of the tested sandwich structures.

Table 1. The eigenfrequencies of the tested sandwich structures.

Scenario	Type of Damage	Mode 1, Hz	Mode 2, Hz	Mode 3, Hz	Mode 4, Hz	Mode 5, Hz
{1}	Partial local lack of the core	326	932	1075	1326	1405
{2}	Full local lack of the core	350		1073.75	1355	1427.5
{3}	Debonding between the core and the upper facing	335		1075	1331.25	1410

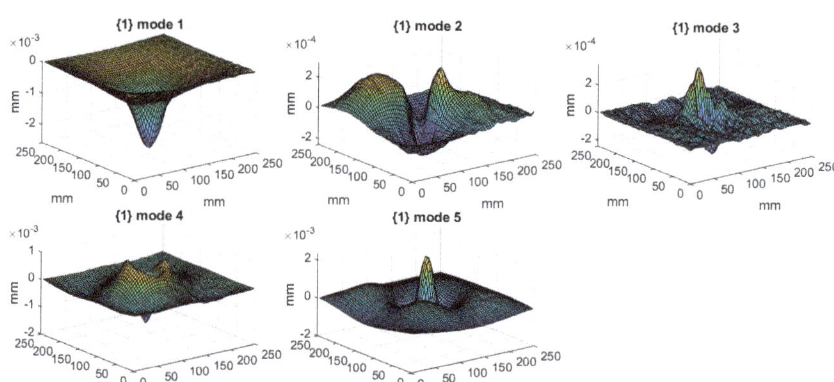

Figure 4. The considered mode shapes for the plate with a partial local lack of the core.

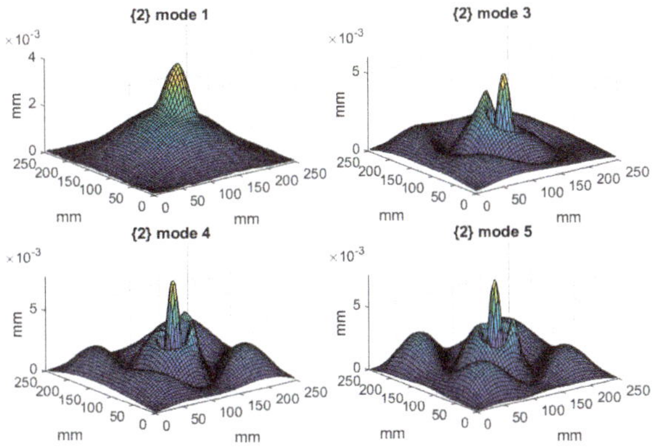

Figure 5. The considered mode shapes for the plate with a full local lack of the core.

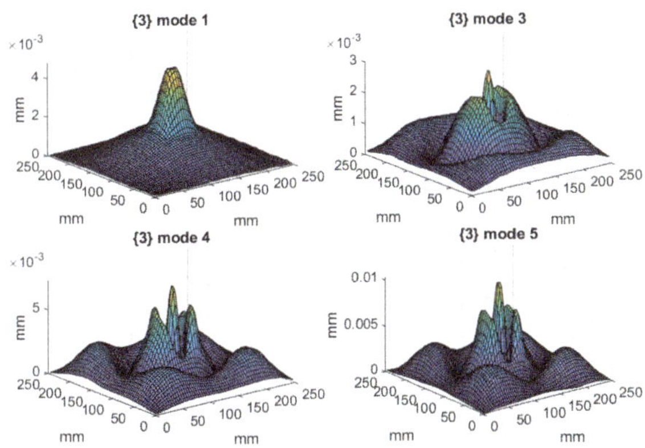

Figure 6. The considered mode shapes for the plate with a debonding between the core and the upper facing.

2.3. Acquisition of Mode Shapes

CT developed by Candès et al. [50] is the multiscale transform with numerous improved properties with respect to traditional wavelet-based multiscale representations. The main advantages of CT cover multidirectionality, thanks to the construction of the pyramidal representation across the scales, a possibility of effective detection of curved anomalies in a signal, and the enhanced filtering capabilities. All of them are of crucial importance in damage identification problems and have a significant influence on damage detectability, which is demonstrated in the next section.

The family of curvelets considered in this transform as basis function are defined as follows [51]:

$$\varphi_{j,l,k}(x) = \varphi_j\left(\mathbf{R}_{\theta_l}\left(x - x_k^{(j,l)}\right)\right), \tag{1}$$

where $x_k^{(j,l)} = R_{\theta_1}^{-1}\left(2^{-j}k_1, 2^{-j/2}k_2\right)$ defines the radial position, and $\mathbf{R}_\theta = \begin{pmatrix} \cos\theta & \sin\theta \\ -\sin\theta & \cos\theta \end{pmatrix}$ defines the angular position of a curvelet at scales of 2^j. The curvelet coefficients $c(\cdot)$ are defined by the inner product of a curvelet and a function $f(x)$:

$$c(j,l,k) = \int_{\mathbb{R}^2} f(x)\varphi_{j,l,k}(x)\mathrm{d}x, \qquad f \in L^2\left(\mathbb{R}^2\right). \tag{2}$$

As one can see, CT is defined in polar coordinates, and the decomposition process is performed using a pair of the window functions: radial $W(r)$ and angular $V(r)$, subject to the admissibility conditions:

$$\sum_{j=-\infty}^{\infty} W^2\left(2^j r\right) = 1, \qquad r \in \left(\frac{3}{4}, \frac{3}{2}\right), \tag{3}$$

$$\sum_{l=-\infty}^{\infty} V^2(t-l) = 1, \qquad t \in \left(-\frac{1}{2}, \frac{1}{2}\right), \tag{4}$$

which lead to the frequency window defined in the Fourier domain as follows [50]:

$$U_j(r,\theta) = 2^{-3j/4} W\left(2^{-j} r\right) V\left(\frac{2^{\lfloor j/2 \rfloor}\theta}{2\pi}\right), \tag{5}$$

where $\lfloor \cdot \rfloor$ denotes integer. Since CT operates in the frequency domain, it is essential to present (2) in terms of frequency ω:

$$c(j,l,k) = \frac{1}{4\pi^2} \int \hat{f}(\omega) U_j(\mathbf{R}_{\theta_l}\omega) \exp\left(\mathrm{i}\langle x_k^{(j,l)}, \omega\rangle\right) \mathrm{d}\omega, \tag{6}$$

where $\hat{\cdot}$ denotes the frequency domain. The representation can be easily transformed into Cartesian coordinates, making it useful for data processing in the form of 2D matrices. Following this, Equation (6) takes a form [50]:

$$c(j,l,k) = \frac{1}{4\pi^2} \int \hat{f}(\mathbf{S}_{\theta_l}\omega) \widetilde{U}_j(\omega) \exp\left(\mathrm{i}\langle 2^{-j}k_1, 2^{-j/2}k_2, \omega\rangle\right) \mathrm{d}\omega, \tag{7}$$

where $\mathbf{S}_\theta = \begin{pmatrix} 1 & 0 \\ -\tan\theta & 1 \end{pmatrix}$ is the shear matrix, and $\widetilde{U}_j(\omega)$ is the frequency window in the Cartesian coordinate system, which changes its form from (5) to the following:

$$\widetilde{U}_j(\omega) = \widetilde{W}_j(\omega)\widetilde{V}_j(\omega), \tag{8}$$

where

$$\widetilde{V}_j(\omega) = V\left(2^{\lfloor j/2 \rfloor} \frac{\omega_2}{\omega_1}\right), \tag{9}$$

$$\widetilde{W}_j(\omega) = \sqrt{\Phi_{j+1}^2(\omega) - \Phi_j^2(\omega)}, \qquad j \geq 0. \tag{10}$$

In (9) and (10), $\Phi_j(\omega_1, \omega_2) = \varphi(2^{-j}\omega_1)\varphi(2^{-j}\omega_2)$ is the product of the 1D windows, $0 \leq \varphi \leq 1$, $\varphi = \begin{cases} 1, & \left[-\frac{1}{2}, \frac{1}{2}\right] \\ 0, & [-2,2] \end{cases}$, $2^j \leq \omega_1 \leq 2^{j+1}$, $-2^{-j/2} \leq \frac{\omega_2}{\omega_1} \leq 2^{-j/2}$.

The discrete version of CT, called by the authors of [50] the digital CT, changes the expression (7) to the following form:

$$c(j,l,k) = \sum_{n_1,n_2} \hat{f}[n_1, n_2 - n_1\tan\theta_l] \widetilde{U}_j[n_1,n_2] \exp\left(\mathrm{i}2\pi\left(\frac{k_1 n_1}{L_{1,j}} + \frac{k_2 n_2}{L_{2,j}}\right)\right), \tag{11}$$

which is one of two possible methods of computing the coefficients called the digital CT via wrapping. More details on the theoretical basis of CT can be found in [50–52].

2.4. Damage Identification Algorithm

The damage identification algorithm was based on the discrete CT implemented by the Curvelet.org team in the form of the CurveLab Toolbox for Matlab® (version 2020b, MathWorks, Natick, MA, USA) The curvelet coefficients were calculated according to (11) for all considered mode shapes (see Figures 4–6) of the tested structures, and the obtained values were raised to the power of 2 in order to separate insignificant low-value coefficients. In the following study, CT via wrapping was used. According to [51], obtaining the coefficients $c(j,l,k)$ consists of four steps:

(1) Calculating 2D Fourier transform (FT) to transform space domain variables $f[s,p]$ (mode shapes, in our case) in the form of the frequency domain variables $\hat{f}[n_1,n_2]$.
(2) Calculating the product of the determined variables with a frequency window: $\hat{f}[n_1,n_2]\widetilde{U}_j[n_1,n_2]$.
(3) Wrapping the above product around the origin: $\widetilde{f}_{j,l}[n_1,n_2] = W\left(U_{j,l}\hat{f}\right)[n_1,n_2]$.
(4) Calculating inverse 2D FT for $\widetilde{f}_{j,l}$ to obtain $c(j,l,k)$.

In practical situations, experimental mode shapes are inevitably biased by measurement noise, which can cause local perturbations in curvelet coefficients. These perturbations could be mistakenly interpreted as damage or they could mask the peaks induced by damage, and thus, lead to false-positive or false-negative damage identification. To overcome this problem, it was proposed to define DI as the summation of the curvelet coefficients for all modes normalized with respect to the largest magnitude value of each mode as follows:

$$\mathrm{DI}(s,p) = \sum_{m=1}^{M} \frac{c(s,p,m)^2}{c(m)^2_{\max}}, \tag{12}$$

where s and p are the coordinates of DIs corresponding to the grid of measurement points, and m stands for the mode number.

The principal goal of the damage identification algorithm was to provide useful information about damage in the sandwich structures by quantitatively evaluating the obtained results after the application of CT. In recent years, statistical interference methods, such as estimation and hypothesis testing have been widely used in the field of damage identification for the evaluation of the mode shape data, among other purposes [33,53]. In this study, a statistical hypothesis testing scheme was followed to classify damaged and healthy elements based on the obtained CT results and pre-defined damage threshold values.

In the first step of the hypothesis testing, the DIs for each element calculated according to (12) were standardized in order to obtain SDI:

$$\mathrm{SDI}(s,p) = \frac{\mathrm{DI}(s,p) - \mu_{\mathrm{DI}}}{\sigma_{\mathrm{DI}}}, \tag{13}$$

where μ_{DI} and σ_{DI} are the mean and the standard deviation of DIs, respectively.

The next step was to define the null and alternative hypotheses for the classification of the elements:

- H_0—the element (*s*,*p*) of a structure is healthy;
- H_1—the element (*s*,*p*) of a structure is damaged.

In general, rejection of the null hypothesis H_0 in the test indicates the presence of damage in the element and vice versa.

The probability density of SDIs that is obtained from a mode shape of a healthy structure would usually have a normal distribution. On the other hand, the probability density of SDI is expected to have non-normal distribution with extreme values in the tails of the density distribution due to a damage presence in a structure. Therefore, by assuming that SDI is normally distributed, the probability density function may be used to classify elements by employing a one-tailed hypothesis test [53]:

$$H_o : \mathrm{SDI}(s,p) < C_r, \tag{14a}$$

$$H_1 : \mathrm{SDI}(s,p) > C_r. \tag{14b}$$

The objective of the hypothesis testing is to decide whether the elements of a structure are damaged based on the significance level—a pre-designed probability threshold, above which the null hypothesis will be rejected. The most commonly used threshold values C_r for the damage identification include 1.28, 2, and 3, which corresponds to 90%, 95%, and 99% confidence levels for the presence of damage, respectively [33,53]. In the final step, the threshold value for the 90% confidence level was selected to classify damaged elements, which resulted from the initial analysis of considered data. For clarity, the above-described damage identification algorithm is presented in the form of a flowchart in Figure 7.

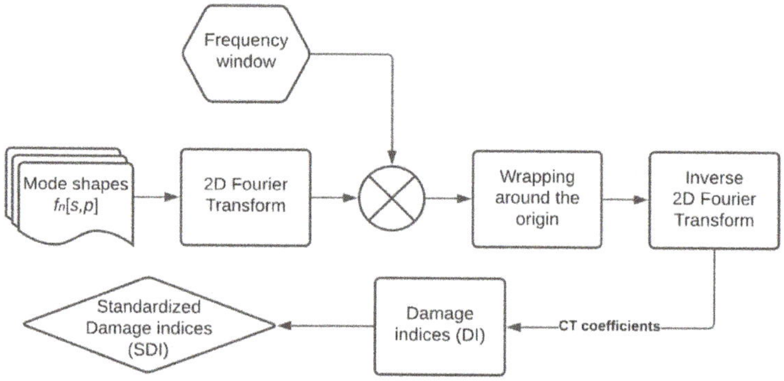

Figure 7. The steps of processing the mode shapes according to the developed damage identification algorithm.

This assumed threshold value provides an advantageous tradeoff between the accuracy of damage detection and localization and the level of noise. The values of SDIs lower than the threshold value of 1.28 were set to zero to filter out smaller peaks, which are not associated with damage.

3. Results and Discussion

The acquired mode shapes of the tested sandwich plates were processed according to the algorithm presented in Section 2.4. The obtained results were focused on the detectability of simulated damage as well as their quantification in terms of their planar shapes, which is presented below.

3.1. *Analysis of Detectability of Various Damage Types*

The determined DIs according to the algorithm presented in Section 2.4 for particular mode shapes are presented in Figures 8–10. According to the results of the previous studies [33], the confidence level for SDI was assumed as 90%, which resolves a compromise between measurement and processing noise and the true-positive DIs representing the damage signatures.

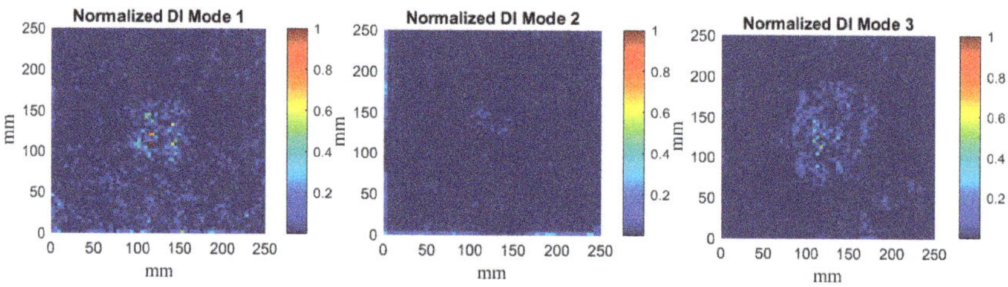

Figure 8. *Cont.*

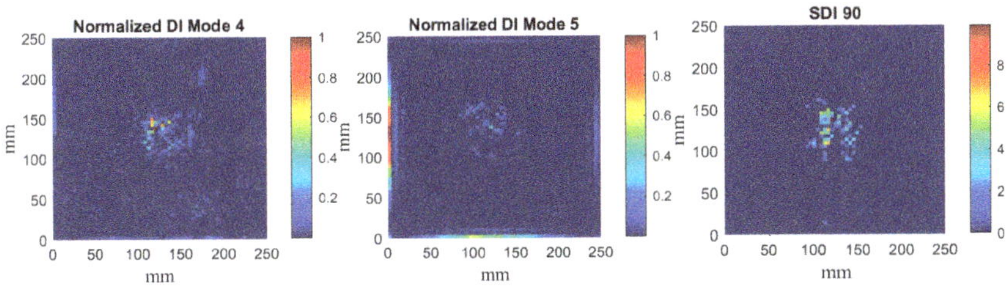

Figure 8. The determined normalized DIs for the mode shapes for the plate {1} and SDIs with the 90% confidence level.

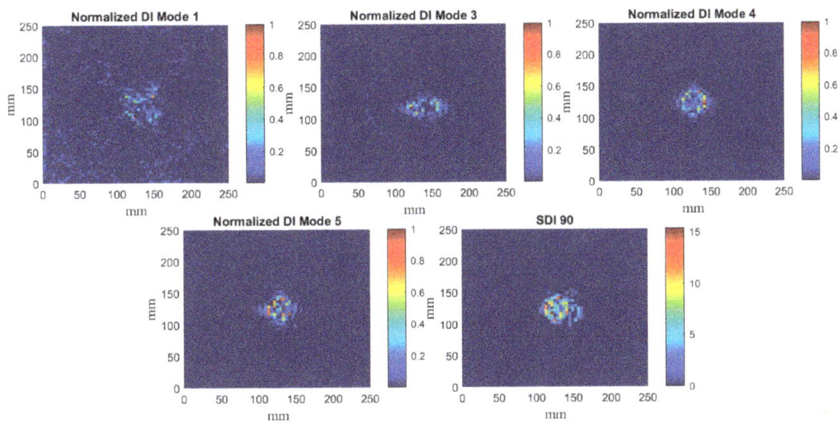

Figure 9. The determined normalized DIs for the mode shapes for the plate {2} and SDIs with the 90% confidence level.

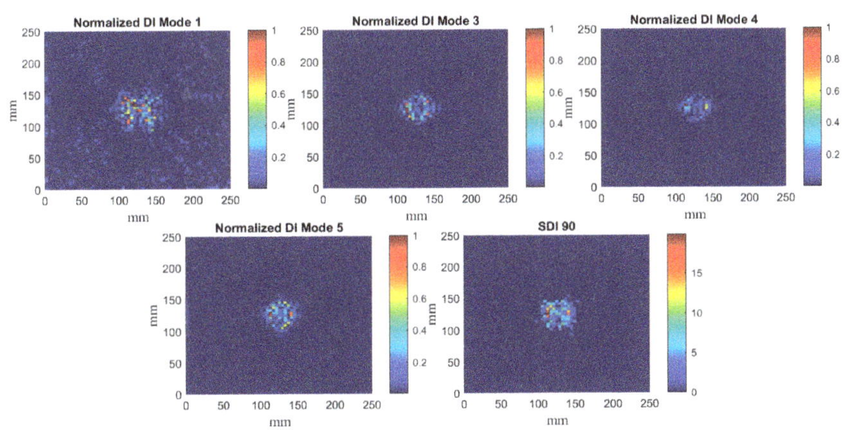

Figure 10. The determined normalized DIs for the mode shapes for the plate {3} and SDIs with the 90% confidence level.

The obtained DIs clearly show that the considered damage scenarios are detectable in every considered case, however, the planar shapes of damage signatures differ from each other for various mode shapes, which can be explained by the variability of local magnitudes of mode shapes and their direct influence on the curvelet coefficients.

It can be noticed that CT demonstrates excellent performance in the filtering of the raw mode shapes obtained from measurements, i.e., in most cases only the damage signatures are visible. Comparing the results presented in Figures 8–10 with the previous results obtained with the fractional discrete WT [28] and with optimization of its WT's parameters [30], one can conclude about good detectability of damage together with an enhanced resolution of CT with respect to the previously applied WT. Comparing the obtained results to those reported in [28,30] one can observe that the proposed algorithm allows for more accurate mapping of a damage shape than the WT-based algorithms used previously. However, in contrast to these WT-based algorithms, the algorithm proposed in this study slightly underestimates the true damage extent. This underestimation is a result of an application of the second step of processing, namely, the SDI procedure. However, the applied SDI procedure (see Section 2.4 for details) allowed to reduce the remaining measurement and processing noise in the resulting sets of DIs, which made it possible to prepare the obtained results for the next processing step, namely the quantification of the shapes of damage signatures. The application of the SDI procedure is a compromise between the proper mapping of a true damage shape and filtering out measurement noise that can be considered as a false indicator of damage. It is also worth mentioning that CT did not produce the boundary effect, i.e., the extremely high values of coefficients at the boundaries of a signal due to the translation of a basis function on this signal, typical for most of WTs.

The calculated DIs for the plate with a partial local lack of the core (Figure 8) confirms the suitability of the proposed SDI procedure for the pursuit of the development of an independent and autonomous damage identification algorithm. One can see that the resulting SDI plot clearly indicates the damage location, while individually only DIs for 3 out of 5 considered mode shapes were able to detect damage and correctly point out its location when a 90% confidence level was selected to classify elements. These results emphasize the necessity to normalize DIs for particular mode shapes in order to obtain equally weighted resulting sets of DIs. This way, the risk of few corrupted measurement data sets causing false or missed damage identification is significantly reduced. Another advantage of the present SDI procedure highlighted by Figure 8, is that mode shape transformation data can simply be fed to the algorithm, and the decision on whether any element of the structure is damaged is acquired automatically based on the selected damage threshold value. Hence, the damage identification algorithm can operate autonomously by minimizing the engagement of the data interpreter in the process.

3.2. Quantification of Damage Shape

To compare the obtained results with the true shapes of damage in the considered damage scenarios, the photographs of damaged sandwich plates presented in Figure 1 were subjected to initial image processing, which was performed to retrieve the true shapes of damage. Firstly, the contours of damage were obtained using boundary detection techniques, and the detected boundaries were appropriately scaled to match true dimensions. Then, the boundaries were transformed to the sets of coordinates of points representing these boundaries and merged with the identified damage for each considered scenario (see Figure 11a). Finally, the resulting sets of coordinates of the detected boundaries were discretized in order to match the defined grid of 64×64 measurement points. The results of merging the discretized boundaries with the identified damage are presented in Figure 11b. The colors on the presented plots were selected in such a way as to obtain good contrast between the identified SDIs and the boundaries of true damage.

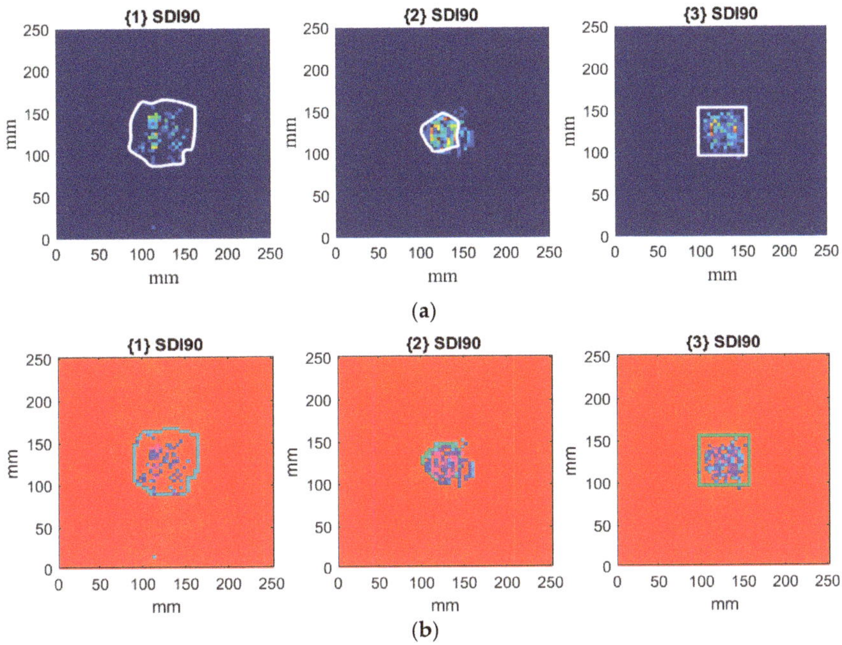

Figure 11. The results of merging of the boundaries of damage with the identified damage for the considered scenarios (**a**) and with the discretized version of the true boundaries of damage (**b**).

As it can be observed, the identified damage corresponds to the true boundaries of damaged regions, however, some misestimations are observable. In particular, the underestimation of damage areas is noticed for scenarios {1} and {3}; while an overestimation is observable for scenario {2}. The observed misestimations may appear due to several reasons: specific location of nodal lines of the considered mode shapes, which made it impossible to obtain positive SDIs in some locations due to low vibration magnitude in these locations; processing procedures, like filtering of CT and rejection of SDIs following the hypotheses (14). Nevertheless, the qualitative analysis of the obtained results shows a good correspondence of the identified damage with the boundaries of true damage for all considered scenarios.

To assess the accuracy of damage quantification quantitatively, the envelopes for identified positive SDIs were calculated. The envelopes were controlled by the shrink factor, which is a scalar value in the range of [0, 1], and takes the value of 1 for a convex hull and the value of 0 for the tight boundary. In these limit cases, one can observe overestimation in the case of application of a convex hull, and underestimation of the results in the case of a tight boundary. After preliminary testing, the value of a shrink factor was assumed as 0.5, which provides a tradeoff between the mentioned limit cases. The obtained results with the determined ratios of the damage areas for the considered damage scenarios are presented in Figure 12.

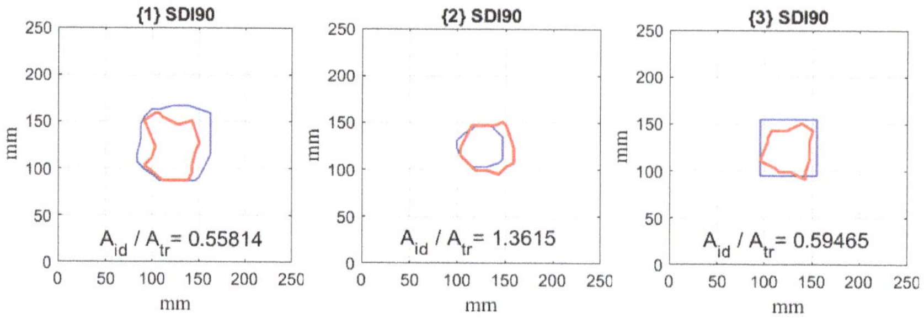

Figure 12. The quantification results for the considered damage scenarios with the ratios of areas of identified A_{id} and true damage A_{tr}.

From the quantification results, one can observe that the identified damage matches well the true boundaries of damage, which confirms the effectiveness of the proposed algorithm in this study. The quantitative analysis resulted in differences in the true and identified damage areas not exceeding 50%, which is comparable with results of quantification, obtained by other NDT methods, e.g., ultrasonic testing in the C-Scan mode.

The obtained results make it possible to formulate a statement that the proposed algorithm can be successfully used in a larger class of damage identification and quantification problems, including various types of damage and structures containing them.

4. Conclusions

The effective hybrid algorithm based on CT and SDI concept, which combines the advantages of CT, namely directional selectivity and excellent filtration ability, and the advantages of SDI, such as selection of only relevant information on damage, is proposed in this study. The algorithm was tested on mode shapes of sandwich structures with various types of damage obtained experimentally during modal analysis testing. The results show a great sensitivity of CT to these types of damage, which resulted in the identification of all considered damage scenarios with higher accuracy in comparison to the previously applied classic WT-based algorithms. The application of the SDI concept improved the quantification of damage sufficiently by discarding the irrelevant curvelet coefficients that do not represent damage. The results of quantification of subsurface damage in the tested sandwich structures show high validity, i.e., in all scenarios the error in comparison with the boundaries of the true damage not exceeding 50%, which is comparable with numerous NDT methods, like ultrasonic testing, widely applied in structural inspections.

Based on the results of the performed studies, it can be concluded that using the proposed approach one can effectively identify and quantify subsurface damage in sandwich structures. The proposed algorithm can be applied to numerous other structural damage identification problems, especially when damage is small, or measurement data is biased by measurement and/or processing noise.

Author Contributions: A.K. and S.R. contributed equally to all stages of preparation of the paper. Both authors have read and agreed to the published version of the manuscript.

Funding: This publication is partially supported within the framework of the Rector's grant no. 10/060/RGJ21/2004 in the area of scientific and development research of the Silesian University of Technology, Poland. The second author would like to acknowledge the support of the European Regional Development Fund within the Activity 1.1.1.2 "Post-doctoral Research Aid" of the Specific Aid Objective 1.1.1 "To increase the research and innovative capacity of scientific institutions of Latvia and the ability to attract external financing, investing in human resources and infrastructure" of the Operational Programme "Growth and Employment" (No.1.1.1.2/VIAA/3/19/414).

Data Availability Statement: The data that support the findings of this study are freely available as a part of the WavStructDamAs benchmark at the website http://kpkm.polsl.pl/wavstructdamas (accessed on 22 June 2021).

Acknowledgments: The authors would like to express their gratitude to the Curvelet.org team for sharing the CurveLab Toolbox.

Conflicts of Interest: The authors declare no conflict of interest. The funders had no role in the design of the study; in the collection, analyses, or interpretation of data; in the writing of the manuscript, or in the decision to publish the results.

References

1. Pieczonka, L.; Ukowski, P.; Klepka, A.; Staszewski, W.J.; Uhl, T.; Aymerich, F. Impact damage detection in light composite sandwich panels using piezo-based nonlinear vibro-acoustic modulations. *Smart Mater. Struct.* **2014**, *23*, 105021. [CrossRef]
2. Manalo, A.; Aravinthan, T.; Fam, A.; Benmokrane, B. State-of-the-art review on FRP sandwich systems for lightweight civil infrastructure. *J. Compos. Constr.* **2017**, *21*, 04016068. [CrossRef]
3. Birman, V.; Kardomateas, G.A. Review of current trends in research and applications of sandwich structures. *Compos. Part B* **2018**, *142*, 221–240. [CrossRef]
4. Castanie, B.; Bouvet, C.; Ginot, M. Review of composite sandwich structure in aeronautic applications. *Compos. Part C* **2020**, *1*, 100004.
5. Yang, J.-S.; Liu, Z.-D.; Schmidt, R.; Schröder, K.-U.; Ma, L.; Wu, L.-Z. Vibration-based damage diagnosis of composite sandwich panels with bi-directional corrugated lattice cores. *Compos. Part A* **2020**, *131*, 105781. [CrossRef]
6. Bulut, M. Low-velocity impact tests on basalt fiber/polypropylene core honeycomb sandwich composites. *Mech. Compos. Mater.* **2020**, *56*, 121–130. [CrossRef]
7. Růžek, R.; Lohonka, R.; Jironč, J. Ultrasonic C-Scan and shearography NDI techniques evaluation of impact defects identification. *NDT&E Int.* **2006**, *39*, 132–142.
8. Revel, G.M.; Pandarese, G.; Cavuto, A. Advanced ultrasonic non-destructive testing for damage detection on thick and curved composite elements for constructions. *J. Sandw. Struct. Mater.* **2012**, *15*, 5–24. [CrossRef]
9. Bělský, P.; Kadlec, M. Capability of non-destructive techniques in evaluating damage to composite sandwich structures. *Int. J. Struct. Integr.* **2019**, *10*, 356–370. [CrossRef]
10. Gillespie, D.I.; Hamilton, A.W.; Atkinson, R.C.; Bellekens, X.; Michie, C.; Adronovic, I.; Tachtatzis, C. Defect detection in aerospace sandwich composite panels using conductive thermography and contact sensors. *Sensors* **2020**, *20*, 6689. [CrossRef] [PubMed]
11. Xie, H.; Fang, H.; Li, X.; Wan, L.; Wu, P.; Yu, Y. Low-velocity impact damage detection and characterization in composite sandwich panels using infrared thermography. *Compos. Struct.* **2021**, *269*, 114008. [CrossRef]
12. He, Y.; Tian, G.Y.; Pan, M.; Chen, D. Non-destructive testing of low-energy impact in CFRP laminates and interior defects in honeycomb sandwich using scanning pulsed eddy current. *Compos. Part B* **2014**, *59*, 196–203. [CrossRef]
13. Mustapha, S.; Ye, L.; Dong, X.; Alamdari, M.M. Evaluation of barely visible indentation damage (BVID) in CF/EP sandwich composites using guided wave signals. *Mech. Syst. Signal Process.* **2016**, *76–77*, 497–517. [CrossRef]
14. Sikdar, S.; Banerjee, S. Guided wave based nondestructive analysis of localized inhomogeneity effects in an advanced sandwich composite structure. *Compos. Part B* **2019**, *176*, 107195. [CrossRef]
15. Zhu, J.; Wang, Y.; Qing, X. Modified electromechanical impedance-based disbond monitoring for honeycomb sandwich composite structure. *Compos. Struct.* **2019**, *217*, 175–185. [CrossRef]
16. Mieloszyk, M.; Jurek, M.; Majewska, K.; Ostachowicz, W. Terahertz time domain spectroscopy and imaging application for analysis of sandwich panel with embedded fibre Bragg grating sensors and piezoelectric transducers. *Opt. Las. Eng.* **2020**, *134*, 106226. [CrossRef]
17. Janeliukstis, R.; Mironovs, D. Smart composite structures with embedded sensors for load and damage monitoring—A review. *Mech. Compos. Mater.* **2021**, *57*, 131–152. [CrossRef]
18. Nsengiyumva, W.; Zhong, S.; Lin, J.; Zhang, Q.; Zhong, J.; Huang, Y. Advances, limitations and prospects of nondestructive testing and evaluation of thick composites and sandwich structures: A state-of-the-art review. *Compos. Struct.* **2020**, *256*, 112951. [CrossRef]
19. Kumar, M.; Shenoi, R.A.; Cox, S.J. Experimental validation of modal strain energies based damage identification method for a composite sandwich beam. *Compos. Sci. Technol.* **2009**, *69*, 1635–1643. [CrossRef]
20. Li, B.; Li, Z.; Zhou, J.; Ye, L.; Li, E. Damage localization in composite lattice truss core sandwich structures based on vibration characteristics. *Compos. Struct.* **2015**, *126*, 34–51. [CrossRef]
21. Lu, L.; Song, H.; Huang, C. Experimental investigation of unbound nodes identification for metallic sandwich panels with truss core. *Compos. Struct.* **2017**, *163*, 248–256. [CrossRef]
22. Zhou, Y.; Xu, Y.; Liu, H.; Guo, Y.; Yi, X.; Jia, Y. Debonding identification of Nomex honeycomb sandwich structures based on the increased vibration amplitude of debonded skin. *Compos. Part B* **2020**, *200*, 108233. [CrossRef]
23. Lestari, W.; Qiao, P. Damage detection of fiber-reinforced polymer honeycomb sandwich beams. *Compos. Struct.* **2005**, *67*, 365–373. [CrossRef]

24. Rucevskis, S.; Wesolowski, M.; Chate, A. Damage detection in laminated composite beam by using vibration data. *J. Vibroeng.* **2009**, *11*, 363–373.
25. Seguel, F.; Meruane, V. Damage assessment in a sandwich panel based on full-field vibration measurements. *J. Sound Vib.* **2018**, *417*, 1–18. [CrossRef]
26. Yang, J.-S.; Ma, L.; Chaves-Vargas, M.; Huang, T.-X.; Schröder, K.-U.; Schmidt, R.; Wu, L.-Z. Influence of manufacturing defects on modal properties of composite pyramidal truss-like core sandwich cylindrical panels. *Compos. Sci. Technol.* **2017**, *147*, 89–99. [CrossRef]
27. Lu, L.; Song, H.; Huang, C. Effects of random damages on dynamic behavior of metallic sandwich panel with truss core. *Compos. Part B* **2017**, *116*, 278–290. [CrossRef]
28. Katunin, A. Vibration-based spatial damage identification in honeycomb-core sandwich composite structures using wavelet analysis. *Compos. Struct.* **2014**, *118*, 385–391. [CrossRef]
29. Zhou, J.; Li, Z. Damage detection based on vibration for composite sandwich panels with truss core. *Compos. Struct.* **2019**, *229*, 111376. [CrossRef]
30. Katunin, A.; Przystałka, P. Automated wavelet-based damage identification in sandwich structures using modal curvatures. *J. Vibroeng.* **2015**, *17*, 2977–2986.
31. Liu, Z.; Ardabilian, M.; Zine, A.; Ichchou, M. Crack damage identification of a thick composite sandwich structure based on Gaussian Processes classification. *Compos. Struct.* **2021**, *225*, 112825. [CrossRef]
32. Katunin, A. Damage identification in composite plates using two-dimensional B-spline wavelets. *Mech. Syst. Signal Process.* **2011**, *25*, 3153–3167. [CrossRef]
33. Janeliukstis, R.; Rucevskis, S.; Wesolowski, M.; Chate, A. Experimental structural damage localization in beam structure using spatial continuous wavelet transform and mode shape curvature methods. *Measurement* **2017**, *102*, 253–270. [CrossRef]
34. Candès, E.J.; Donoho, D.L. Curvelets: A surprisingly effective nonadaptive representation of objects with edges. In *Curve and Surface Fitting*; Cohen, A., Rabut, C., Schumaker, L.L., Eds.; Vanderbilt University Press: Nashville, TN, USA, 2000.
35. Wang, L.; Yang, Z.; Waters, T.P. Structural damage detection using cross correlation functions of vibration response. *J. Sound Vib.* **2010**, *329*, 5070–5086. [CrossRef]
36. Katunin, A. Damage identification and quantification in beams using Wigner-Ville distribution. *Sensors* **2020**, *20*, 6638. [CrossRef]
37. Katunin, A. Identification of structural damage using S-transform from 1D and 2D mode shapes. *Measurement* **2021**, *173*, 108656. [CrossRef]
38. Welland, G.V. Beyond Wavelets. In *Studies in Computational Mathematics 10*; Chui, C.K., Monk, P., Wuytack, L., Eds.; Academic Press: San Diego, CA, USA, 2003.
39. Mumtaz, R.; Mumtaz, M.; Mansoor, A.B.; Masood, H. Computer aided visual inspection of aircraft structures. *Int. J. Image Process.* **2012**, *6*, 38–53.
40. Xu, K.; Liu, S.; Ai, Y. Application of Shearlet transform to classification of surface defects for metals. *Image Vis. Comput.* **2015**, *35*, 23–30. [CrossRef]
41. Umamaheswara Raju, R.S.; Ramachandra Raju, V.; Ramesh, R. Curvelet transform for estimation of machining performance. *Optik* **2017**, *131*, 615–625. [CrossRef]
42. Tzanis, A. A versatile tuneable curvelet-like directional filter with application to fracture detection in two-dimensional GPR data. *Signal Proc.* **2017**, *132*, 243–260. [CrossRef]
43. Anandan, P.; Sabeenian, R.S. Fabric defect detection using Discrete Curvelet Transform. *Proc. Comput. Sci.* **2018**, *133*, 1056–1065.
44. Bagheri, A.; Ghodrati Amiri, G.; Seyed Razzaghi, S.A. Vibration-based damage identification of plate structures via curvelet transform. *J. Sound Vib.* **2009**, *327*, 593–603. [CrossRef]
45. Nicknam, A.; Hosseini, M.H.; Bageri, A. Damage detection and denoising in two-dimensional structures using curvelet transform by wrapping method. *Arch. Appl. Mech.* **2011**, *81*, 1915–1924. [CrossRef]
46. De Marchi, L.; Baravelli, E.; Ruzzene, M.; Speciale, N.; Masetti, G. Guided wave expansion in warped curvelet frames. *IEEE Trans. Ultrason. Ferroelectr.* **2012**, *59*, 949–957. [CrossRef] [PubMed]
47. Vafaie, S.; Salajegheh, E. Comparisons of wavelets and contourlets for vibration-based damage identification in the plate structures. *Adv. Struct. Eng.* **2019**, *22*, 1672–1684. [CrossRef]
48. Ansari, R.A.; Budhhiraju, K.M. A comparative evaluation of denoising of remotely sensed images using wavelet, curvelet and contourlet transforms. *J. Ind. Soc. Remote Sens.* **2016**, *44*, 843–853. [CrossRef]
49. Katunin, A.; John, M.; Joszko, K.; Kajzer, A. Characterization of quasi-static behavior of honeycomb core sandwich structures. *Model Eng.* **2014**, *22*, 78–84.
50. Candès, E.; Demanet, L.; Donoho, D.; Ying, L. Fast discrete curvelet transforms. *Multiscale Model. Simul.* **2006**, *5*, 861–899. [CrossRef]
51. Gao, R.X.; Yan, R. Beyond wavelets. In *Wavelets. Theory and Applications for Manufacturing*; Gao, R.X., Yan, R., Eds.; Springer: Boston, MA, USA, 2011.
52. Donoho, D.L.; Duncan, M.R. Digital curvelet transform: Strategy, implementation, and experiments. *Proc. SPIE* **2000**, *4056*, 12–30. [CrossRef]
53. Bayissa, W.L.; Haritosa, N.; Thelandersson, S. Vibration-based structural damage identification using wavelet transform. *Mech. Syst. Signal Process.* **2008**, *22*, 1194–1215. [CrossRef]

materials

Article

The Correlation between Shrinkage and Acoustic Emission Signals in Early Age Concrete

Magdalena Bacharz *, Kamil Bacharz * and Wiesław Trąmpczyński

Department of Concrete Construction and Geotechnics, Faculty of Civil Engineering and Architecture, Kielce University of Technology, 25-314 Kielce, Poland
* Correspondence: mbacharz@tu.kielce.pl (M.B.); kbacharz@tu.kielce.pl (K.B.); Tel.: +48-41-34-24-579 (M.B.); +48-41-34-24-752 (K.B.)

Abstract: This study analysed the processes of damage formation and development in early age unloaded concrete using the acoustic emission method (IADP). These are of great importance in the context of the durability and reliability of a structure, as they contribute to reducing its failure-free operation time. Concrete made with basalt aggregate and Portland or metallurgical cement cured under different conditions after demoulding was the test material. The obtained damage values were compared with the measured concrete shrinkage, and a shrinkage strain–acoustic emission signal (resulting from damage) correlation was found. The correlation allows easy measurement of damage level in the early period of concrete hardening, and consequently can be the basis of a non-destructive method.

Keywords: new diagnostic method; concrete; early age damage; acoustic emission; shrinkage; NDT

Citation: Bacharz, M.; Bacharz, K.; Trąmpczyński, W. The Correlation between Shrinkage and Acoustic Emission Signals in Early Age Concrete. *Materials* **2022**, *15*, 5389. https://doi.org/10.3390/ma15155389

Academic Editors: Stanisław Radkowski and Szymon Gontarz

Received: 6 July 2022
Accepted: 2 August 2022
Published: 5 August 2022

Publisher's Note: MDPI stays neutral with regard to jurisdictional claims in published maps and institutional affiliations.

1. Introduction

Concrete, especially reinforced concrete (RC), is currently the most commonly used material for the construction of buildings and engineering structures. It is subjected to various effects, i.e., mechanical loads (static and dynamic) [1,2], variable temperature [3,4] and humidity, as well as chemical and biological attacks, contributing to the corrosion of these objects [5,6].

An important aspect is not only the current strength of elements or entire structures but also their durability [7].

Increasingly complex and ambitious design challenges and sustainable development goals require using concrete and steel with new properties that significantly impact the behaviour of structures [8,9]. Tight construction schedules, which is another factor influencing structural durability, have increased the attractiveness of prefabrication. The practice of assembling components in a factory has helped improve the quality and durability of the structures being built. Since prefabricated components require a high degree of manufacturing accuracy, poor supervision may affect the durability of mass-produced elements, as experience shows [10]. Manufacturing errors in newly constructed buildings [11,12] impact their performance [13,14], leading to extensive and costly repairs.

Undoubtedly, the width and growth of cracks are crucial parameters in diagnosing buildings [15]. For this reason, cracks in concrete [16,17] and reinforced concrete [18–20] are still extensively studied and analysed. Furthermore, the description of cracks is constantly modified and developed due to the heterogeneity of concrete and complex states of stress and strain that accompany cracking and microcracking [21–23].

The cracks observed in the structures are the final result of a more complex process of the formation and development of damage in early age concrete [24]. When hardening, concrete reduces its volume due to moisture loss and chemical processes, referred to as shrinkage. Shrinkage is often discussed in the literature [25] not only as a phenomenon that takes place in early-age concrete [14,25] but also as a factor impacting structural durability [26]. Various calculation models have been developed to determine shrinkage strain as another type of load acting on a structure [27–30]. Shrinkage is responsible for cracking that may lead to fatigue, shortened service life, or compromised load capacity. Extensive research is performed to mitigate early microcracking in concrete by adding fly ash or slag cement and using lightweight aggregate, as described in [31–33]. Another approach to mitigate microcracks is to incorporate nanomaterials. For example, carbon nanotubes were found to reduce the microcracks and shrinkage [34].

Stress concentration, caused by heterogeneous and often excessive temperature and humidity fields, is also a factor in microcracking [14,35]. Microcracks can transform into cracks observed on the surface of structures [36], thereby contributing to the ingress of water and other aggressive substances that have an adverse effect on durability. A possible solution to this issue may be the use of carbon nanotubes, although it requires additional analysis regarding structural concretes [37].

Non-destructive or semi-destructive methods [38–41] are essential in diagnosing concrete. Chemical (qualitative and quantitative) analyses, physical (optical, thermographic, radiological, acoustic, electromagnetic) methods, or biological methods (macroscopy, microscopy) allow the assessment of strength and homogeneity (sclerometric and acoustic methods), location of defects and damage (acoustic and radiological methods), location of reinforcement, corrosion assessment (electromagnetic, radiological, electrical methods), and the evaluation of humidity and temperature distribution (indirect—physical, chemical).

One of the non-destructive methods (used in this paper) is the acoustic emission (AE) method—IADP (identification of active destructive processes), which has been successfully used to analyse the development of cracks resulting from loads on the structure, including the service load [42–47]. AE methods are applied in analysing the parameters of elastic (acoustic) waves generated in the material during the cracking process [48–51].

The objective of this paper is to analyse AE signals caused by destructive processes in unloaded early age concrete, detected using the AE IADP technique, and their correlation with the measured shrinkage strains.

In the papers [52,53], AE signals were assigned to the destructive processes identified and tracked using the IADP technique. The authors of [54,55] verified the suitability of IADP for testing destructive processes in unloaded concrete hardening at various temperatures and under different maintenance conditions, taking into account the type of cement and aggregate and the presence of reinforcement and admixtures.

This paper attempts to determine the correlation between the strain values and the destructive processes recorded in the hardening concrete. The tests were performed on unloaded concrete that varied in the type of cement and maintenance conditions after removal from the moulds. The strains and the course of the destructive processes in concrete were tested for 56 days. Furthermore, the shrinkage strains were estimated according to three standard approaches: Eurocode 2 (EC2) [27], Model Code 90–99 (MC90–99) [28], and Bazant Baweja—model B3 (B3) [29]. The influence of maintenance conditions on the level of predicted strains was analysed. A robust correlation was observed and described between the number of destructive processes, their energy, and the shrinkage strains. This means that by knowing (measuring) early age concrete shrinkage strains, one can determine the AE signal range as the basis for determining whether in addition to the basic processes and other processes indicative of progressive internal micro-damage occurring in concrete, thereby reducing structural durability and reliability. The tests and their analyses are a continuation of the work aimed at developing a universal non-destructive method for assessing early age concrete.

2. Materials and Methods

Destructive processes in the material (microcracks, crack propagation, and dislocations) are accompanied by a rapid release of energy, generating elastic (acoustic) waves in the material. These waves (Figure 1) gradually disappear as a result of energy absorption in the thermal process.

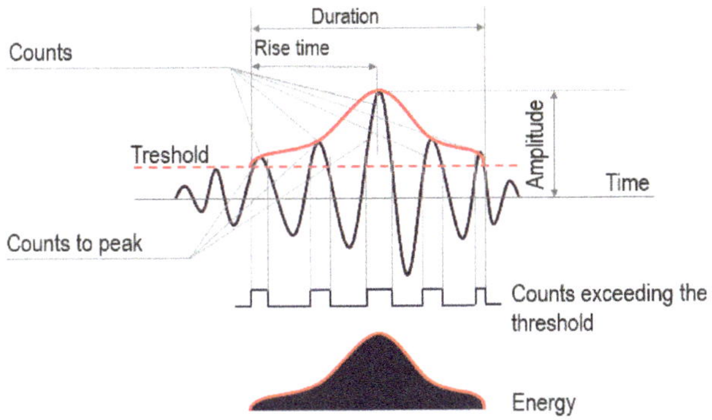

Figure 1. Acoustic signal parameters.

In this study, the IADP method was used to measure destructive processes. The method does not rely on the selected parameters of the AE signal, but is based on the reference signals created with the use of 12 parameters of the recorded electric signal: counts, counts to peak, duration, rise time, amplitude in mV or dB, energy, strength, root mean square, mean level, mean frequency, reverberation frequency, and initial frequency. The use of this method for the evaluation of reinforced concrete structures (bridges) is presented in [45–47].

The signal is measured when an active destructive process occurs in a given element during the measurement, e.g., a crack is formed or propagated [56]. Damage in the element but not developing does not generate AE signals. Acoustic waves generated in the material can be recorded using acoustic sensors (usually piezoelectric). Their proper arrangement enables the localisation of their source. The selected signal parameters are also analysed [57] and can be used to determine the type of failure [49–51].

Using AE signals, it is also possible to analyse other destruction processes. In the case of prestressed elements, these are [42,43]: microcracking, friction between crack faces, initiation and growth of cracks, cracking at the concrete reinforcement interface, concrete spalling, friction at the concrete reinforcement interface, corrosion, plastic deformation, and cracking of cables and other reinforcement. This method was successfully used in diagnosing prestressed concrete elements and structures [44].

In [53,55], the IADP method was used to analyse the failure processes in the early phase of concrete hardening. Destructive (12 signal parameters) were assigned to following classes of reference signals [47,52]:

- microcracks in the cement paste and at the aggregate-paste interface (Class 1),
- internal propagation of microcracks (Class 2),
- formation of microcracks on the concrete surface (Class 3), and
- growth of microcracks (Class 4).

A destructive process in concrete (for example, microcracks in the cement paste) is an acoustic wave sources. Twelve parameters of the AE wave (signal) are recorded by the AE sensors and compared with the base of reference signals, which allows us to determine the destructive process (in this case, Class 1) and its location based on the arrival time of the AE signal.

Although microcracks and damage in early age concrete do not have a direct impact on the safety of the structure [21] (unless their size exceeds a certain level [26,58]), they affect the durability of the structure because they become the sites of future crack initiation.

2.1. Test Elements

Nine concrete samples (three series of three samples) with dimensions of $150 \times 150 \times 600$ mm were used in the tests. The samples varied in terms of selected parameters described below and summarised in Table 1. The samples were made with basalt aggregate 2–16 mm from the Gracze quarry and different cement types. In sample C−I (MC) CEM I 42.5N MSR/NA, Portland cement from the WARTA cement plant was used, while in samples C−III (MC) and C−III (AC) CEM III/A 42.5N—LH/HSR/NA, metallurgical cement from the Górażdże cement plant was used. After removal from the moulds, C−III (MC) samples were cured at 100% humidity for 10 days and then subjected to proper tests, i.e., measurements of strains and volume change, and AE signal recording. The C−III (AC) samples were tested without prior curing in water. The moisture-cured samples (C−III (MC) and C−I (MC)) and the non-moist-cured sample (C−III (AC)) hardened at a constant temperature of 22 ± 2 °C for 56 days.

Table 1. Characteristics of concrete samples.

Symbol	Aggregate	Cement	Hardening Conditions	Temperature Condition
C−I (MC)	Basalt	CEM I	10 days of wet curing	Constant
C−III (MC)	Basalt	CEM III	10 days of wet curing	Constant
C−III (AC)	Basalt	CEM III	drying in air	Constant

The composition of individual concrete mixes is shown in Table 2.

Table 2. Composition of concrete mixtures [kg/m^3].

Symbol	Basalt 2–8	Basalt 8–16	Sand 0–2	CEM I	CEM III	Water
C−I (MC)	581	731	691	360	x	180
C−III (MC)	581	731	691	x	360	180
C−III (AC)	581	731	691	x	360	180

2.2. Research Methods

2.2.1. Strain Measurements and Prediction

An 8 inch (~20 cm) demountable mechanical strain gauge was used for the strain test (Figure 2b). The measurement points were steel elements glued to the four walls of the sample, as shown in Figure 2a.

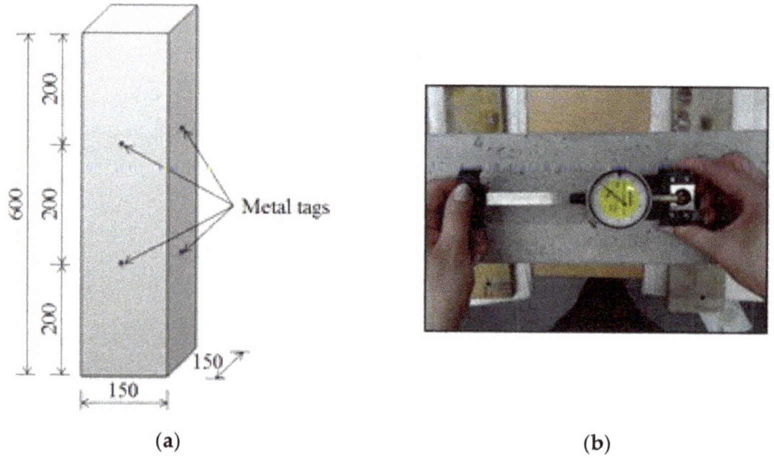

Figure 2. (**a**) Spacing of metal tags for strain measurements, (**b**) measurement of strain with an extensometer.

Strains were measured on four faces of each specimen and the basic statistical parameters were calculated, i.e., mean value, coefficient of variation, and variance. The strain results obtained for three series of specimens at varied curing conditions are shown in Figure 3, where the vertical axis shows the results of shrinkage strains and the horizontal axis shows the time, in days, for which the measurements were made. It also should be noted that during the test, the samples were not additionally loaded.

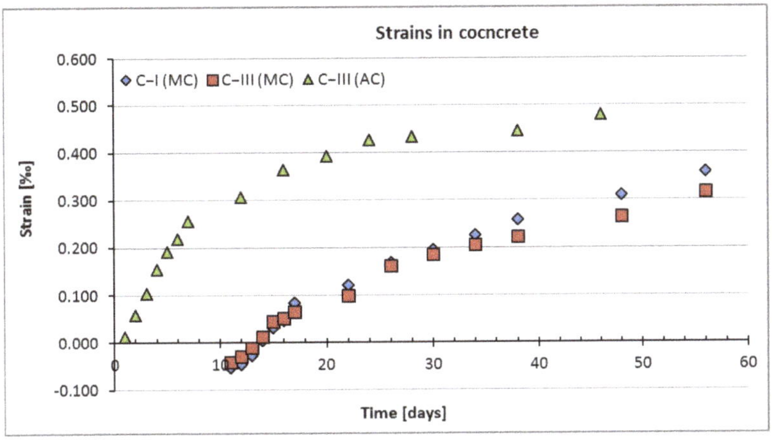

Figure 3. The result of the strain test of specimens from series C−I (MC), C−III (MC), and C−III (AC).

The strains (shrinkage) from the tests were compared with the results predicted according to the following standards: Eurocode 2 (EC2)—PN-EN−1992−1−1 [27], Model Code 90–99 (MC) [28], and the Bazant-Baweja approach (B3) [29]. The adopted assumptions are given in Figure 4.

The strain values obtained in the C−I (MC) and C−III (MC) concrete (subjected to 10-day curing after removal from the moulds) are shown in Figure 5a,b, and for C−IIII (AC) (no curing after demoulding) are shown in Figure 6.

The strains estimated according to the EC2, MC90–99, and B3 approaches in water-cured samples C−I (MC) (with Portland cement) and C−III (MC) (with metallurgical cement) fell within the range of +/− 20% in relation to the values obtained in the laboratory tests. In the first week of the test, swelling of concrete C−I (MC) and C−III (MC) was observed. The phenomenon was most likely due to the higher humidity in the thermal chamber than the ambient humidity in the laboratory hall where the samples were prepared for testing.

Higher ambient humidity in the climatic chamber resulted from samples with higher humidity being placed there. With the same number of samples and the volume of the climatic chamber, the result was a significant increase in humidity inside the chamber. As a result, the humidity of the samples themselves increased, and, as a result, their volume increased. Over time, the humidity decreased, contributing to the recorded shrinkage deformations, not swelling. This was largely due to the low humidity of the environment in which the chamber itself was located, which contributed to the final decrease in the level of humidity.

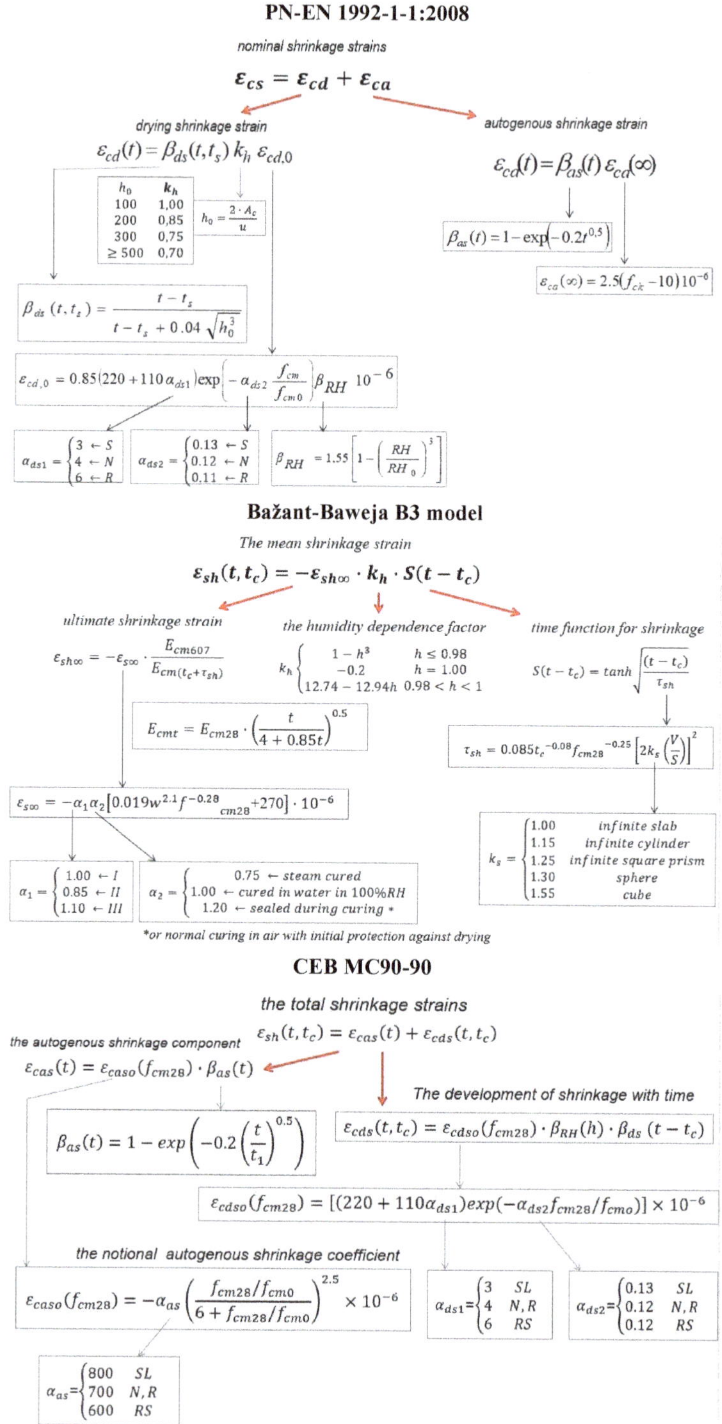

Figure 4. Procedures for predicting shrinkage strain according to selected standards.

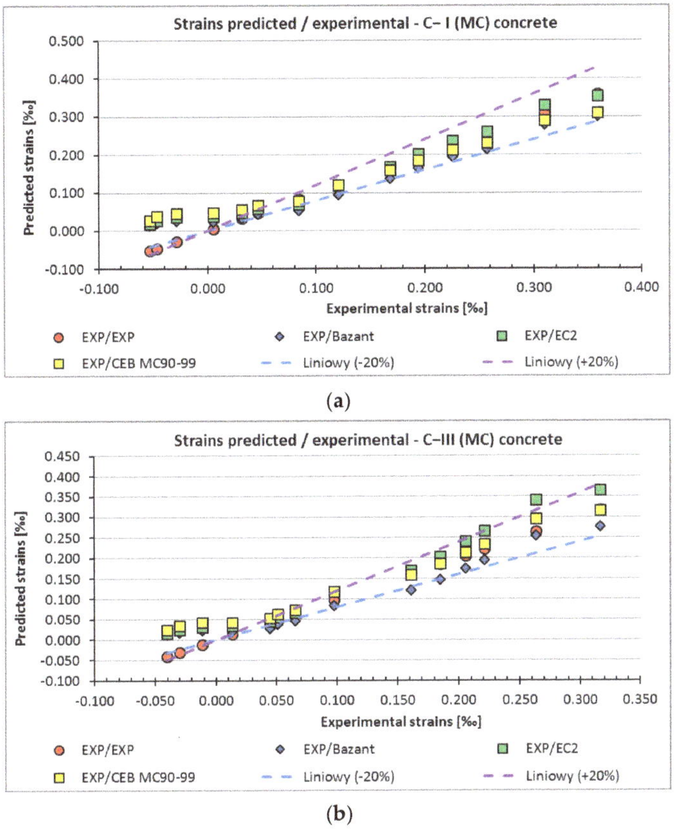

Figure 5. The strains in the pre-cured samples: (**a**) C−I (MC), (**b**) C−III (MC).

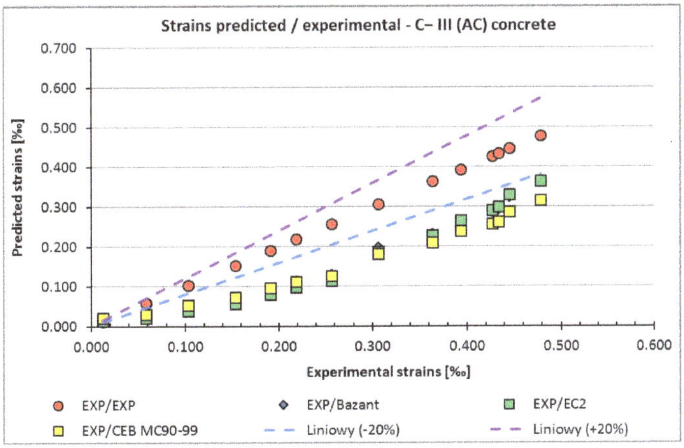

Figure 6. The strain results of non-cured C−III (AC) samples.

In the case of the non-moisture-cured C−III (AC) concrete with metallurgical cement, the strain values predicted according to the standards were lower than the values obtained from laboratory tests.

The impact of not curing and curing errors in the initial stage of hardening on the shrinkage strains are not taken into account in the adopted standards [27–29], which may be the reason for the differences (Figure 6).

2.2.2. Measurements of Destructive Processes—IADP Method

The tests were carried out using the IADP acoustic emission method, validated, among others, in [53]. The flow chart of the method is shown in Figure 7.

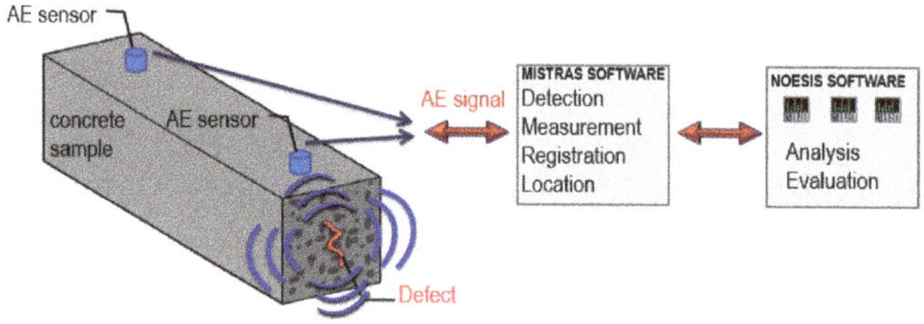

Figure 7. The concept of the IADP method for early-age concrete.

Concrete samples C−I (MC) and C−III (MC) were removed from the moulds after 10 days of cure. C−III (AC) samples were tested without curing. Two Vallen VS30-V sensors were placed on one side of each sample. The concrete surface was cleaned before the sensors were attached to the surface with a thermally conductive silicone paste and elastic rubber.

The calibration before the start of the test consisted of checking the amplitude of the recorded AE signal from breaking the graphite of a pencil with a hardness of 2H, diameter 0.3 mm, length 3 mm (Hsu Nielsen source [59,60]) inclined to the surface of the sample at an angle of 30° (Figure 8).

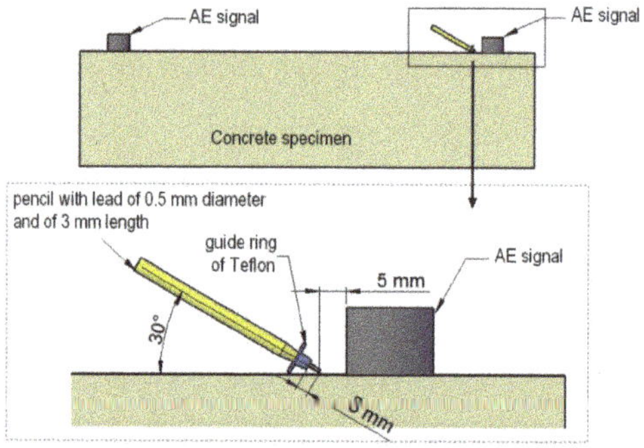

Figure 8. Sensor calibration before testing—Hsu Nielsen source.

If the amplitude of the signal excited by breaking the pencil recorded on each sensor reached 100 dB, calibration was performed successfully. If a lower amplitude was recorded, the amount of thermal paste was added to improve the contact between the sensor and the concrete surface.

Before the test, an input data file was created in the Mistras program. The dimensions of the tested element, the spacing, and number of AE sensors as well as measurement parameters were set. AE signals were recorded in the samples in 12 h cycles preceding the measurement at 1–8, 12, 16, 20, 24, 28, 38, and 46 days at the points where strain values were taken, the samples were weighed, and cracks were observed.

The test results were analysed using Noesis software and the reference signal database [53]. Based on that, the recorded signals were assigned to three classes that defined individual processes using the supervised approach.

The classification involved comparing the registered AE signals with the previously prepared reference base. It was developed in the research described in [54].They were based on the division of a broad database into 3 groups of signals (classes) on the basis of the analysis of 12 parameters describing the AE signal. The exact process of creating databases is presented in [43,49,54].

3. Results

3.1. Number of AE Signals Analysis

The correlation between the strains obtained from the tests and the number of AE signals (destructive processes such as microcracks in the cement paste and at the aggregate-paste interface (Class 1), the internal microcrack propagation (Class 2), and the formation of microcracks on the concrete surface (Class 3) was recorded in concrete series C−I (MC)–C−III (AC), and as shown in Figure 9a–c.

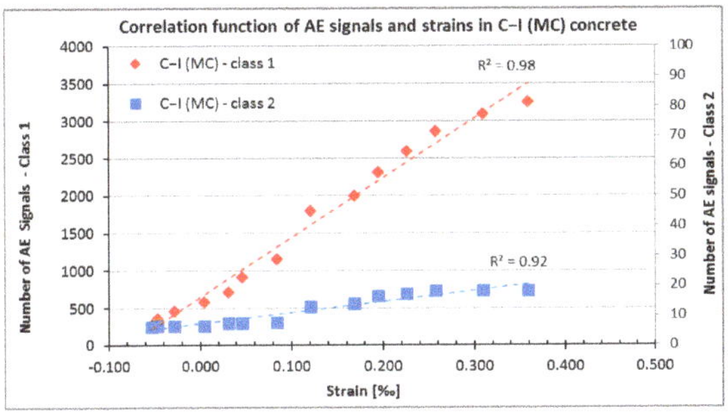

(a)

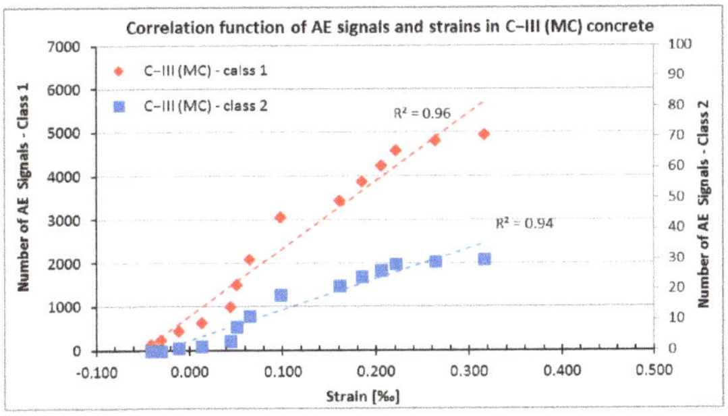

(b)

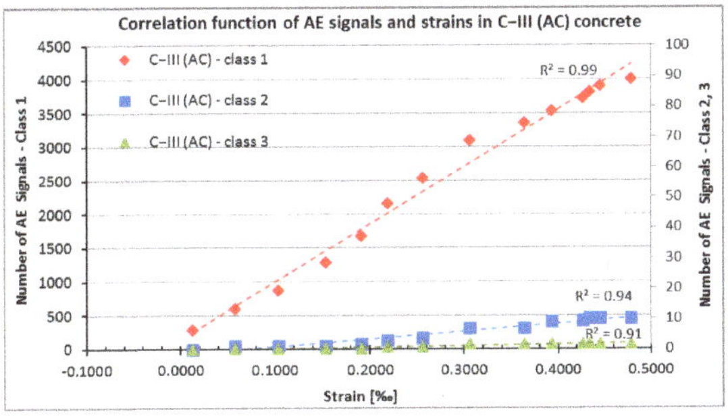

(c)

Figure 9. Correlation functions of AE signals and strains for the C−I (MC) (**a**) and C−III (MC) concrete series (subjected to curing) (**b**) and C−III (AC) (without curing) (**c**).

In the presented graphs on the horizontal axis, the values of shrinkage strains are shown, while the vertical axes show the number of class 1 signals on the left axis, and the number of class 2 and 3 signals on the right axis. The use of two vertical axes is due to significant differences between the quantities of class 1, 2, and 3 signals. Presenting them on one axis would make their analysis impossible. However, if class 3 signals were not recorded in a sample, no additional vertical axis was added.

From the results in Figure 9, it follows that there is a very strong correlation, practically linear, between the strain values and the number of destructive processes recorded in the sample: microcracks in the cement paste and at the grain boundaries (Class 1), the development of microcracks in the cement paste (Class 2), and the formation of microcracks on the concrete surface (Class 3).

Equations of linear functions shown in Figure 9 are summarised in Table 3.

Table 3. Equations of linear functions of the correlation between strains and number of hits.

Concrete Series	Class 1	Class 2	Class 3
C−I (MC)	y = 8000x + 650	y = 35x + 7	
C−III (MC)	y = 4200x + 210	y = 1800x + 55	
C−III (AC)	y = 8800x + 110	y = 25x − 1.5	y = 5x − 0.5

3.2. AE Signal Energy Analysis

In a paper [53], the unit energy values of the signals of individual classes (1, 2, and 3) were determined for concrete series C−III (MC) and C−III (AC). These values, summarised in Table 4, were used to determine an increase in the signal energy in C−I (MC)-C−III (AC).

Table 4. The unit energy values.

Unit Energy	C−I (MC)	C−III (MC)	C−III (AC)
Class 1	0.34	0.27	0.23
Class 2	14.05	17.57	18.99
Class 3			113.65

The obtained correlation between the unit energy of the signals of individual classes (representing damage) and the strains is shown in Figure 10.

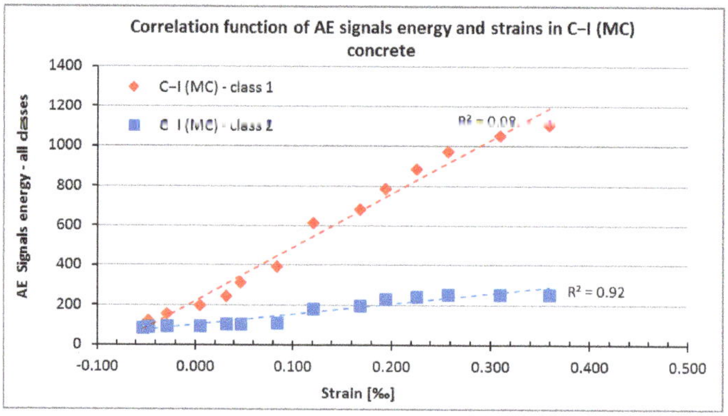

(a)

Figure 10. *Cont.*

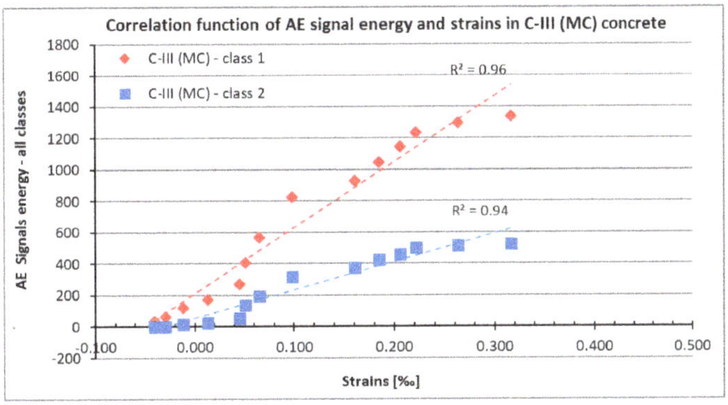

(**b**)

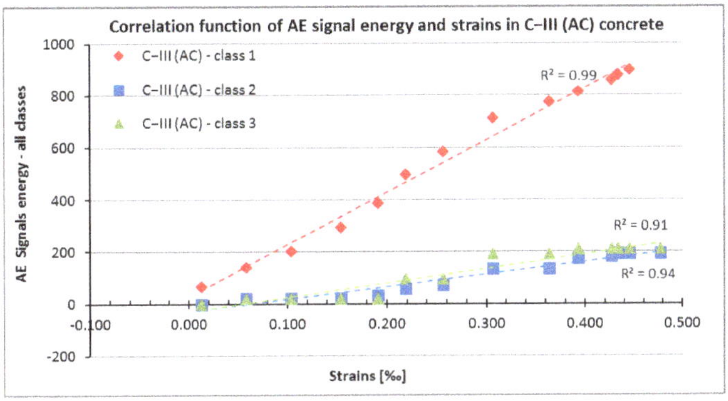

(**c**)

Figure 10. Correlation functions of AE signals energy and strains in samples subjected to curing (**a**) C−I (MC), (**b**) C−III (MC), and (**c**) C−III (AC) (without prior curing).

From the results in Figure 9, it follows that there is a very strong linear correlation between the increase in strains and energy of destructive processes recorded for the samples.

The equations of linear functions shown in Figure 10 are summarised in Table 5.

Table 5. Equations of linear functions of the correlation between strains and energy of AE signals.

Concrete Series	Class 1	Class 2	Class 3
C−I (MC)	y = 2700x + 220	y = 515x + 100	
C−III (MC)	y = 15,600x + 785	y = 100x + 3	
C−III (AC)	y = 2000x + 25	y = 480x − 30	y = 555x − 30

3.3. *Correlation Function of Strains and Destructive Processes in Unloaded Concrete*

One of the objectives of the study was an attempt to develop a relationship that would allow the prediction of the number of AE signals due to damage and processes in time.

On the basis of the trend lines in the graphs showing the increase in both shrinkage strains and acoustic emission signals, a natural logarithm function was adopted as the base function modified with adjustment coefficients.

$$y = \alpha \cdot A_{SK} \cdot \ln(x) + \beta \cdot B_{SK} \quad (1)$$

where:

α, β—estimated correction factors,

A_{sk}, B_{sk}—coefficients of the logarithmic trend line of the additional shrinkage strain for the tested sample, and

x—time in days.

Correction factors, α and β, were adopted to better match the results. The coefficients were estimated as the ratios of the A and B coefficients read from the trend line (logarithm functions) of the increase in shrinkage strains (A_{SK} and B_{SK}) and the acoustic emission signals (A_{AE} and B_{AE}).

$$\alpha = \frac{A_{AE}}{A_{SK}}, \beta = \frac{B_{AE}}{B_{SK}} \tag{2}$$

The values of the α and β coefficients were estimated with two methods (API and APII) based on the results obtained from the concrete series. In the first approach (API), the A_{AE}/A_{SK} and B_{AE}/B_{SK} ratios were determined for each concrete series (C−I (MC) to C−III (AC) separately, and then the values were averaged, obtaining α and β coefficients, respectively. The second approach (APII) consisted of determining the average A_{AE}, A_{SK}, B_{AE}, and B_{SK} coefficients from the C−I (MC) to C−III (AC) concrete series. Then, α and β coefficients were calculated from A_{AE}/A_{SK} and B_{AE}/B_{SK} relationships. The parameters adopted to estimate the theoretical values of AE signal increase are presented in Table 6.

Table 6. Parameters determined with two methods.

Concrete Series	A_{SK}	B_{SK}	A_{EA}	B_{EA}	API		APII	
					α_I	β_I	α_{II}	β_{II}
C−I (MC)	0.024	−0.027	909	−388	34.614	9151	34.555	18.314
C−III (MC)	0.018	−0.023	612	−270				
C−III (AC)	0.034	0.015	1129	18				

The initial verification of the results was carried out based on the calculated coefficients (α and β) and the values of experimental shrinkage strains from the C−I (MC) to C−III (AC) concrete series. The course of the predicted increase in acoustic emission signals over time in relation to their experimental values is shown in Figure 11. The graphs show the increase in the number of AE signals for all three classes of AE signals.

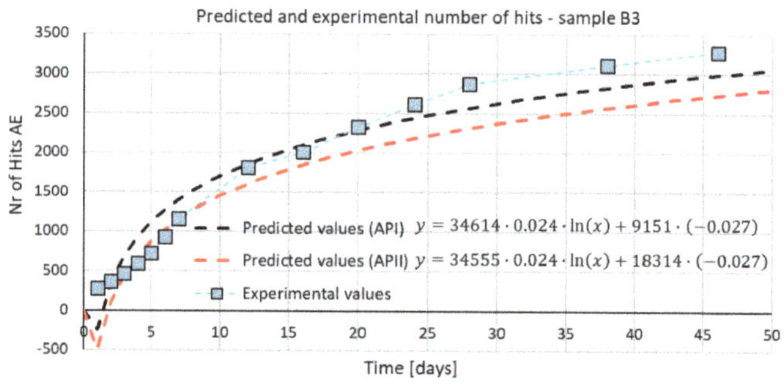

(a)

Figure 11. *Cont.*

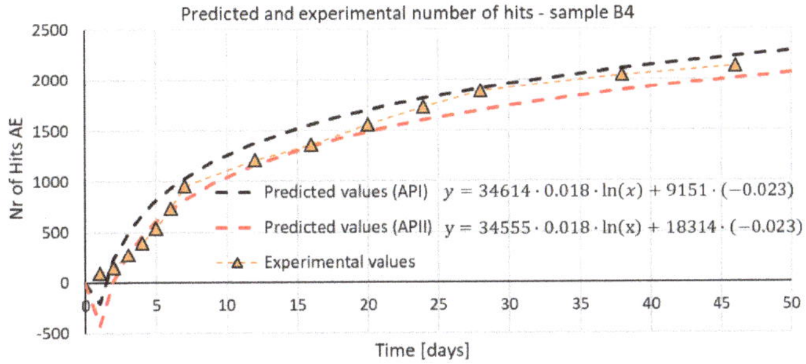

(b)

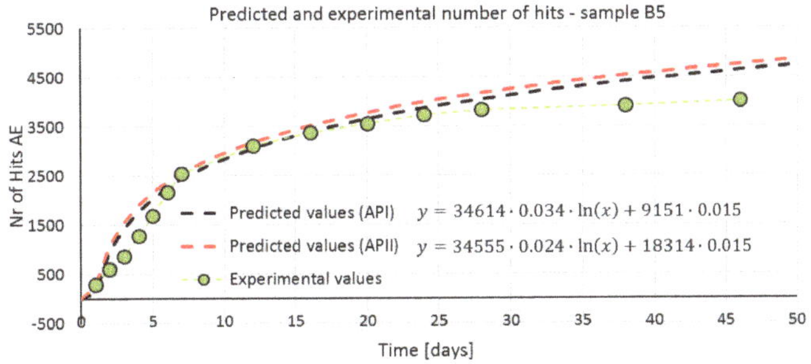

(c)

Figure 11. The results of the number of hits predicted (based on strain measurements) and recorded in samples: (**a**) C−I (MC), (**b**) C−III (MC) (series subjected to curing) and (**c**) C−III (AC).

The dotted lines (with markers at measurement points) show the real increase in AE signals for the C−I (MC) to C−III (AC) concrete series over time, while the dashed lines show the increase in predicted course of the AE signals. The black dashed line represents the theoretical gain estimated from the α_I and β_I coefficients, while the dashed red line represents the theoretical gain of the AE signals estimated from the α_{II} and β_{II} correction factors. Both lines were developed on the basis of the shrinkage strains of the examined concrete, i.e., the trend line in the diagram of the actual strain increment. The analysis showed that the predicted values of the increase in the number of AE signals were consistent with those obtained experimentally and, therefore, the adopted algorithm of the procedure can be considered properly matched.

As has been shown, it is possible to describe the number of AE signals as a function of shrinkage strains, which is the basis for the wide testing program taking into account various parameters that affect the shrinkage in concrete.

4. Discussion

Using the AE technique, a very strong correlation, in the range of 0.91 to 0.99, between the strain values and the number of AE signals related to destructive processes in early age concrete was found. Additionally, the recorded processes were divided into three classes, i.e., microcracks in the cement paste and at the grain boundaries in the cement paste (Class 1), the formation of microcracks in the cement paste (Class 2), and the formation of microcracks on the concrete surface (Class 3). The indicated correlation was described using a simple logarithm function.

Additionally, a very strong correlation was found between the increase in strain and the increase in energy of the AE signals, which can be described using a similar logarithm function.

Hence, by measuring shrinkage strains and knowing experimentally defined factors, one can estimate the theoretical range of AE signals. On this basis, one can determine whether, in addition to the basic processes, there are also those indicative of progressive internal micro-damage. This is important, as they can reduce the structure's durability and reliability.

Additionally, a comparative analysis of experimental shrinkage strains and strains estimated according to selected calculation models, including those used in the relative standards shown in Figures 5 and 6, was made. The values estimated according to approaches EC2, MC 90–99, and B3 are within the +/− 20% range of these obtained in laboratory tests for concrete subjected to 10-day water curing after demoulding (C−I (MC)) with Portland cement and with metallurgical cement (C−III (MC)). However, in the case of concrete with metallurgical cement hardening at the declared temperature without prior curing in water (C−III (AC)), the strain values predicted according to the standards are lower than those obtained from the laboratory tests. This may be due to the fact that no guidelines are available in the adopted standards for estimating the impact of curing errors or lack of curing in the initial period of concrete hardening on the values of shrinkage strains.

5. Conclusions

Presented results show that:

- The increase in the number of AE signals resulting from damage:
    - Class 1—microcracks in the cement paste and at the aggregate-paste interface,
    - Class 2—internal propagation of microcracks, and
    - Class 3—formation of microcracks on the concrete surface

is strongly correlated with the increase in strain over time and can be described by logarithmic function. Simple logarithm functions describing this relation were developed for C−I (MC), C−III (MC), and C−III (AC) concrete with specific parameters and variables.

- The increase in the energy of Class 1, 2, and 3 signals over time is strongly correlated with the increase in strain over time and functions describing this relation were developed.
- The results presented may form the basis for simple diagnostics of new elements. It means that by knowing the early age shrinkage strains (easy to measure), one can estimate the early age damage.
- More concrete tests should be performed to optimise the function formula and other variables should be added, e.g., aggregate, admixtures, and concrete strength.
- It was also shown that:
- Experimental shrinkage strains and those estimated according to selected standards show a strong correlation in the case of cured concrete C−I (MC) and C−III (MC)).
- In the case of non-cured concrete (C−III (AC)), the shrinkage strains estimated according to the standards are lower than those measured in the laboratory tests.

The obtained results are very promising, in particular due to the diagnostic possibilities they offer. For this reason, an extended program to verify these findings on other specimens made with a different concrete composition is being prepared.

Author Contributions: Conceptualization, W.T., M.B.; methodology, W.T., M.B., K.B.; software, M.B. and K.B.; validation, M.B. and K.B.; formal analysis, W.T., M.B., K.B.; investigation, M.B.; resources, W.T.; data curation, M.B. and K.B.; writing—original draft preparation, M.B. and W.T.; writing—review and editing, W.T., M.B., K.B.; visualization, M.B. and K.B.; supervision, W.T.; project administration, W.T. and M.B.; funding acquisition, W.T. All authors have read and agreed to the published version of the manuscript.

Funding: The work is supported by the program of the Minister of Science and Higher Education under the name: Regional Initiative of Excellence in 2019–2022 project number 025/RID/2018/19 financing amount PLN 12,000,000.

Institutional Review Board Statement: Not applicable.

Informed Consent Statement: Not applicable.

Data Availability Statement: The data presented in this study are available in: M. Bacharz, Wykorzystanie metody emisji akustycznej do badania procesów destrukcyjnych w betonie nieobciążonym, Ph.D. Thesis, Politechnika Swietorzyska, Kielce, Poland, 2016. (In Polish).

Acknowledgments: The authors would like to thank Anton K. Schindler from Auburn University, for cooperation and helpful comments during the article preparation.

Conflicts of Interest: The authors declare no conflict of interest.

References

1. Oh, B.H.; Kim, S.H. Advanced Crack Width Analysis of Reinforced Concrete Beams under Repeated Loads. *J. Struct. Eng.* **2007**, *133*, 411–420. [CrossRef]
2. Bacharz, K.; Goszczyńska, B. Shear Capacity of Reinforced Concrete Beams under Monotonic and Cyclic Loads: Experiments and Computational Models. *Materials* **2021**, *14*, 4092. [CrossRef] [PubMed]
3. Wang, X.; Chen, Q.; Tao, J.; Han, R.; Ding, X.; Xing, F.; Han, N. Concrete thermal stress analysis during tunnel construction. *Adv.in Mech. Eng.* **2019**, *11*, 1–15. [CrossRef]
4. Linek, M. Airport Cement Concrete with Ceramic dust of increased thermal resistance. *Materials* **2022**, *15*, 3673. [CrossRef] [PubMed]
5. Oh, B.H.; Kim, K.H.; Jang, B.S. Critical Corrosion Amount to Cause Cracking of Reinforced Concrete Structures. *ACI Mater. J.* **2009**, *106*, 333–339.
6. Raczkiewicz, W.; Wójcicki, A. Temperature Impact on the Assessment of Reinforcement Corrosion Risk in Concrete by Galvanostatic Pulse Method. *Appl. Sci.* **2020**, *10*, 1089. [CrossRef]
7. Li, K.; Li, L. Crack-altered durability properties and performance of structural concretes. *Cem. Concr. Res.* **2019**, *124*, 105811. [CrossRef]
8. Pawelska-Mazur, M.; Kaszyńska, M. Mechanical Performance and Environmental Assessment of Sustainable Concrete Recycled End-of-Life Tyre Fibers. *Materials* **2021**, *14*, 256. [CrossRef]
9. Zegardło, B.; Szeląg, M.; Ogrodnik, P. Ultra-high strength concrete made with recycled aggregate from sanitary ceramic wastes—The method of production and the interfacial transition zone. *Constr. Build. Mater.* **2016**, *122*, 736–742. [CrossRef]
10. Raczkiewicz, W.; Bacharz, K.; Bacharz, M.; Grzmil, W. Manufacturing Errors of Concrete Cover as a Reason of Reinforcement Corrosion in a Precast Element—Case Study. *Coatings* **2019**, *9*, 702.
11. Buczkowski, W. Zagrożenie Awarią Żelbetowego Zbiornika Prostopadłościennego. In Proceedings of the Materiały XII Konferencji Naukowo-Technicznej pt.: Żelbetowe i Sprężone Zbiorniki na Materiały Sypkie i Ciecze, Cracow, Poland, 19–21 November 2003.
12. Hadidi, R.; Saadeghvaziri, M.A. Transverse Cracking of Concrete Bridge Decks:State-of-the-Art. *J. Bridge Eng.* **2005**, *10*, 503–510. [CrossRef]
13. Godycki-Ćwirko, T.; Korzeniowski, P.; Piotrkowski, P. Awaryjne zarysowanie ścian zbiornika spowodowane skurczem. *Inżynier Budownictwa* **2009**, *10*, 67–71.
14. Safiuddin, M.; Kaish, A.B.M.A.; Woon, C.-O.; Raman, S.N. Early-Age Cracking in Concrete: Causes, Consequences, Remedial Measures, and Recommendations. *Appl. Sci.* **2018**, *8*, 1730. [CrossRef]
15. Kossakowski, P.; Ślusarczyk, J. A case study of pre-service cracks in the concrete decks of a two-level basement car park. *Arch. Civ. Eng.* **2017**, *63*, 79–97. [CrossRef]
16. Sadowski, T.; Golewski, G. Effect of Aggregate Kind and Graining on Modeling of Plain Concrete under Compression. *Comput. Mater. Sci.* **2008**, *43*, 119–126. [CrossRef]
17. Yuan, M.; Qiang, S.; Xu, Y.; Li, Y.; Xu, W. Research on Cracking Mechanism of Early-Age Restrained Concrete under High-Temperature and Low-Humidity Environment. *Materials* **2021**, *14*, 4084. [CrossRef]
18. Subramanian, N. Controlling the Crack Width of Flexural RC Members. *Indian Concr. J.* **2005**, *79*, 31–36.
19. Giry, C.; Oliver-Leblond, C.; Dufour, F.; Ragueneau, F. Cracking analysis of reinforced concrete structures. *Eur. J. Environ. Civ. Eng.* **2014**, *18*, 724–737. [CrossRef]
20. Goszczyńska, B.; Trąmpczyński, W.; Tworzewska, J. Analysis of Crack Width Development in Reinforced Concrete Beams. *Materials* **2021**, *14*, 3043. [CrossRef]
21. Neville, A.M. *Properties of Concrete*; Pearson Education Limited: London, UK, 2012; p. 846.
22. Kurdowski, W. *Cement and Concrete Chemistry*; Springer: London, UK, 2014; p. 699.
23. Mehta, P.K.; Monteiro, P.J.M. *Concrete: Microstructure, Properties, and Materials*; McGraw-Hill: New York, NY, USA, 2014.
24. Klun, M.; Bosiljkov, V.; Bokan-Bosiljkov, V. The Relation between Concrete, Mortar and Paste Scale Early Age Properties. *Materials* **2021**, *14*, 1569. [CrossRef]
25. Kovler, K.; Zhutovsky, S. Overview and Future Trends of Shrinkage Research. *Mater. Struct.* **2006**, *39*, 827–847. [CrossRef]
26. Flaga, K. The influence of concrete shrinkage on durability of reinforced structural members. *Bull. Pol. Acad. Sci. Tech. Sci.* **2015**, *63*, 15–22. [CrossRef]

27. *PN-EN-1992-1-1:2008*; Eurokod 2: Projektowanie Konstrukcji z Betonu-Część 1: Reguły Ogólne i Reguły dla Budynków. PKN: Warszawa, Poland, 2008. (In Polish)
28. *Structural Concrete–Textbook on Behaviour, Design and Performance*; Updated Knowledge of the CEB/FIP Model Code 1990, fib Bulletin 2; Federation Internationale du Beton: Lausanne, Switzerland, 1999; Volume 2.
29. Zdenêk, P.; Bažant, Z.P.; Baweja, S. *Creep and Shinkage Characterization for Design of Concrete Structures-Model B3*; Structural Engineering Report 94-10/603c; Northwestern University: Evanston, IL, USA, 1994.
30. *ACI 231R-10:2010*; Report on Early—Age Cracking: Causes, Measurement, and Mitigation. Report by ACI Committee 231. American Concrete Institute: Farmington Hills, MI, USA, 2010.
31. Schindler, A.; Byard, B.E. Mitigation of early-age cracking in concrete structures. *MATEC Web Conf.* **2019**, *284*, 07005. [CrossRef]
32. Byard, B.E.; Schindler, A.K.; Barnes, R.W. Early-Age Cracking Tendency and Ultimate Degree of Hydration of Internally Cured Concrete ASCE. *J. Mat. Civ. Eng.* **2012**, *24*, 1025–1033. [CrossRef]
33. Golewski, G.L.; Sadowski, T. The fracture toughness the KIIIc of concretes with F fly ash (FA) additive. *Constr. Build. Mater.* **2017**, *143C*, 444–454. [CrossRef]
34. Ramezani, M.; Dehghani, A.; Sherif, M.M. Carbon nanotube reinforced cementitious composites: A comprehensive review. *Constr. Build. Mater.* **2022**, *315*, 125100. [CrossRef]
35. Raoufi, K.; Pouhr-Gaz, M.; Poursaee, A.; Weiss, J.; Asce, M. Restrained Shrinkage Cracking in Concrete Elements: Role of Substrate Bond on Crack Development. *J. Mater. Civ. Eng.* **2011**, *23*, 895–908. [CrossRef]
36. Ghorbani, R.; Matta, F.; Sutton, M. Full-Field Deformation Measurement and Crack Mapping on Confined Masonry Walls Using Digital Image Correlation. *Exp. Mech.* **2015**, *55*, 227–243. [CrossRef]
37. Ramezani, M.; Kim, Y.H.; Sun, Z.; Sherif, M.M. Influence of carbon nanotubes on properties of cement mortars subjected to alkali-silica reaction. *Cem. Concr. Compos.* **2022**, *131*, 104596. [CrossRef]
38. Schabowicz, K. Testing of Materials and Elements in Civil Engineering. *Materials* **2021**, *14*, 3412. [CrossRef]
39. Hoła, J.; Schabowicz, K. State-of-the-art non-destructive methods for diagnostics testing of building structures—Anticipated development trends. *Arch. Civ. Mech. Eng.* **2010**, *10*, 5–18. [CrossRef]
40. Domaneschi, M.; Niccolini, G.; Lacidogna, G.; Cimellaro, G.P. Nondestructive Monitoring Techniques for Crack Detection and Localization in RC Elements. *Appl. Sci.* **2020**, *10*, 3248. [CrossRef]
41. Iatsko, O.; Anjan, R.B.; Stallings, J.M.; Nowak, A.S. Weigh-in-Motion-Based Fatigue Damage Assessment. *Transp. Res. Rec. J. Transp. Res. Board* **2020**, *2674*, 036119812091975. [CrossRef]
42. Gołaski, L.; Świt, G.; Kalicka, M.; Ono, K. Acoustic Non-Destructive Techniques as a New Method for Evaluation of Damages in Prestressed Concrete Structures: Failure of Concrete Structures. *J. Acoust. Emiss.* **2006**, *24*, 187–195.
43. Świt, G. *Predicting Failure Processes for Bridge—Type Structures Made of Prestressed Concrete Beams Using the Acoustic Emission Method*; Wydawnictwo Politechniki Świętokrzyskiej: Kielce, Poland, 2011; pp. 1–179.
44. Gołaski, L.; Goszczyńska, B.; Świt, G.; Trąmpczyński, W. System for the global monitoring and evaluation of damage processes developing within concrete structures under service loads. *Balt. J. Road Bridge Eng.* **2012**, *7*, 237–245. [CrossRef]
45. Goszczyńska, B.; Świt, G.; Trąmpczyński, W.; Krampikowska, A.; Tworzewska, J.; Tworzewski, P. Experimental validation of concrete crack identification and location with the acoustic emission method. *Arch. Civ. Mech. Eng.* **2012**, *12*, 23–28. [CrossRef]
46. Goszczyńska, B.; Świt, G.; Trąmpczyński, W. Monitoring of Active Destructive Processes as a Diagnostic Tool for the Structure Technical State Evaluation. *Bull. Pol. Acad. Sci. Tech. Sci.* **2013**, *61*, 97–108. [CrossRef]
47. Goszczyńska, B. Analysis of the process of crack initiation and evolution in concrete with acoustic emission testing. *Arch. Civ. Mech. Eng.* **2014**, *14*, 134–143. [CrossRef]
48. Ohtsu, M. Estimation of Crack and Damage Progression in Concrete by Quantitative Acoustic Emission Analysis. *Mater. Eval.* **1999**, *57*, 521–525.
49. Ohno, K.; Ohtsu, M. Crack classification in concrete based on acoustic emission. *Constr. Build. Mater.* **2010**, *24*, 2339–2364. [CrossRef]
50. Ohno, K. Application of acoustic emission for structure diagnosis. *Diagnosis* **2011**, *2*, 3–18.
51. Ohtsu, M.; Isoda, T.; Tomoda, Y. Acoustic emission techniques standardized for concrete structures. *J. Acoust. Emission* **2007**, *25*, 21–32.
52. Bacharz, M.; Goszczyńska, B.; Trąmpczyński, W. Analysis of destructive processes in unloaded early-age concrete with the acoustic emission method. *Procedia Eng.* **2015**, *108*, 245–253. [CrossRef]
53. Bacharz, M.; Trąmpczyński, W. Identification of Active Destructive Processes in Unloaded Early-Age Concrete with the Use of the Acoustic Emission Method. In Proceedings of the 2016 Prognostics and System Health Management Conference (PHM-Chengdu), Chengdu, China, 19–21 October 2016.
54. Bacharz, M. Application of the Acoustic Emission Method to the Evaluation of Destructive Processes in Unloaded Concrete. Ph.D. Dissertation, Kielce University of Technology, Kielce, Poland, 27 June 2016.
55. Trąmpczyński, W.; Goszczyńska, B.; Bacharz, M. Acoustic Emission for Determining Early Age Concrete Damage as an Important Indicator of Concrete Quality/Condition before Loading. *Materials* **2020**, *13*, 3523. [CrossRef] [PubMed]
56. Królikowski, J.; Witczak, Z. Wzorcowe Źródła Emisji Akustycznej. In *Emisja Akustyczna, Źródła, Metody, Zastosowania*; Malecki, I., Ranachowski, J., Eds.; Polska Akademia Nauk, Instytut Podstawowych Problemów Techniki: Warszawa, Poland, 1994; pp. 467–485.

57. Zhang, Z.-H.; Deng, J.-H. A new method for determining the crack classification criterion in acoustic emission parameter analysis. *Int. J. Rock Mech. Min. Sci.* **2020**, *130*, 104323. [CrossRef]
58. Flaga, K. *Concrete Shrinkage and Its Influence on the Capacity, Serviceability of RC and Prestressed Structures*; Monograph No. 73; Cracow University of Technology: Cracow, Poland, 2002; pp. 1–67. (In Polish)
59. Hsu, N.; Breckenridge, F. Characterization and calibration of acoustic emission sensors. *Mater. Eval.* **1981**, *39*, 60–68.
60. RILEM Technical Committee (Masayasu Ohtsu). Recommendation of RILEM TC 212-ACD: Acoustic emission and related NDE techniques for crack detection and damage evaluation in concrete/Measurement method for acoustic emission signals in concrete. *Mater. Struct.* **2010**, *43*, 1177–1181. [CrossRef]

materials

MDPI

Article

Mechanical Properties Study of Miniature Steel Specimens Based on the Small Punch Test and Simulation Methods

Jingwei Zhang *, Zijian Guo and Kanglin Liu

College of Chemical Engineering, Fuzhou University, Fuzhou 350108, China
* Correspondence: zhjw@fzu.edu.cn

Abstract: The small punch test (SPT) can be very convenient to obtain mechanical properties due to its unique advantages from small-volume samples, and has gained wide popularity and appreciation among researchers. In this paper, the SPT test and finite element (FE) simulations were performed for three alloys, and the yield stresses (σ_{YS}) and ultimate tensile strengths (σ_{UTS}) from the uniaxial tensile test (UTT) were correlated with the yield force (F_y) and maximum force (F_m) of the small punch test (SPT) before and after compliance calibration. Finally, the effect of specimen size on the SPT curves was discussed. The results showed that the deviation between SPT test and FE simulation was due to the loading system stiffness, which was confirmed by the loading system compliance calibration test. The SPT curves before and after calibration have less influence on the empirical correlation results for σ_{UTS}, while the correlation results for σ_{YS} depend on the method used to determine F_y in the SPT curve. Finally, the simulation results indicated that the effect of specimen size on the force–displacement curve in the SPT is slight. This work also provides a reference for subsequent researchers to conduct empirical correlation studies using different specimen sizes.

Keywords: small punch test; finite element simulation; material properties; compliance calibration; specimen size

Citation: Zhang, J.; Guo, Z.; Liu, K. Mechanical Properties Study of Miniature Steel Specimens Based on the Small Punch Test and Simulation Methods. *Materials* **2022**, *15*, 6542. https://doi.org/10.3390/ma15196542

Academic Editors: Stanisław Radkowski and Szymon Gontarz

Received: 18 August 2022
Accepted: 19 September 2022
Published: 21 September 2022

1. Introduction

In the field of nuclear and petrochemical industries, a large number of facilities are affected by high-temperature environments, neutron irradiation, corrosion, and other harsh environments. After a period of service, the material properties of the equipment will deteriorate. How to obtain real-time mechanical properties of materials without damaging the structural integrity of in-service components has become the focus of many researchers.

As one of the small sample testing techniques (SSTT), the small punch test (SPT) has attracted extensive attention in the field of nuclear and petrochemical industry due to its unique advantages in obtaining material mechanical properties from small volume samples [1–10]. The SPT method, which originated in the early 1980s, was originally developed to test the changes in material properties caused by tempering embrittlement or irradiation embrittlement of in-service nuclear materials [1]. Now it has gradually developed to evaluate the mechanical property parameters of materials such as tensile properties [2–5], brittle–ductile transition temperature [6], fracture toughness [7], fatigue property [8–10], and creep properties [11–13]. In addition, the SPT method is also used to investigate creep crack propagation by some researchers [14–16]. The SPT test method is also used in the field of biomechanics [17]. At the same time, because of its advantages in the characteristics of "micro-damage" and "tests", SPT is particularly useful for some parts that cannot be tested by standardized tests (such as welded joints [18], thermal barrier coatings [19] and functionally gradient materials [20]).

However, despite the many advantages of SPT mentioned above, it faces many problems and challenges nowadays. One of these important issues is how to ensure the accuracy, reliability and comparability of SPT data from different laboratories. Although

SPT is widely developed in the last decades, the standardization of this method is still in progress [21–25]. An important factor affecting the results that need to be considered is the accuracy of the correlation results between SPT and the uniaxial tensile test (UTT), which is highly dependent on the sensitivity of the test equipment [26–28]. This result makes it difficult to transfer the procedures related to the evaluation of mechanical parameters (σ_{YS} and σ_{UTS}) by different researchers [29,30].

To solve this problem, a number of researchers have extensively discussed the different experimental parameters that affect the reliability of SPT results. Lucas et al. [31] first studied the effects of specimen thickness, hole diameter of the lower die, and punch size on force–displacement curves with different materials by experiments. Campitelli et al. [32] studied the effects of specimen thickness and friction coefficient on the force–displacement curves of AISI 316L- and F82H-MOD-tempered martensitic steels by numerical simulation. Xu et al. [33] used the FE method to discuss the influence of the punch diameter and hardness, hole diameter of lower die and chamfer radius, distance of punch from the center and specimen thickness on SPT results. Zhou et al. [34] studied the effects of friction coefficient, specimen thickness, stamping rate and a lower die aperture on force–displacement curves of SUS304 stainless steel by the GTN model; Andrés et al. [35] studied the influence of different clamping conditions on the upper die on the SPT and the small punch creep test (SPC). Peng et al. [36] systematically studied the influence of small deviations of various test parameters on the results of SPT, from the aspects of specimen geometric deviation, material mechanical properties, damage parameters, pre-tightening condition of the upper die and the friction coefficient between the punch or die and the specimen. And finally, the sensitivity of each test parameter in the five stages of the force–displacement curve was summarized.

However, in addition to the factors mentioned above, there is an important factor whose influence is often ignored, and this is the way in which the value of the displacement used for representing the force-displacement response is determined [37,38]. Different researchers have used different SPT displacement measurement systems, resulting in differences in loading system compliance, which in turn may cause their measured displacement values to shift in the direction of displacement between the SPT test and the FE simulation [39]. This deviation may be more obvious for some non-contact displacement sensors or COD-type extensometers that measure the displacement through the punch in the experimental equipment [40]. Moreno et al. [39] performed a thorough analysis in order to clarify this matter and the results showed that the different force–displacement curves in SPT can be obtained by the above different measurement methods. Hähner et al. [40] also pointed out that the initial stage curves of force–displacement in SPT are affected by using the non-contact displacement transducers. The results of Ávila et al. [27] showed that this factor leads to a deviation between SPT tests and FE simulations, and attributed this uncertainty to the large effect of elastic displacement associated with the low stiffness of the device setup. Although the LVDT sensor is recommended to obtain relatively accurate displacement results through direct contact with the bottom of the specimen in the current standard [21], this displacement measurement system is not used by all researchers. Therefore, it is a matter of discussion whether the direct use of the SPT curve before compliance calibration to correlate it with the material properties produces significantly different results from the empirical correlation of the SPT curve after calibration. In addition, the diameter of a standard small specimen is usually 8 mm or 10 mm, and the specimen's shape is either round or square, which depends on different countries and regions. As in European Union, the round specimen of Φ 8 mm × 0.5 mm is adopted. In China, round specimens with Φ 10 mm × 0.5 mm are usually used. In contrast, a 10 mm × 10 mm × 0.5 mm or 8 mm × 8 mm × 0.5 mm square specimen can be conveniently sampled from the sharp impact specimen compared with a round specimen [33]. Therefore, it is necessary to investigate the effect of different size tests on the SPT results.

In this paper, the mechanical properties of 316L, 347L stainless steels, and a new high-entropy alloy, $Co_{32}Cr_{28}Ni_{32.94}Al_{4.06}Ti_3$, were investigated by SPT and UTT. The effect of loading system compliance on the empirical correlation results between SPT and UTT was systematically investigated. Finally, the effect of specimen sizes on the SPT curves was investigated by the (3D) FE model. The objective of this paper was to evaluate the mechanical properties of steels for pressure vessels by the SPT method and to discuss the results of the empirical correlation between the characteristic parameters (F_y and F_m) on the force–displacement curves before and after the compliance calibration and the material properties (σ_{YS} and σ_{UTS}) of the standardized tests. These findings will surely contribute to the future standardization of SPT and provide a reference for subsequent researchers to conduct empirical correlation studies using different specimen sizes.

2. Materials and Methods

2.1. Materials

The materials used in this paper were 316L, 347L stainless steels and $Co_{32}Cr_{28}Ni_{32.94}Al_{4.06}Ti_3$ high-entropy alloys, which are generally used for pressure vessels. 316L and 347L stainless steel are widely used in petrochemical and pharmaceutical fields, and $Co_{32}Cr_{28}Ni_{32.94}Al_{4.06}Ti_3$ is a new high-entropy alloy with promising applications in nuclear power. More details about $Co_{32}Cr_{28}Ni_{32.94}Al_{4.06}Ti_3$ are introduced in the previous article [41]. The main chemical components were analyzed by EDS (Energy Dispersive Spectroscopy, BRUKER corp., Karlsruhe, Germany). The results are shown in Table 1. The composition (atomic fraction and weight ratio, %) of $Co_{32}Cr_{28}Ni_{32.94}Al_{4.06}Ti_3$ is shown in Table 2. The optical microscope images of the metallographic microstructure of the above materials are shown in Figure 1.

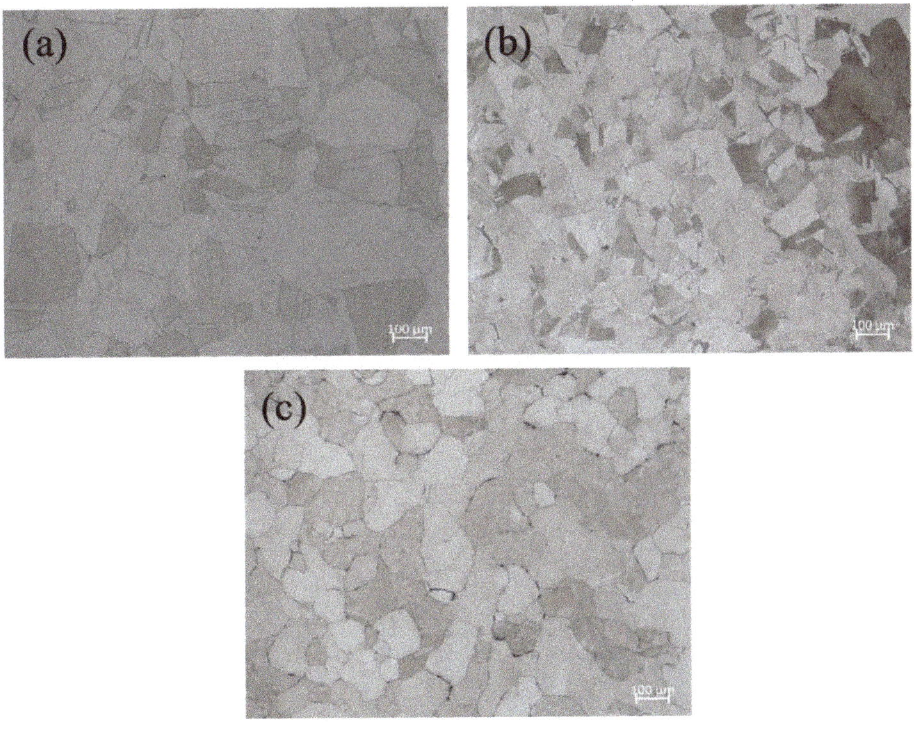

Figure 1. Optical microscope images of the metallographic microstructure of three materials. (**a**) 316L stainless steel; (**b**) 347L stainless steel; (**c**) $Co_{32}Cr_{28}Ni_{32.94}Al_{4.06}Ti_3$.

Table 1. Chemical composition of stainless steel (wt%).

Steel	Chemical Composition (wt%)								
	C	Si	Mn	P	S	Cr	Mo	Ni	Nb
316L	0.015	0.70	0.54	0.02	0.007	16.53	2	11.55	—
347L	0.04	0.75	1.97	0.02	0.01	18.5	—	11.0	0.95

Table 2. Atomic and weight ratios of the principal elements of $Co_{32}Cr_{28}Ni_{32.94}Al_{4.06}Ti_3$.

Composition	Co	Cr	Ni	Al	Ti
Atomic ratio (%)	32	28	32.94	4.06	3
Weight ratio (%)	34.11	26.34	34.97	1.98	2.60

2.2. Uniaxial Tensile Test

In order to obtain the mechanical properties of the materials, uniaxial tensile tests were carried out at room temperature using AG-X plus a universal electronic testing machine (SHIMADZU corp., Kyoto, Japan), as shown in Figure 2. A 52 mm × 14 mm × 2.5 mm plate-shaped tensile test specimen was used (Figure 3), and tensile tests were carried out at a strain rate of 1×10^{-3} mm/s. Finally, the yield stress (σ_{YS}), ultimate tensile strength (σ_{UTS}), and uniform elongation (ε) of three materials are presented in Table 3.

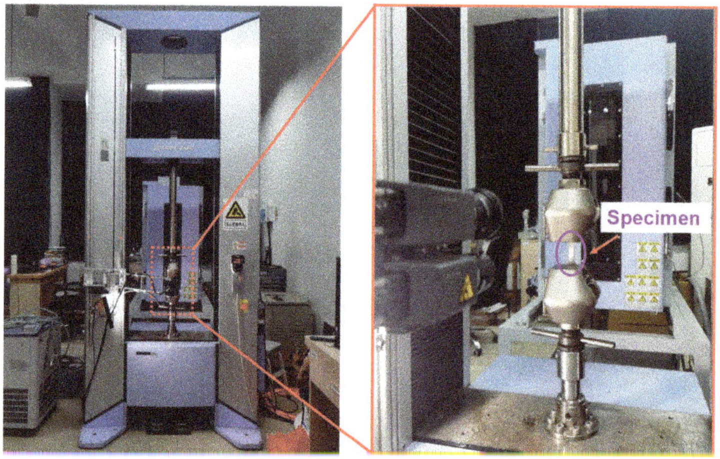

Figure 2. Overview of uniaxial tensile test.

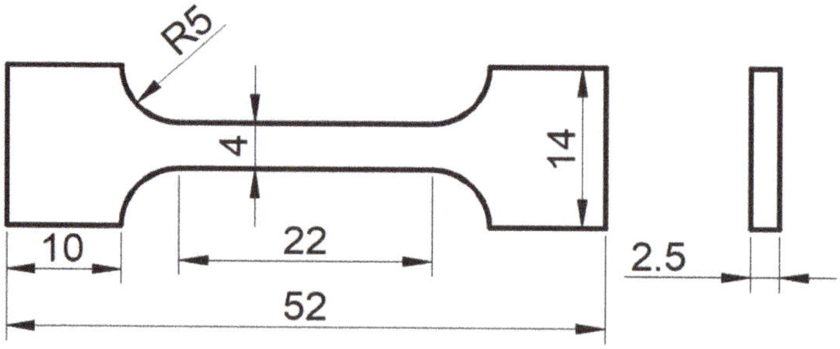

Figure 3. Dimensions of the tensile specimen (mm).

Table 3. Mechanical properties of three materials at room temperature.

Materials	Yield Stress (σ_{YS})/MPa	Ultimate Tensile Strength (σ_{UTS})/MPa	Uniform Elongation (ε)/%
316L	350	635	58.1
347L	320	540	9.5
$Co_{32}Cr_{28}Ni_{32.94}Al_{4.06}Ti_3$	530	985	37.2

2.3. Small Punch Test

SPT specimens with a diameter of 10 mm and a thickness of 0.8 mm were cut from the plate-shaped tensile specimens after the experiment. Then, the specimens were polished to 0.55 mm on both sides by 600, 1200, and 1500 grit size papers, and finally, the specimens were finely polished to a mirror brightness of 0.5 mm by diamond lapping paste to meet the standard requirements for the surface roughness of the specimens. Figure 4 shows the macroscopic morphology of the small punch specimen before and after the experiment.

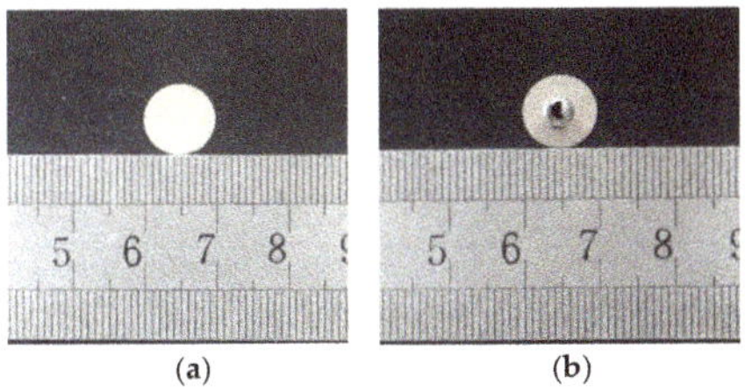

Figure 4. Macro morphology of a small punch specimen. (**a**) before test; (**b**) after test.

In this paper, the small punch test apparatus, modified based on the small punch creep test machine, was used to conduct SPT. Its force sensor and displacement sensor are coupled to the loading brake, and the force and displacement changes are recorded by loading the punch. Once the force is received on the force sensor, the displacement sensor begins to record the punch displacement data. This displacement sensor can be used to obtain relatively accurate displacement data of the punch on the top. The schematic diagram of the apparatus is shown in Figure 5. The apparatus for clamping the specimen consists of an upper die, and a circular lower die with a hole diameter of 4 mm. The specimen is placed horizontally and perpendicular to the direction of the force. During the test, the loading is applied to the specimen by means of a punch and a ball with a diameter of 2.5 mm. The test was conducted in a quasi-static condition with a displacement rate of 0.5 mm/min.

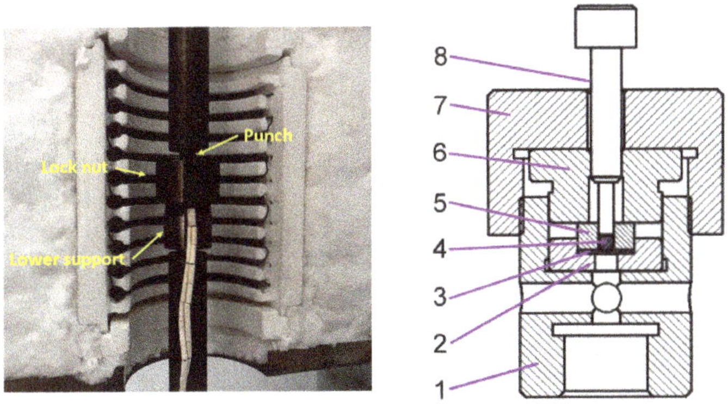

Figure 5. Schematic diagram of small punch test. 1. Lower support 2. Lower die 3. Specimen 4. Ball 5. Guide block 6. Upper die 7. Lock nut 8. Punch.

2.4. FE Model and Numerical Simulation

The FE model used in this paper is shown in Figure 6. Both a two-dimensional (2D) axisymmetric model and quarter-symmetric three-dimensional (3D) model were used to simulate the SPT by ABAQUS large commercial FE software [42]. A 2D model with 2000 axisymmetric 4-node square elements, homogenized mesh, and reduced integration with hourglass control (CAX4R) was mainly used to investigate the effect of equipment compliance on experimental data, while a 3D model with about 64,000 8-node linear brick elements, homogenized mesh, and reduced integration with hourglass control (C3D8R) was used to study the effect of specimen size on experimental data. Most researchers widely use these two models in SPT simulations [5,43]. They are suitable for analyzing cases involving large stress and strain gradients, as well as for studying complex contact problems. The agreement between the 2D and 3D model results was confirmed in the article [29,30] by Altstadt et al. Isotropic elastic–plastic materials are subject to the von Mises yield criterion and the corresponding J_2 flow theory [44,45], which takes the form as follows,

$$2\sigma_{YS}^2 = (\sigma_1 - \sigma_2)^2 + (\sigma_2 - \sigma_3)^2 + (\sigma_3 - \sigma_1)^2 \quad (1)$$

where σ_1, σ_2, and σ_3 are the three principal stresses respectively. The material properties' inputs in ABAQUS were imported from real stress–strain data evaluated by uniaxial tensile tests of related materials, as shown in Table 3. The constitutive equation can be expressed in the following form,

$$\sigma = \sigma_{YS} + \sigma_P(\varepsilon_p) \quad (2)$$

where the $\sigma_p(\varepsilon_p)$ equation is derived from the uniaxial tensile behavior [32]. Generally, the FE model constructed consists of four parts, namely the (1) punch, (2) upper die, (3) lower die, and (4) specimen. Unlike the test, the FE model perfectly simulates the test setup under ideal conditions. The upper and lower dies and punch are defined as rigid bodies, and the specimen is defined as a deformable body. The upper and lower dies were completely fixed to simulate the clamping condition during the experiment, while the displacement constraint was applied to the punch until the specimen fails. The friction coefficient between the punch and specimen was set to 0.2, which is a typical friction coefficient for steel-steel contact in the absence of lubrication [46]. The equipment dimensions for the SPT test and FE simulation were kept consistent. The relevant parameters of specimens and apparatus for different models are summarized in Table 4. The mesh refinement was performed to ensure the accuracy of simulation results. The mesh sizes of square and round specimens in the 3D model were consistent to avoid the impact of mesh sensitivity on simulation results [47,48].

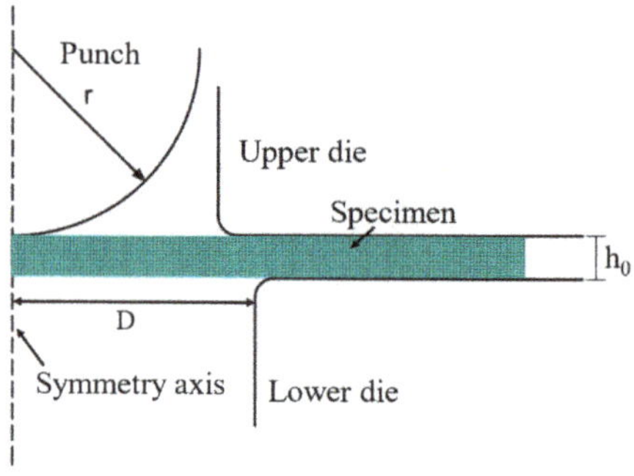

(a)

Figure 6. *Cont.*

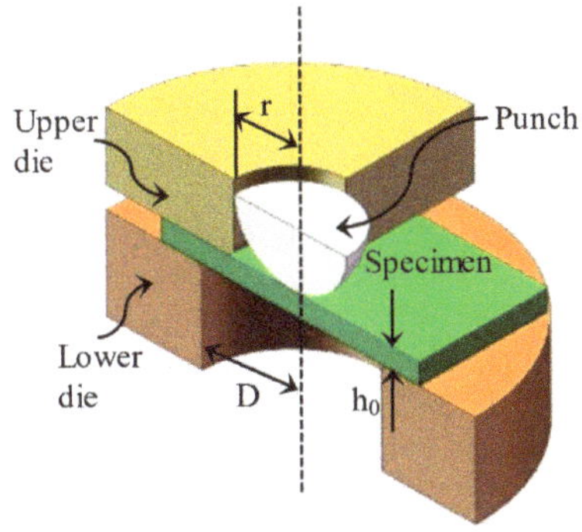

(b)

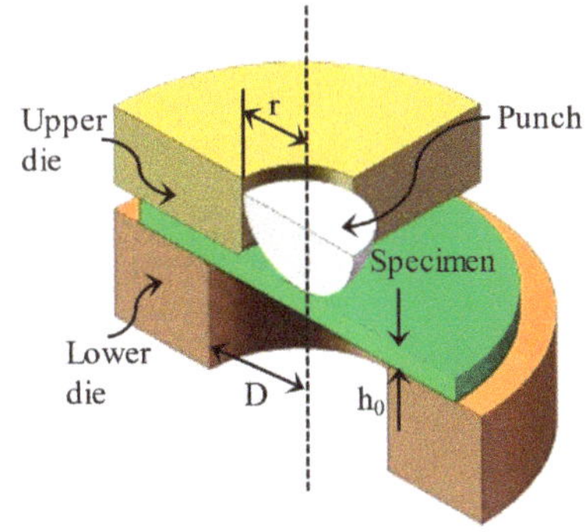

(c)

Figure 6. Schematic diagram of FE model of SPT. (**a**) 2D model; (**b**) 3D model with square specimen; (**c**) 3D model with round specimen.

Table 4. Geometric parameters of SPT for 2D and 3D models.

Set No.	Specimen Type	Specimen Diameter d (mm)	Punch Radius r (mm)	Receiving Hole Radius D (mm)	Specimen Thickness h_0 (mm)	Edge Type
2D	round	10 mm	1.25 mm	4 mm	0.5 mm	Full clamped
3D	round	10 mm/8 mm	1.25 mm	4 mm	0.5 mm	Full clamped
3D	square	10 mm/8 mm	1.25 mm	4 mm	0.5 mm	Full clamped

3. Results

3.1. SPT Experiment Results

The typical force–displacement responses of the three materials are shown in Figure 7a. Three test groups were analyzed for each material, and the results show a high degree of coincidence of the force–displacement curves. The force–displacement data of the SPT experiment curve was recorded until the force exceeded the maximum force point and after the specimen ruptured.

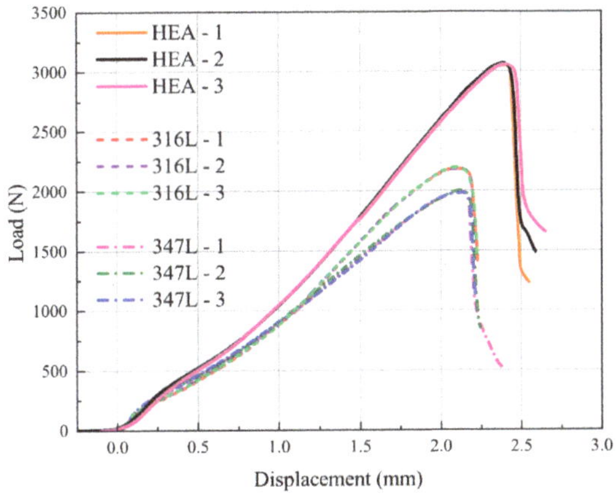

(a)

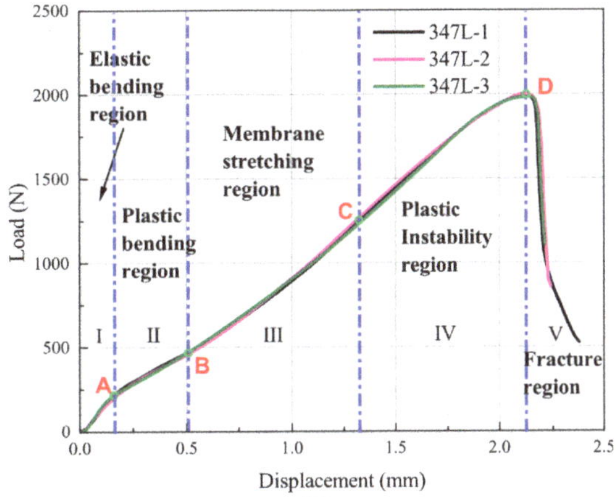

(b)

Figure 7. Force–displacement curve of SPT. (**a**) The three materials; (**b**) representation of the different stages.

As shown in Figure 7b, The force–displacement curve can be divided into five different regions: elastic bending (region I), plastic bending (region II), membrane-stretching (region III), plastic instability (region IV), and unstable fracture (region V). However, the boundaries between these five regions are not clearly defined, and the delineation of the different regions is highly artificial. The force in the initial elastic stage (region I) and plastic stage (region II) increases almost linearly, and an obvious inflection point can be seen between the two regions. This inflection point is called the elastic–plastic transition point. The ordinate value corresponding to this point is defined as the elastic–plastic transition force or yield force (F_P) (corresponding to point A in Figure 7b) by most researchers, which is usually used to correlate with the yield strength (R_P) and regarded as the dividing point between region I and region II. The plastic bending region (region II) ends when the slope of the force–displacement curve increases significantly. In region III, it can be seen that the slope of the force–displacement curve gradually increases. The reason is that after the plastic bending stage, the effect of strain hardening in the material overcomes the reduction of the specimen thickness caused by the punch and enables it to bear the force at an increased rate. This situation is similar to the strain-strengthening stage of the uniaxial tensile test, and the difference is that the stress condition is a biaxial stress state. Therefore, the dividing point between region II and region III (corresponding to point B in Figure 7b) is regarded as the specimen entering the membrane-stretching stage from bending deformation.

The dividing point (corresponding to point C in Figure 7b) between membrane stretching (region III) and plastic instability (region IV) can be regarded as a balance point between the material force bearing and deformation resisting. When the value of force exceeds the value of Point C, due to the decrease in the local thickness of the specimen, the microvoids in the material continuously nucleate, grow, and coalesce, thus inducing microcracks on the specimen surface. On the macro level, the specimen will become softened, and the slope of the force–displacement curve gradually decreases until the maximum force (corresponding to point D in Figure 7b) is reached and the stage of plastic instability (region IV) ends.

Figure 8 shows the SEM images of the three materials observed after the specimens fractured (region V). It can be observed that a large number of dimples formed in the material, and a large number of microvoids coalesced. Then, the force of the specimens begins to drop sharply, which is called an unstable fracture region (region V).

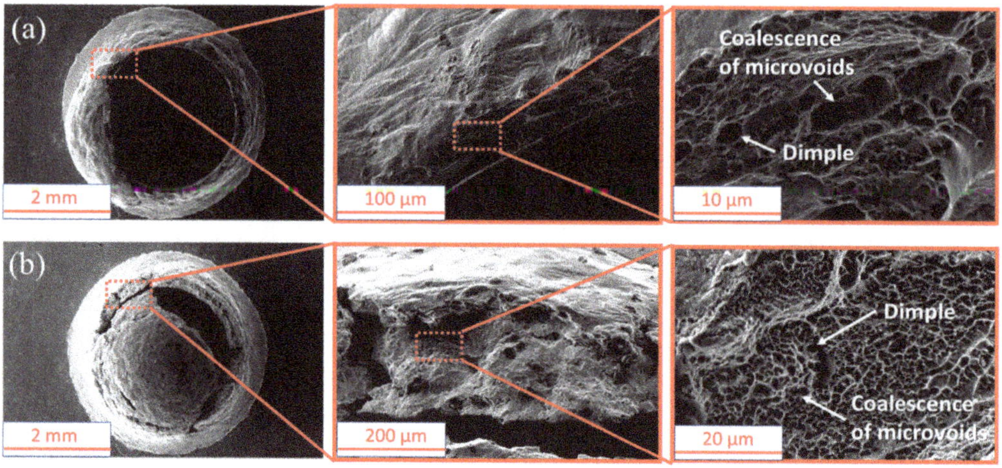

Figure 8. *Cont.*

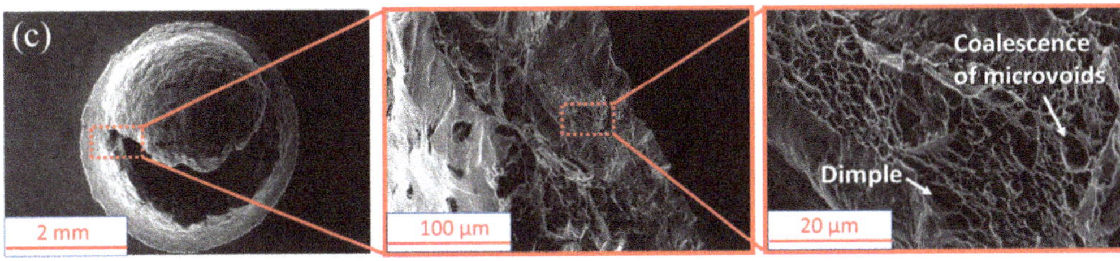

Figure 8. SEM image for three materials. (**a**) 316L stainless steel; (**b**) 347L stainless steel; (**c**) $Co_{32}Cr_{28}Ni_{32.94}Al_{4.06}Ti_3$.

3.2. SPT Simulation Results

Figure 9a shows the experiment and simulation results of the three materials used in this study. When compared with the experiment results, the dividing points in the force–displacement curve of simulation are more pronounced. The general trends of the experiment and simulation responses are approximately similar in the initial four regions but different in the slopes. This deviation is more pronounced at the initial stage of the force–displacement curve, as shown in Figure 9b. The reason can be explained as that, contrary to the experiment, the specimen dies and punch are defined as rigid bodies in the simulation process, but in fact are elastic deformers. In the practical data recording, different displacement measurement methods lead to the displacement sensor at the top of the punch recording additional displacement due to the elastic deformation of the test frame [39] during the loading process, so that the displacement data recorded by the displacement sensor at the top of the punch (corresponding to SPT test) are slightly higher than the displacement data on the upper face of the specimen (corresponding to SPT simulation). The above situation leads to deviations of the force–displacement curves between the test and simulation. This deviation is also reported in literature [28,32].

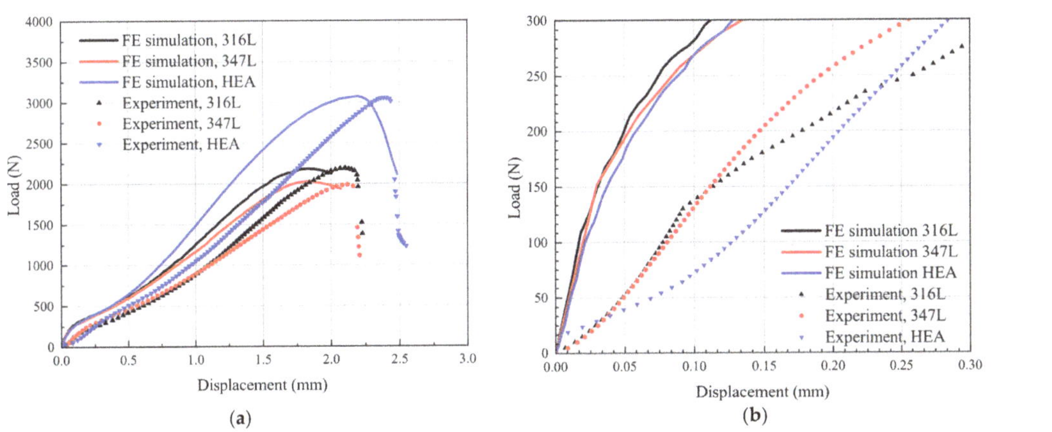

Figure 9. Comparison of force–displacement curve between the experiment and FE simulation. (**a**) at all stages; (**b**) at the initial stage.

3.3. Calibration of Loading System Compliance

In the previous section, we analyzed and explained the reasons for the deviations between the experimental and simulation results. However, the reasonableness of the analytical results for the curves in Figure 9 needs to be further investigated by applying the compliance calibration of the loading system to the experimentally obtained force–displacement curves. It is worth noting that regarding the compliance calibration of the loading system, different compliance calibration methods are used by researchers [27,28,32,39,40,46,49]. The displacement (δ_{upper}) on the upper face of the specimen can be derived from the displacements (δ_{ext}) through an appropriate correction of the respective elastic compliances involved, referred to as C_{ext} [28,32,39,40]. Some researchers [27,43,46] believe that a linear calibration of the force–displacement (δ_{ext}) response can be achieved with the compliance calibration curve obtained by this method. However, other researchers have argued that the effective contact area with the punch, the applied stress, and the strain field obtained during the SPT are neither constant nor uniform and continuously vary during the test. At the same time, as the punch displacement increases, the deformation of the specimen and the effective contact area of the material increases, and the specimen stiffness gradually increases [27]. Thus, the nonlinear calibration for the force–displacement (δ_{ext}) response should be considered in the actual compliance calibration to obtain the best match between the test and simulation curves. Meanwhile, Ávila et al. [27] also pointed out that if only the initial stiffness is used to correct the original curve, a large overcorrection may be obtained for high displacement values.

Therefore, a nonlinear compliance calibration curve was used in this paper. The compliance calibration of the loading system was performed using a cylindrical tungsten carbide specimen with a diameter of 10 mm and a thickness of 5 mm instead of the small circular piece in the test. The calibration method is similar to the reference [43]. In the first loading step, the maximum force was obtained, The subsequent force should not be exceeded in the SPT, and some unloading–loading cycles were performed until a steady state of the force–displacement (δ_{ext}) curve was reached. The last loading step of this calibration test was recorded, and a fifth-order polynomial regression was established from this data as a calibration equation. This curve was used to calibrate the δ_{ext} obtained from the SPT test, which resulted in a new displacement (δ_{upper}) equal to the displacement of the upper surface of the specimen. It should be noted that the high sensitivity of the calibration process on the test apparatus makes the loading–unloading process unstable. Therefore, in addition to the compliance calibration curves obtained in this study, the correction results for AISI 304L by Moreno et al. [39] and the three calibrations for JRQ by Lucon et al. [49] are also shown in Figure 10.

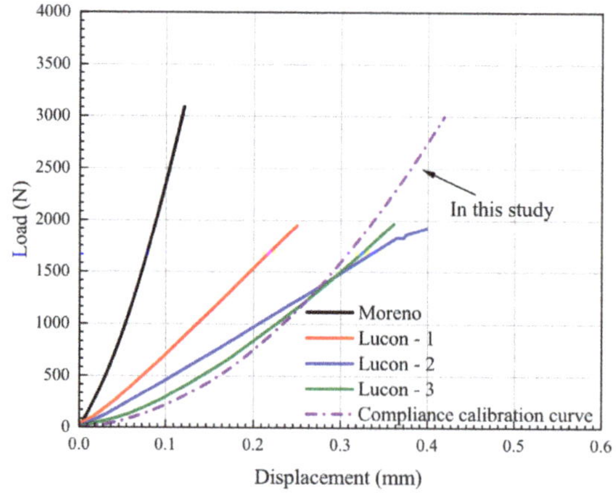

Figure 10. Compliance calibration curve by different researchers.

As can be seen from the results in Figure 10, the final calibration curves are different due to the displacement measurement systems, test apparatus, and punch materials used by the researchers. It can also be found that the results in the reference [47] show that there are still slight deviations in the three compliance calibration results for the same material with the same test apparatus. Therefore, the compliance calibrations in Appendix A of EN 10,371 [21] are given as a recommended range.

The compliance curve in Figure 10 can be expressed in the following form:

$$\delta_N{}^{WC} = P_1(F) \quad (3)$$

where $\delta_N{}^{WC}$ is the elastic deformation of the test apparatus and punch, F is the corresponding force. The force–displacement curve in Figure 7 can also be expressed in the following form:

$$\delta_{exp} = P_2(F) \quad (4)$$

where δ_{exp} includes the elastic bending deformation due to specimen deflection plastic indentation of the punch and elastic deformation of the test frame. Thus, the actual displacement of the specimen can be expressed as:

$$\delta_{upper} = \delta_{exp} - \delta_N{}^{WC} = P_2(F) - P_1(F) \quad (5)$$

where δ_{upper} is the displacement of the upper face of the specimen; the corrected results for the above three materials by Equation (5) are shown in Figure 11. As can be seen, the corrected curves obtained using the above calibration method generally agree well with the simulated curves, with only slight deviations at the initial stages of the curves. Therefore, the deviation between the experiment and simulation results can be attributed to the influence of loading system compliance.

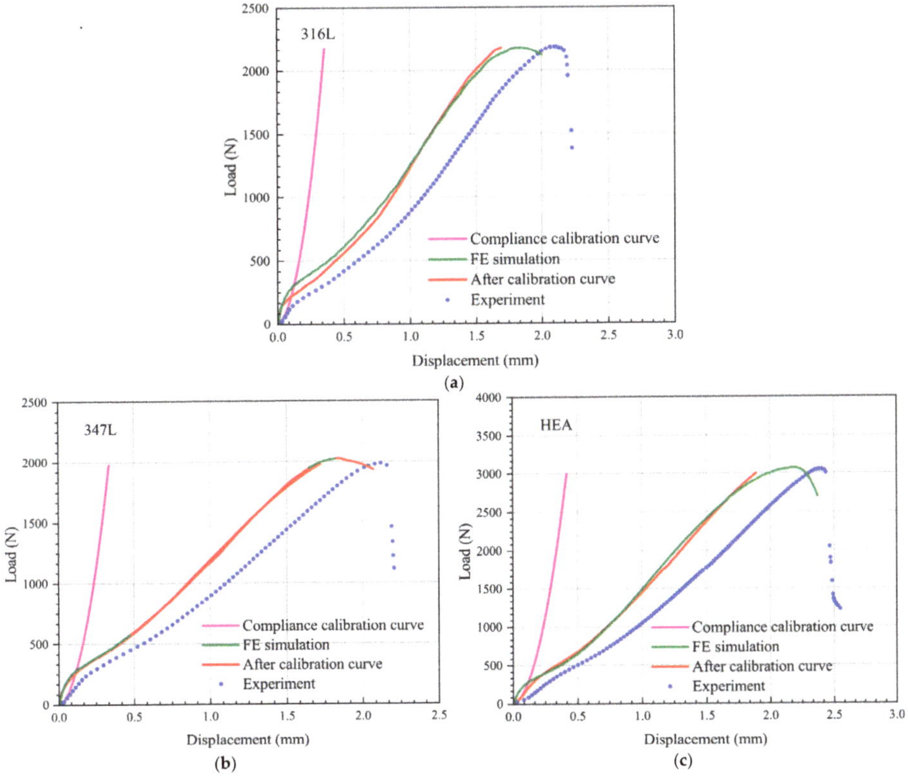

Figure 11. Force–displacement curve of SPT before and after compliance correction. (**a**) 316L; (**b**) 347L; (**c**) HEA.

4. Discussion

4.1. Correlation between SPT Curve and Tensile Properties before and after Compliance Calibration

The above results confirm that the deviation between experimental and FE results in the displacement is caused by the compliance of the testing machine configuration and the displacement measurement method. It is important to note that due to the high sensitivity of the SPT, the disturbance generated in the process of correcting the compliance of the loading system may lead to the failure of the calibration process, and extra care should be taken in this process. In fact, due to the different displacement measurement methods and materials in the SPT equipment, the results of the compliance calibration are very different. One of the best methods to avoid this problem is to use LVDT contact displacement measurements directly. This method can directly obtain accurate displacement data at the bottom of the specimen and prevent the possible failure of compliance calibration. However, due to the conditions in different laboratories, direct contact displacement measurements sometimes cannot be used. Thus, the effect of testing machine configuration compliance on the empirical correlation results is worth discussing.

4.1.1. Correlation with Ultimate Tensile Strength (σ_{UTS})

Researchers widely use the maximum force (F_m) as the characteristic force associated with the ultimate tensile strength (σ_{UTS}). The primary empirical correlation forms are as follows [50]:

$$\sigma_{UTS} = \alpha_1 \cdot \frac{F_m}{h_0^2} + \alpha_2 \tag{6}$$

$$\sigma_{UTS} = \alpha_1' \cdot \frac{F_m}{(h_0 \cdot u_m)} + \alpha_2' \tag{7}$$

where h_0 is the specimen's initial thickness, F_m is the maximum force used to determine the force characteristic value of σ_{UTS}, and u_m is the value of displacement corresponding to the force characteristic F_m. α_1, α_2, α_1', and α_2' are the correlation coefficients related to the material. The results before and after compliance calibration for the three materials are presented in Table 5.

Table 5. Characteristic parameters (maximum force) of the SPT curve before and after compliance calibration.

Compliance	Steel	u_m (mm)	F_m/h_0^2 (MPa)	$F_m/(h_0 \cdot u_m)$ (Mpa)
Before	316L	1.70	8720	2565
	347L	2.14	8000	1869
	$Co_{32}Cr_{28}Ni_{32.94}Al_{4.06}Ti_3$	2.40	12,212	2544
After	316L	2.10	8748	2573
	347L	1.83	8088	2573
	$Co_{32}Cr_{28}Ni_{32.94}Al_{4.06}Ti_3$	2.19	12,268	2800

Equations (6) and (7), respectively, were used to empirically correlate the maximum force (F_m) and the ultimate tensile strength (σ_{UTS}) of the three materials before and after compliance calibration. Equations (8) and (9) are the empirical correlation equations before compliance calibration and Equations (10) and (11) are the after-compliance calibration equations. The relation between the SPT characteristic force (F_m) and UTT strength property (σ_{UTS}) can then be expressed as follows:

Before compliance calibration:

$$\sigma_{UTS} = 0.10 \cdot \frac{F_m}{h_0^2} - 282, R^2 = 0.99 \tag{8}$$

$$\sigma_{UTS} = 0.67 \cdot \frac{F_m}{(h_0 \cdot u_m)} - 727, R^2 = 0.99 \tag{9}$$

After compliance calibration:

$$\sigma_{UTS} = 0.10 \cdot \frac{F_m}{h_0^2} - 727, R^2 = 0.99 \quad (10)$$

$$\sigma_{UTS} = 0.72 \cdot \frac{F_m}{(h_0 \cdot u_m)} - 1090, R^2 = 0.99 \quad (11)$$

Comparing Equation (8) with Equation (10), it can be seen that the correlation coefficients in the empirical correlation equations obtained from Equation (6) are less affected before and after the compliance calibration and differ only in the correlation coefficient α_2. However, it can be seen from the results of the compared Equation (9) with Equation (11) that the correlation coefficients are slightly different before and after compliance calibration. The reason for the above situation is that empirical correlation equation, Equation (6), for each material is only concerned with F_m, which hardly changes before and after the calibration, and thus the difference is not significant. For the same reason, the empirical correlation equation, Equation (7), for each material is not only relevant to F_m but to u_m, and u_m changed significantly before and after the calibration, so the data obtained after the compliance calibration of Equation (7) are different from before. Suppose the uncorrected curve is used to evaluate the material ultimate tensile strength (σ_{UTS}); an overestimation of results may be obtained (i.e., the larger value of σ_{UTS} will be obtained by the uncorrected curve compared with the corrected curve with the same $F_m/(h_0 \cdot u_m)$), and therefore unsafe results will be obtained.

4.1.2. Correlation with Yield Stress (σ_{YS})

The empirical correlation equation of yield stress is usually expressed in the following form,

$$\sigma_{YS} = \beta_1 \cdot \frac{F_y}{h_0^2} + \beta_2 \quad (12)$$

where β_1 and β_2 are the empirical correlation constants, h_0 is the initial thickness of SPT specimen, and F_y is the elastic–plastic transition force in the SPT curve. The key to determining yield stress (σ_{YS}) lies in how to determine the corresponding characteristic force, i.e., elastic–plastic transition force (F_y). Different methods for determining F_y are proposed, as described in references [28,50]. In all the current assessments, F_{y_Mao} [51], F_{y_CEN} [22], $F_{y\_t/10}$ [52], and $F_{y\_t/100}$ [6] methods are adopted by most researchers, as shown in Figure 12.

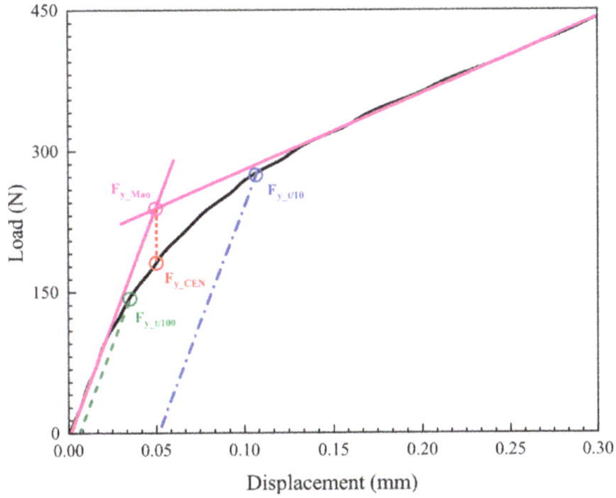

Figure 12. Different methods of determining the force F_y on SPT curve.

The *Mao* method [51] minimizes the error between the linear function and the SPT initial stage curve by establishing two linear procedures, and the resulting intersection point is $F_{y\_Mao,}$ and the intersection point projected vertically onto the SPT curve is F_{y_CEN}. The $t/10$ and $t/100$ methods were performed by drawing a parallel line tangent to the elastic region I of the SPT curve. The line was translated along the deflection axis to the points whose values are $t/10$ and $t/100$, respectively. The intersection value between the parallel line tangent and SPT curve was determined as the characteristic force F_y. The maximum slope of the tangent line corresponding to the slope at the inflection point is suggested in Hähner et al. [40] as the slope of the elastic region I of the SPT curve before compliance calibration, and this method was used in this paper to determine the yield stress (σ_{YS}) before compliance calibration.

The characteristic forces (F_y) associated with the yield stress (σ_{YS}) before and after the compliance calibration are listed in Table 6. The relation between the SPT characteristic force (F_y) and UTT material property (σ_{YS}) can then be expressed as follows:

Table 6. Characteristic force of the SPT curve determined by different methods before and after compliance calibration (yield force).

Compliance	Steel	F_{y_Mao}/h_0^2 (MPa)	F_{y_CEN}/h_0^2 (MPa)	$F_{y\_t/10}/h_0^2$ (MPa)	$F_{y\_t/100}/h_0^2$ (MPa)
Before	316L	768	684	868	548
	347L	956	928	1173	845
	$Co_{32}Cr_{28}Ni_{32.94}Al_{4.06}Ti_3$	1410	1372	1710	1321
After	316L	760	668	788	560
	347L	895	797	1141	756
	$Co_{32}Cr_{28}Ni_{32.94}Al_{4.06}Ti_3$	1323	1253	1652	1202

Before compliance calibration:

$$\sigma_{YS}= 0.36 \cdot \frac{F_{y\_Mao}}{h_0^2} + 41.58, R^2 = 0.87 \tag{13}$$

$$\sigma_{YS}= 0.35 \cdot \frac{F_{y\_CEN}}{h_0^2} + 84.80, R^2 = 0.89 \tag{14}$$

$$\sigma_{YS}= 0.22 \cdot \frac{F_{y\_t/10}}{h_0^2} + 134.07, R^2 = 0.73 \tag{15}$$

$$\sigma_{YS}= 0.31 \cdot \frac{F_{y\_t/100}}{h_0^2} + 137.23, R^2 = 0.82 \tag{16}$$

After compliance calibration

$$\sigma_{YS}= 0.34 \cdot \frac{F_{y\_Mao}}{h_0^2} + 56.83, R^2 = 0.85 \tag{17}$$

$$\sigma_{YS}= 0.32 \cdot \frac{F_{y\_CEN}}{h_0^2} + 97.23, R^2 = 0.86 \tag{18}$$

$$\sigma_{YS}= 0.25 \cdot \frac{F_{y\_t/10}}{h_0^2} + 171.74, R^2 = 0.76 \tag{19}$$

$$\sigma_{UTS}= 0.10 \cdot \frac{F_{y\_t/100}}{h_0^2} - 282, R^2 = 0.85 \tag{20}$$

where Equations (13)–(16) as well as Equations (17)–(20) correspond to the four methods for determining F_y, respectively. A comparison of the empirical correlation coefficients in the equations above intuitively shows that the differences before and after the compliance correction are obvious for different methods of determining the yield force (F_y). The empirical correlation equations before and after compliance calibration determined using F_{y_Mao}, F_{y_CEN}, and $F_{y\_t/10}$ methods showed small deviations, and the empirical correlation coefficients β_1 and β_2 do not differ significantly. In contrast, the empirical correlation equations determined by the $F_{y\_t/100}$ method showed larger deviations in the empirical correlation coefficients β_1 and β_2 before and after compliance calibration, which indicates that the F_{y_Mao}, F_{y_CEN}, and $F_{y\_t/10}$ methods are less affected by the compliance calibration and are more applicable to the characteristic forces of the above-mentioned study materials.

It can also be seen that the regression coefficient R^2 of the empirical correlation equation for yield strength (σ_{YS}) is relatively small compared to the empirical correlation equation for ultimate tensile strength (σ_{UTS}), indicating that its empirical correlation equation has a large scattering, which may be caused by the small number of data points. Table 6 illustrates the values of the empirical correlation constant β_1 from the literature of different researchers. It should be noted that for the empirical correlation equation of yield strength, the coefficient β_2 may not necessarily exist, depending on the differences in the mechanical properties of the material and the influence of the test apparatus. Table 7 combined with Equations (17) and (18) shows that the values of β_1 determined by the *Mao* and *CEN* methods are close to the values in the literature [5,32,41–53]. Therefore, by combining the two perspectives above, F_{y_Mao} and F_{y_CEN} may be the more desirable characteristic forces for the three materials of interest in this study.

Table 7. Recommended values of parameters β_1 and β_2 in different literature.

β_1	β_2	Materials	Method	Reference
0.36	-	SUS316, PCA, HT-60, A533B, HT-9	F_{y_Mao}	Mao and Takahashi [51]
0.39	-	316L and F82H	F_{y_CEN}	Campitelli et al. [32]
0.38	-	HAZ of 30CrMo5-2	F_{y_CEN}	Rodriguez et al. [52]
0.364	-	Different steels and Al alloy	$F_{y\_t/10}$	Garcia et al. [50]
0.442	-	Different steels and Al alloy	F_{y_CEN}	Garcia et al. [50]
0.476	-	Different steels and Al alloy	F_{y_Mao}	Garcia et al. [50]
0.349	133.48	Ti-6Al-4V, Stainless Steel, Cu, Al, In718	F_{y_Mao}	Lancaster et al. [53]

Figure 13 compares the different methods of determining the yield force (F_y) for the three materials studied in this paper. It can be seen that the difference in the methods of determining the yield force (F_y) before and after compliance correction is slight. This result indicates that the determination of the yield stress (σ_{YS}) is not affected by the loading system stiffness. It can also be seen that if the empirical correlation coefficient in Equation (8) is constant, the $t/10$ method obtains the highest yield strength of the material, while the other three methods obtain the yield strength of the material relatively close. However, this result was demonstrated only for the three materials studied in this paper, and its applicability to other materials remains further confirmed.

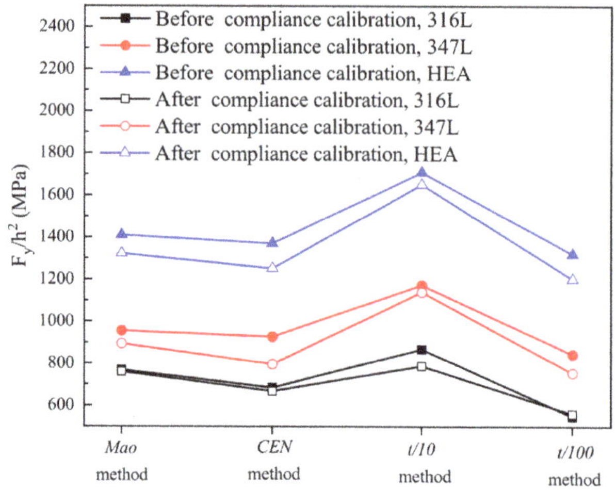

Figure 13. Different methods for determining the F_y.

4.2. Effect of Different Specimen Sizes on SPT Curve

In order to avoid the influence of some neglected factors on the SPT results during the experiments, the subsequent studies of different specimen sizes were based on FE analysis. The different specimen sizes generally used by most researchers are presented in Table 8. The data of 316L stainless steel mentioned above were used for FE simulations. The mesh sizes of square specimens and round specimens in the 3D model were identical to avoid the impact of mesh sensitivity on simulation results.

Table 8. FE model parameters of 316L stainless steel with different specimen sizes.

Material	Specimen Diameter (mm)		Specimen Shape	
316L	8	10	Round specimen	Square specimen

Figure 14 shows the results of the SPT curves for all cases in Table 7. It can be seen that the FE simulation results for round and square specimens with the same diameter exhibit almost identical SPT curves. This result shows that the SPT curves were not influenced by the specimen's shape when the specimen periphery was in a fully clamped condition. Moreover, for specimens with diameters of 8 mm and 10 mm, respectively, there is a high degree of coincidence in the first four regions of the SPT curves. Still, the two curves begin to diverge near the maximum force, and the general downward trend in force remains consistent beyond the maximum force. In general, the effect of specimens' diameter on the force–displacement curve is not apparent, and a slight difference induced by the specimen diameter can be found beyond the maximum force point.

Figure 15 shows the equivalent stress distribution of four different types of specimens of 316L stainless steel when the punch displacement was 1.7 mm. It can be seen from the figure that although the overall stress distribution of square specimens and round specimens is different, the stress distribution of the contact part with the punch at the center of the specimen is roughly the same. The difference between round or square specimens with different diameters is only evident in the non-contact area with the punch.

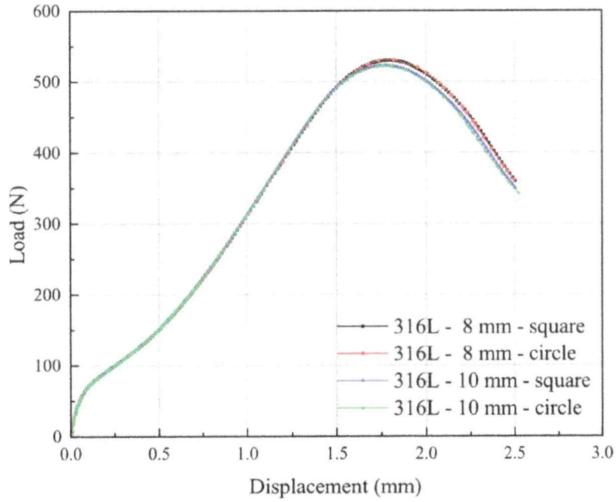

Figure 14. Force–displacement curve of SPT test for specimens in Table 7.

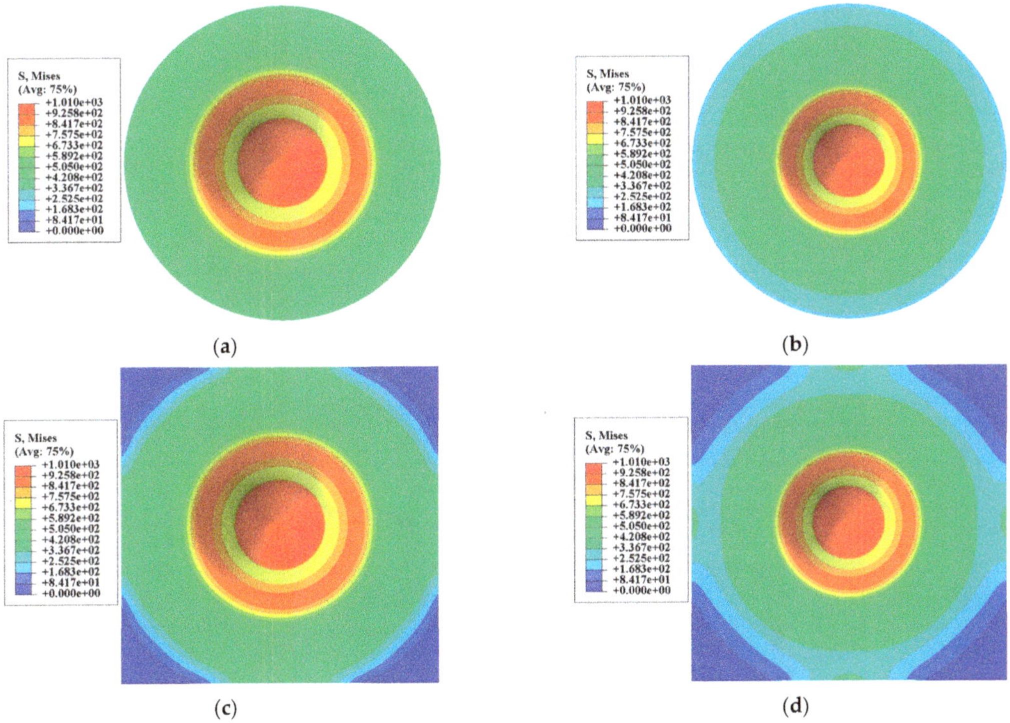

Figure 15. Contour plots for Von Mises stress on FE model of SPT. (**a**) round specimen with a diameter of 8 mm; (**b**) round specimen with a diameter of 10 mm; (**c**) square specimen with a diameter of 8 mm; (**d**) square specimen with a diameter of 10 mm.

Figure 16 shows the equivalent plastic strain distribution on the bottom surface of the specimen at the maximum force in the path from the center of the specimen to the edge. Combined with the contour plots of the equivalent plastic strain distribution for specimens of different diameters and shapes in Figure 15, it can be more intuitively seen that the different specimen sizes have almost the same equivalent plastic strain distributions for the SPT deformation process.

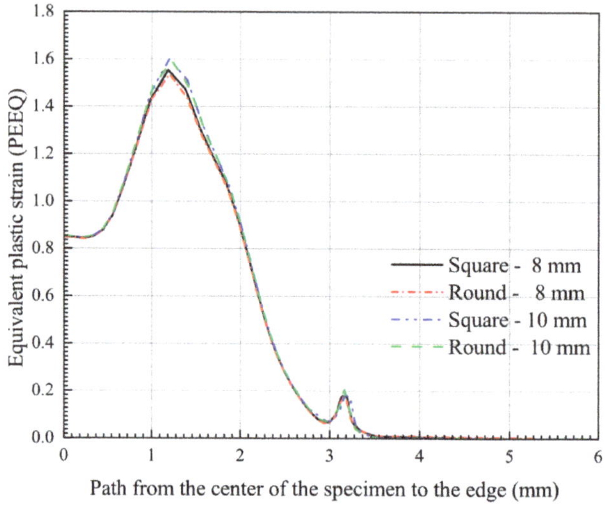

Figure 16. PEEQ distribution on the bottom surface of the specimen at the maximum force in the path from the center of the specimen to the edge.

Based on the above results, it can be seen that different specimen sizes, i.e., round and square specimens with different diameters kept clamped around, have little effect on the SPT curve results. This implies that the results obtained are theoretically transferable and comparable when SPT tests are performed by different researchers using square and circular specimens, respectively. However, this conclusion must be carried out under the condition that the periphery of the specimen is wholly clamped in the experimental results.

5. Conclusions

In this paper, SPT experiments and simulations were carried out on 316L, 347L stainless steels, and a new high-entropy alloy $Co_{32}Cr_{28}Ni_{32.94}Al_{4.06}Ti_3$, and the effect of specimen sizes on the SPT curve were investigated by simulation. The main conclusions are as follows:

(1) The discrepancy of the force–displacement of SPT between the test and the simulation results is mainly due to the loading system stiffness.

(2) The empirical correlation results for the ultimate tensile strength (σ_{UTS}) by Equation (6) and for the yield strength (σ_{YS}) by Equation (12) have little difference before and after the loading system compliance calibration, but the correlation results by Equation (7) have a greater difference because of the changes in u_m before and after the loading system compliance calibration.

(3) Simulation results on different specimen shapes and sizes show that the effect of specimen shapes and sizes on the SPT results can be ignored. This conclusion was confirmed by the 316L stainless steel simulation results. Therefore, the experimental results obtained based on the above specimen size provide a reference for different researchers even if their specimen shapes and sizes differ.

This investigation shows that insufficient stiffness of a loading system can lead to an inaccurate material properties by the small punch test, but the results after correction are still trustworthy. It also provides a reference for subsequent researchers to conduct empirical correlation studies using different specimen sizes.

Author Contributions: Conceptualization, J.Z. and Z.G.; methodology, J.Z. and Z.G.; software, Z.G.; validation, J.Z. and Z.G.; formal analysis, Z.G.; investigation, J.Z., Z.G. and K.L.; data curation, Z.G.; writing—original draft preparation, Z.G.; writing—review and editing, J.Z. and Z.G.; visualization, J.Z. and K.L.; supervision, K.L.; project administration, J.Z. and K.L.; funding acquisition, J.Z. All authors have read and agreed to the published version of the manuscript.

Funding: This research was funded by National Natural Science Foundation of China (No.51705079), Natural Science Foundation of Fujian Province (No.2018J01767).

Institutional Review Board Statement: Not applicable.

Informed Consent Statement: Not applicable.

Data Availability Statement: Not applicable.

Conflicts of Interest: The authors declare no conflict of interest.

References

1. Manahan, M.P.; Argon, A.S.; Harling, O.K. The development of a miniaturized disk bend test for the determination of postirradiation mechanical properties. *J. Nucl. Mater.* **1981**, *104*, 1545–1550. [CrossRef]
2. Nowik, K.; Oksiuta, Z. Experimental and numerical small punch tests of the 14Cr ODS ferritic steel. *Acta Mechanica et Automatica* **2022**, *16*, 225–232. [CrossRef]
3. Bruchhausen, M.; Holmström, S.; Simonovski, I.; Austin, T.; Lapetite, J.M.; Ripplinger, S.; de Haan, F. Recent developments in small punch testing: Tensile properties and DBTT. *Theor. Appl. Fract. Mech.* **2016**, *86*, 2–10. [CrossRef]
4. Cheng, Z.Y.; Sun, J.R.; Tai, P.F.; Zhang, L.Q.; Wei, Y.T.; Chang, H.L.; Thuku, R.; Gichuhi, K.M. Comparative Study between Small Punch Tests and Finite Element Analysis of Miniature Steel Specimens. *J. Mater. Eng. Perform.* **2021**, *30*, 9094–9107. [CrossRef]
5. Cao, Y.; Zu, Y.; Zhen, Y.; Li, F.; Wu, G. Determination of the true stress-strain relations of high-grade pipeline steels based on small punch test correlation method. *Int. J. Pres. Ves. Pip.* **2022**, *199*, 104739. [CrossRef]
6. Contreras, M.A.; Rodriguez, C.; Belzunce, F.J.; Betegon, C. Use of the small punch test to determine the ductile-to-brittle transition temperature of structural steels. *Fatigue Fract. Eng. Mater. Struct.* **2008**, *31*, 727–737. [CrossRef]
7. Ha, J.S.; Fleury, E. Small punch tests to estimate the mechanical properties of steels for steam power plant: II. Fracture toughness. *Int. J. Pres. Ves. Pip.* **1998**, *75*, 707–713. [CrossRef]
8. Lancaster, R.J.; Jeffs, S.P.; Illsley, H.W.; Argyrakis, C.; Hurst, R.C.; Baxter, G.J. Development of a novel methodology to study fatigue properties using the small punch test. *Mater. Sci. Eng. A* **2019**, *748*, 21–29. [CrossRef]
9. Zhao, L.; Wang, X.; Xu, L.Y.; Han, Y.D.; Jing, H.Y. Fatigue performance of Hastelloy X at elevated temperature via small punch fatigue test. *Theor. Appl. Fract. Mech.* **2021**, *116*, 103118. [CrossRef]
10. Lewis, D.T.S.; Lancaster, R.J.; Jeffs, S.P.; Illsley, H.W.; Davies, S.J.; Baxter, G.J. Characterising the fatigue performance of additive materials using the small punch test. *Mater. Sci. Eng. A* **2019**, *754*, 719–727. [CrossRef]
11. Yang, Z.; Wang, Z.W. Relationship between strain and central deflection in small punch creep specimens. *Int. J. Pres. Ves. Pip.* **2003**, *80*, 397–404. [CrossRef]
12. Rouse, J.P.; Cortellino, F.; Sun, W.; Hyde, T.H.; Shingledecker, J. Small punch creep testing: Review on modelling and data interpretation. *Mater. Sci. Technol.* **2013**, *29*, 1328–1345. [CrossRef]
13. Dymáček, P. Recent developments in small punch testing: Applications at elevated temperatures. *Theor. Appl. Fract. Mech.* **2016**, *86*, 25–33. [CrossRef]
14. Yoon, K.B.; Nguyen, T.T. Estimation of high-temperature fracture parameters for small punch specimen with a surface crack. *Fatigue Fract. Eng. Mater. Struct.* **2018**, *41*, 1224–1236. [CrossRef]
15. Nguyen, T.T.; Yoon, K.B. Fully plastic J-integral and C* equations for small punch test specimen with a surface crack. *Int. J. Pres. Ves. Pip.* **2020**, *188*, 104214. [CrossRef]
16. Li, Y.Z.; Dymacek, P.; Hurst, R.; Stevens, P. A novel methodology for determining creep crack initiation and growth properties using FEM with notched small punch specimens. *Theor. Appl. Fract. Mech.* **2021**, *116*, 103112. [CrossRef]
17. Chittibabu, V.; Rao, K.S.; Rao, P.G. Factors affecting the mechnical properties of compact bone and miniature specimen test techniques: A review. *Adv. Sci. Technol. Res. J.* **2016**, *10*, 169–183. [CrossRef]
18. Gulcimen, B.; Hahner, P. Determination of creep properties of a P91 weldment by small punch testing and a new evaluation approach. *Mater. Sci. Eng. A* **2013**, *588*, 125–131. [CrossRef]
19. Wen, W.; Jackson, G.A.; Li, H.; Sun, W. An experimental and numerical study of a CoNiCrAlY coating using miniature specimen testing techniques. *Int. J. Mech. Sci.* **2019**, *157*, 348–356. [CrossRef]
20. Soltysiak, S.; Selent, M.; Roth, S.; Abendroth, M.; Hoffmann, M.; Biermann, H.; Kuna, M. High-temperature small punch test for mechanical characterization of a nickel-base super alloy. *Mater. Sci. Eng. A* **2014**, *613*, 259–263. [CrossRef]
21. *EN 10371*; Metallic Materials–Small Punch Test Method. European Committee for Standardization: Brussel, Belgium, 2021.

22. *CWA 15627*; Small Punch Test method for Metallic Materials. European Committee for Standardization: Brussel, Belgium, 2007.
23. *GB/T 29459.1*; Small Punch Test Methods of Metallic Materials for In-Service Pressure Equipments—Part 1: General Requirements. Chinese Standard: Beijing, China, 2012.
24. *GB/T 29459.2*; Small Punch Test Methods of Metallic Materials for In-Service Pressure Equipments—Part 2: Method of Test for Tensile Properties at Room Temperature. Chinese Standard: Beijing, China, 2012.
25. *ASTM E3205*; Standard Test Method for Small Punch Testing of Metallic Materials. ANSI: New York, NY, USA, 2022.
26. Kalidindi, S.R.; Abusafieh, A.; El-Danaf, E. Accurate characterization of machine compliance for simple compression testing. *Exp. Mech.* **1997**, *37*, 210–215. [CrossRef]
27. Sánchez-Ávila, D.; Orozco-Caballero, A.; Martinez, E.; Portoles, L.; Barea, R.; Carreno, F. High-accuracy compliance correction for nonlinear mechanical testing: Improving Small Punch Test characterization. *Nucl. Mater. Energy* **2021**, *26*, 100914. [CrossRef]
28. Janca, A.; Siegl, J.; Hausild, P. Small punch test evaluation methods for material characterisation. *J. Nucl. Mater.* **2016**, *481*, 201–213. [CrossRef]
29. Altstadt, E.; Houska, M.; Simonovski, I.; Bruchhausen, M.; Holmström, S.; Lacalle, R. On the estimation of ultimate tensile stress from small punch testing. *Int. J. Mech. Sci.* **2018**, *136*, 85–93. [CrossRef]
30. Altstadt, E.; Ge, H.E.; Kuksenko, V.; Serrano, M.; Houska, M.; Lasan, M.; Bruchhausen, M.; Lapetite, J.M.; Dai, Y. Critical evaluation of the small punch test as a screening procedure for mechanical properties. *J. Nucl. Mater.* **2016**, *472*, 186–195. [CrossRef]
31. Lucas, G.E.; Okada, A.; Kiritani, M. Parametric analysis of the disc bend test. *J. Nucl. Mater.* **1986**, *141*, 532–535. [CrossRef]
32. Campitelli, E.N.; Spatig, P.; Bonade, R.; Hoffelner, W.; Victoria, M. Assessment of the constitutive properties from small ball punch test: Experiment and modeling. *J. Nucl. Mater.* **2004**, *335*, 366–378. [CrossRef]
33. Xu, T.; Guan, K.S.; Wang, Z.W. Study on standardization of small punch test (1)—general requirements. *Press. Vessel. Technol.* **2010**, *27*, 37–46. (In Chinese)
34. Zhou, Z.X.; Ling, X. Ductile Damage Analysis for Small Punch Specimens of Type 304 Stainless Steel Based on GTN Model. *J. Test. Eval.* **2009**, *37*, 538–544.
35. Andrés, D.; Dymacek, P. Study of the upper die clamping conditions in the small punch test. *Theor. Appl. Fract. Mech.* **2016**, *86*, 117–123. [CrossRef]
36. Peng, J.; Vijayanand, V.D.; Knowles, D.; Truman, C.; Mostafavi, M. The sensitivity ranking of ductile material mechanical properties, geometrical factors, friction coefficients and damage parameters for small punch test. *Int. J. Pres. Ves. Pip.* **2021**, *193*, 104468. [CrossRef]
37. Moreno, M.F. Effects of thickness specimen on the evaluation of relationship between tensile properties and small punch testing parameters in metallic materials. *Mater. Des.* **2018**, *157*, 512–522. [CrossRef]
38. Peng, J.; Zhang, H.; Wang, Y.Q.; Richardson, M.; Liu, X.D.; Knowles, D.; Mostafavi, M. Correlation study on tensile properties of Cu, CuCrZr and W by small punch test and uniaxial tensile test. *Fusion Eng. Des.* **2022**, *177*, 113061. [CrossRef]
39. Moreno, M.F.; Bertolino, G.; Yawny, A. The significance of specimen displacement definition on the mechanical properties derived from Small Punch Test. *Mater. Des.* **2016**, *95*, 623–631. [CrossRef]
40. Hähner, P.; Soyarslan, C.; Cakan, B.G.; Bargmann, S. Determining tensile yield stresses from Small Punch tests: A numerical-based scheme. *Mater. Des.* **2019**, *182*, 107974. [CrossRef]
41. Guo, J.Q.; Tang, C.Z.Z.; Lai, H.S. Microstructure and Mechanical Properties of $Co_{32}Cr_{28}Ni_{32.94}Al_{4.06}Ti_3$ High-Entropy Alloy. *Materials* **2022**, *15*, 1444. [CrossRef]
42. *ABAQUS Users Manual*, Version 6.10-1; Dassault Systemes Simulia Corp.: Providence, RI, USA, 2010.
43. Chica, J.C.; Diez, P.M.B.; Calzada, M.P. Development of an improved prediction method for the yield strength of steel alloys in the Small Punch Test. *Mater. Des.* **2018**, *148*, 153–166. [CrossRef]
44. Barsanescu, P.D.; Comanici, A.M. von Mises hypothesis revised. *Acta. Mech.* **2017**, *228*, 433–446. [CrossRef]
45. Von Mises, R. Mechanik der festen Körper im plastisch deformablen Zustand. *Nachrichten von der Gesellschaft der Wissenschaften zu Göttingen, Mathematisch-Physikalische Klasse* **1913**, *1913*, 582–592.
46. Chica, J.C.; Diez, P.M.B.; Calzada, M.P. Improved correlation for elastic modulus prediction of metallic materials in the Small Punch Test. *Int. J. Mech. Sci.* **2017**, *134*, 112–122. [CrossRef]
47. Alshboul, O.; Almasabha, G.; Shehadeh, A.; Al Hattamleh, O.; Almuflih, A.S. Optimization of the Structural Performance of Buried Reinforced Concrete Pipelines in Cohesionless Soils. *Materials* **2022**, *15*, 4051. [CrossRef]
48. Almasabha, G.; Alshboul, O.; Shehadeh, A.; Almuflih, A.S. Machine Learning Algorithm for Shear Strength Prediction of Short Links for Steel Buildings. *Buildings* **2022**, *12*, 775.
49. Lucon, E.; Benzing, J.; Hrabe, N. *Development and Validation of Small Punch Testing at NIST*; US Department of Commerce National Institute of Standards and Technology: Gaithersburg, MD, USA, 2020.
50. Garcia, T.E.; Rodriguez, C.; Belzunce, F.J.; Suarez, C. Estimation of the mechanical properties of metallic materials by means of the small punch test. *J. Alloys Compd.* **2014**, *582*, 708–717. [CrossRef]
51. Mao, X.Y.; Takahashi, H. Development of a Further-Miniaturized Specimen of 3 Mm Diameter for Tem Disk (Phi-3 Mm) Small Punch Tests. *J. Nucl. Mater.* **1987**, *150*, 42–52. [CrossRef]

52. Rodriguez, C.; Cabezas, J.G.; Cardenas, E.; Belzunce, F.J.; Betegon, C. Mechanical Properties Characterization of Heat-Affected Zone Using the Small Punch Test. *Weld. J.* **2009**, *88*, 188–192.
53. Lancaster, R.J.; Jeffs, S.P.; Haigh, B.J.; Barnard, N.C. Derivation of material properties using small punch and shear punch test methods. *Mater. Des.* **2022**, *215*, 110473. [CrossRef]

materials

MDPI

Article

Identification of the Domain Structure Defects of a Radially Magnetized Rubber–Ferritic Conglomerate

Karolina Popowska , Szymon Gontarz * and Przemysław Szulim

Institute of Vehicles and Construction Machinery, Warsaw University of Technology, Narbutta 84, 02-524 Warsaw, Poland

* Correspondence: szymon.gontarz@pw.edu.pl

Abstract: Modern solutions in materials engineering are designed not just for the improvement in the mechanical or electromagnetic properties of materials but also to begin to fulfill specific functional roles. A good example of such a modern solution is a composite made of steel and rubber–ferritic conglomerate, which is the research object of the article. The composite, when properly magnetized, can act as a magnetic encoder ring for reading the angular displacement, speed, or acceleration parameter. The paper addresses the problem of identifying and assessing the defects of the magnetic encoder ring domain structure in the form of a radially magnetized ring. It discusses the essential types of the ring's degradation, such as mechanical, thermal, and magnetic, and presents problems related to the identification of emerging defects. The conducted research allows a better understanding of the degradation process in the context of magnetic encoder ring reliability. Based on the conducted research on the proposed test stand, it is possible to track the progressive degradation related to each effect. These degradation case analyses consider both quantitative and qualitative changes in the encoder ring's domain structure. The proposed parameters show the possibilities and perspectives for detecting the ring's defects in the early stage of its development. Solely such an approach will allow for proper exploitation and extension of the applicability of this kind of 'intelligent material'. Additionally, the developed parameters for the encoder ring's defects detection can support the progress of rapidly evolving methods for diagnosing mechanical systems based on a signal from such an element.

Keywords: diagnostic parameters; magnetic-based diagnostics; magnetic encoder ring; magnetic domain structure

Citation: Popowska, K.; Gontarz, S.; Szulim, P. Identification of the Domain Structure Defects of a Radially Magnetized Rubber–Ferritic Conglomerate. *Materials* **2023**, *16*, 3487. https://doi.org/10.3390/ma16093487

Academic Editors: Antoni Planes and Jana Bidulská

Received: 13 February 2023
Revised: 25 April 2023
Accepted: 26 April 2023
Published: 30 April 2023

1. Introduction

Modern technology development almost excludes the existence of technical objects developed within one engineering discipline. At present, these are complex objects that can generally be called mechatronics. Interferences and permeation of various disciplines cause objects to acquire new properties and functionalities. This direction is being pursued in parallel with a clear tendency toward the miniaturization of technical solutions. When considering modern machines whose primary task is performing a rotational movement, it is clear that systems for measuring the parameters of this motion have become necessary. In the beginning, they were independent, with various measurement paths, such as RLC bridges, strobotron circuits, and tachometric generators [1–3], and later more integrated devices, such as in the form of discs rotating with a shaft and inductive sensors [4]. The latest solutions use magnetic encoder rings cooperating with appropriately selected magnetometers [5]. Considering the integration and miniaturization mentioned above [5,6], steel and rubber–ferritic conglomerate composites were created. After appropriate magnetization, it can act as a magnetic encoder for reading parameters, such as angular position, velocity, or angular acceleration [7]. Moreover, such an element may be a rolling bearing sealing (Figure 1b), which means that only minimal interference in the object's structure is enough to obtain information about the rotational movement.

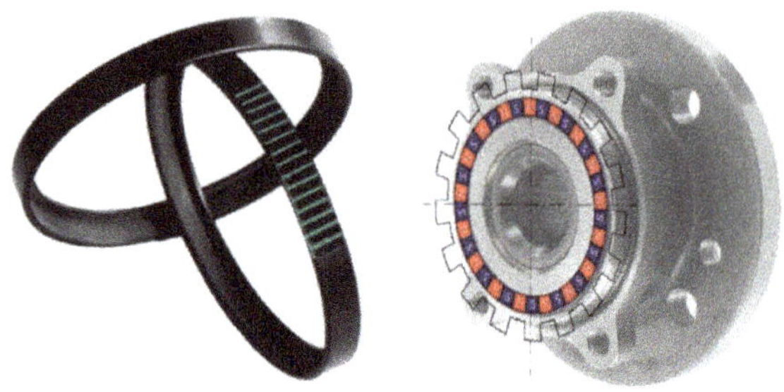

(a) (b)

Figure 1. Selected types of magnetic encoder rings: (**a**) Circumferentially magnetized encoder ring and (**b**) radially magnetized encoder ring [8,9].

In science and industry, magnetic encoders are applied to various topics. The information about rotational speed or angular displacement obtained through them is used for real-time machine monitoring, correcting the operation of machines, and controlling regulation systems, or, for example, ensuring the continuity and maintenance of the required parameters of production processes [10,11]. All these functions will be fulfilled provided that the magnetic encoder ring and magnetometer [12,13] are well maintained. Recently, it has been indicated that magnetic encoder rings can also act as a source of diagnostic information by providing a signal describing the unevenness of the rotary motion correlated with the technical condition of the rotary machine [5]. In this case, the quality and reliability of the signal from the encoder ring are even more critical, which, as a specific material engineering product, is subjected to many specific influences that may cause its degradation. Notably, imperfections not only result in the production process but also may arise during operation. Then, the range of factors influencing the encoder is much wider. Therefore, the behavior of the domain structure is diverse.

We can see from this that the magnetic encoder is the key element for the correct operation of the entire system. Assuming that the relevant standard guidelines are met [10], such as assembly criteria and characteristic parameters of the system elements, it is the encoder quality that can lead to a situation where the obtained information is erroneous. For encoders in automotive applications, standard suppliers [10–12] do not clearly indicate the parameters for assessing the quality of the magnetic field distribution and the amplitudes of the magnetic induction domains. Additionally, with regard to quality assessment requirements, suppliers are not completely consistent with each other. As a result of the analysis of that standard, three measures can be indicated: the number of domain pairs Ndp; their dimensions (e.g., Dw-domain width); and the minimum limit value of domain amplitudes measured from a certain distance $|B_{min}|$. These parameters have their reference values specified in the standards. There are some gaps in terms of measurement and diagnostic parameters. For example, it seems advisable to consider the width of the domains measured on a circumference with a nominal radius R to read the cumulative error of the width of the domains or to analyze the waveform of the magnetic induction value measured on that circumference.

Meeting the parameters for assessing the quality of encoders is necessary to obtain a reliable and repeatable measurement of rotational speed. Therefore, meeting this parameter type in more advanced applications, such as measuring the speed and angular position in devices such as wind turbines or fast and precise industrial robots, will be all the more important. One should also consider the signal from the encoder as a diagnostic signal of the rotating machine, the diagnosis of which will not be reliable without certainty as to the encoder–magnetometer measuring system. Understanding the nature of degradation, together with the possibility of their identification, can be an important contribution wherever encoder rings are used as a source of information about rotational motion parameters. Therefore, the aim of this article is to both identify and assess the defects of a magnetic encoder ring domain structure. Section 2 describes a proposal for a universal measurement stand and experiment conducted with the use of that stand. The next chapter presents the results of the experiments. The results include the use of the proposed diagnostic parameters, the task of which is the reliable identification of the defects at an early stage of their development. The last chapter discusses the possibility of presenting diagnostically oriented analysis of magnetic encoder rings.

2. Measurement Method

Magnetic encoder defects can be broadly categorized as those caused by manufacturing or operation. To make the topic more specific, the paper refers to the faults that could occur during operation. Magnetic encoders work with elements and mechanical systems. Due to a malfunction on the mechanical side, the encoder can be subjected to abrasion. The ring, in addition to the aforementioned function of providing information on the rotational speed of the shaft, also acts as a seal for the hub bearing in the machine. Due to insufficient grease in the bearing, the mating elements may wear and overheat. When we consider the magnetic encoder in a vehicle's ABS system, degradation may occur as a result of an increase in temperature in the drive system. Rapid braking or braking in the event of a high load on the drive system (driving down a steep slope) results in an increase in temperature to above 673 K. It is also worth mentioning that the number of electric cars is currently increasing. In the drive system of such a vehicle, in the event of engine damage (both mechanical and electric), the phenomenon of magnetic flux leakage may occur, which may lead to the demagnetization of the encoder domain structure [14].

An appropriate measurement station and methodology are proposed to detect the defects described in this chapter. The proprietary stand, in conjunction with the parametric evaluation, became the basis for a novel measurement method used in this research. The schematic diagram is shown in Figure 2.

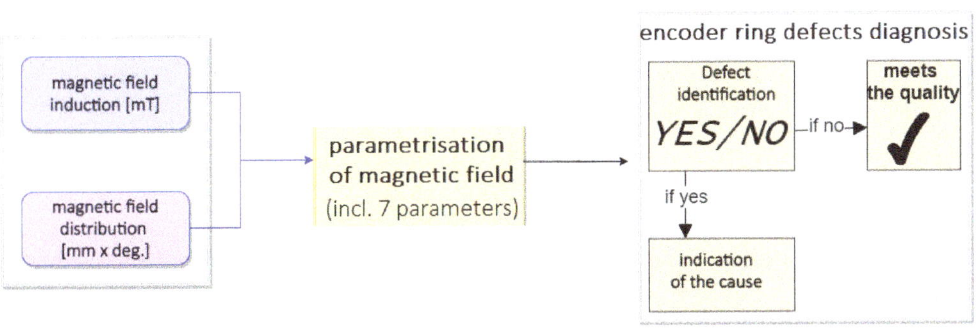

Figure 2. The novel method measurement procedure.

2.1. Hardware and Software Architecture of the Stand

The measuring device had the form of a 4-axis scanner, with the possibility of making 3 translational movements (X, Y, Z) and one rotary movement. During the scanning of the research object, the stand combines rotary and translational motion. Due to such a combination of movements, the distribution of the magnetic field of the encoder is determined. The advantage of introducing a rotational motion in the encoder measurement is the maintenance of a polar coordinate system reflecting the natural working conditions of the target encoder and sensor system. The test stand had the ability to accurately position the sensor on the vertical axis (Figure 3a) to maintain the same distance from the encoder surface throughout the entire measurement. The scanner resolution for the range of translational movements was at the level of 0.1 mm, while for the rotational movement, it was at 0.023 degrees. The measurement of the displacement on the X, Y, and Z axes was based on counting the impulses of the stepper motors (with zeroing after reaching the extreme position detected by optical limit sensors) only, while the axis of rotation was equipped with an additional absolute magnetic encoder with a resolution of 14 bits. Figure 3 shows the measuring stand with the tested object.

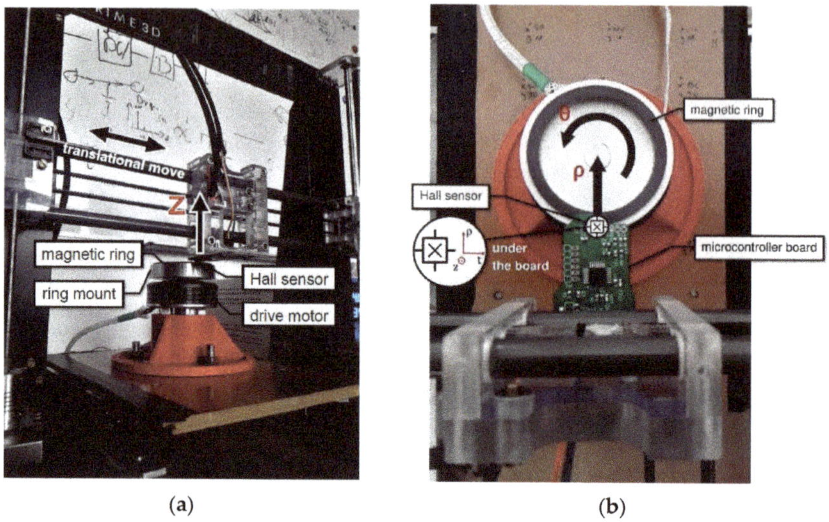

Figure 3. Measurement stand: (**a**) front view and (**b**) top view.

As a measuring element, a Hall magnetometer was used in the stand (Figure 3b) [15]. The measuring axis (X-axis) of the sensor was tangent to the curvature of the magnetic encoder, and the Y-axis was radial and followed the trajectory of the head—ρ (Figure 3b), while the Z-axis was directed vertically upward (Figure 3a).

By keeping the distance constant throughout the measurement, it is possible to obtain a repeatable measurement result of the magnetic field. In the case of measuring encoders, the use of a precise angular measurement significantly increases the result accuracy because the parameters of the number of domain pairs Ndp, domain widths Dw, and the minimum limit value of domain amplitudes $|B_{min}|$ are determined in the circumferential direction. Measurements carried out in this way provide signals that have been subjected to a detailed diagnostically oriented analysis.

2.2. Experiment

Based on the described measuring station, experiments were planned in which magnetic encoders were subjected to various types of degradation. Figure 4 shows the object of the experiment—a magnetic encoder. The encoder ring samples were obtained as a courtesy by the encoder ring's manufacturer. It is a standard magnetic ring, which is one of the seals of the rolling bearing. Such constructions are currently used, for example, in the automotive industry, and are a key element in ABS/ASR systems to determine efficiency.

(a) (b)

Figure 4. Object of the experiment. i.e., the magnetic encoder ring: (**a**) encoder with marked characteristic dimensions—internal and external diameter and (**b**) copies of encoder rings are marked for testing purposes.

The nominal magnetic properties and geometric dimensions of the shown encoder ring (Figure 4), according to the standard [10], are listed as follows:

- Number of domain pairs $Ndp = 48$;
- Inner diameter $Di = 48$ mm and outer diameter $Do = 58$ mm (Figure 4a);
- Diameter of the nominal reading circumference of the measurements $D_{mean} = 53$ mm; hence, the width of a single domain $Dw = 1.64$ mm;
- The minimum, limiting value of domain magnetic induction amplitudes $|B_{min}|$ dependent on the measurement distance [10].

For the purposes of the experiment, the degradation of magnetic encoders was reproduced in laboratory conditions. Subsequent magnetic encoders were subjected to the following impacts: 1—temperature; 2—magnetic field; and 3—mechanical abrasion. The controlled increase in the intensity of these interactions resulted in successive degrees of degradation of the magnetic information contained in the rings. The table below (Table 1) shows the degree of the severity caused by various types of damage to the encoders.

Table 1. Types and intensities of magnetic encoder ring degradations.

		Type of Degradation		
	No.	**A. Mechanical Abrasion [mm] ***	**B. Magnetic Field [mT] ****	**C. Temperature [K]**
INTENSITY	**1.**	0.00	0	293 ***
	2.	0.23	45	603
	3.	0.25	75	653
	4.	0.30 (+ cavities)	100	713
	5.	0.35	120	753
	6.	0.50	50	783

*—max. depth of magnetic layer—0.70 mm, **—encoder ring nominal magnetic field—10 mT, *** ambient temperature.

The proposed experiment has a useful purpose because thermal, mechanical, or magnetic degradation may occur during the stage of operation or even production.

3. Measurement Results

Each measurement of the 15 rings was made on the proprietary measuring stand (Section 2.1), taking into account 55 waveforms of the induction value recorded on successive encoder circuits with given radii. This measurement method makes it possible to obtain the distribution of the magnetic field over the encoder surface. The measurements were carried out at a fixed distance of 0.3 mm from the magnetometer's sensitive element to the encoder surface. A constant distance between the sensor and the encoder surface throughout the measurement allows the control of the distance from the signal source (encoder ring's surface) to the measured magnetic field induction value. The above conditions ensure reliable magnetic induction measurements, which decrease with increasing distance from the signal source [13].

3.1. Mechanical Degradation

Five magnetic encoder ring samples were subjected to five different mechanical degradation intensities. The rubber–ferritic layer of the encoder ring, with a nominal thickness of 0.70 mm, was abraded from a depth of 0.23 mm to 0.50 mm (Table 1). In addition, specific real operating conditions were simulated, i.e., tearing of the magnetic layer, which may occur during the mechanical abrasion process (sample "0.30 + cavities"). The encoder rings were abraded perpendicular to their surface due to mechanical friction.

Preliminary research has shown that mechanical degradation can be observed in the change in the magnetic field of the damaged encoder ring that is directly proportional to the abrasion depth. Additionally, there is a decrease in the domain amplitudes and a flattening of the distribution of magnetic induction values on the encoder ring radial waveform.

Figure 5 shows 3D visualizations of magnetic field induction distributions of the encoder ring subjected to regular mechanical degradation (Figure 5a) and its undamaged prototype (Figure 5b). The data labels show the magnetic induction peak values of the encoder ring domains. The magnetic induction *Bz* peaks of the mechanically abraded ring domains had a flattened shape (Figure 5a), while the domain peaks of the undamaged encoder ring had a slender shape (Figure 5b), which can be observed in the domain magnifications on the graphs.

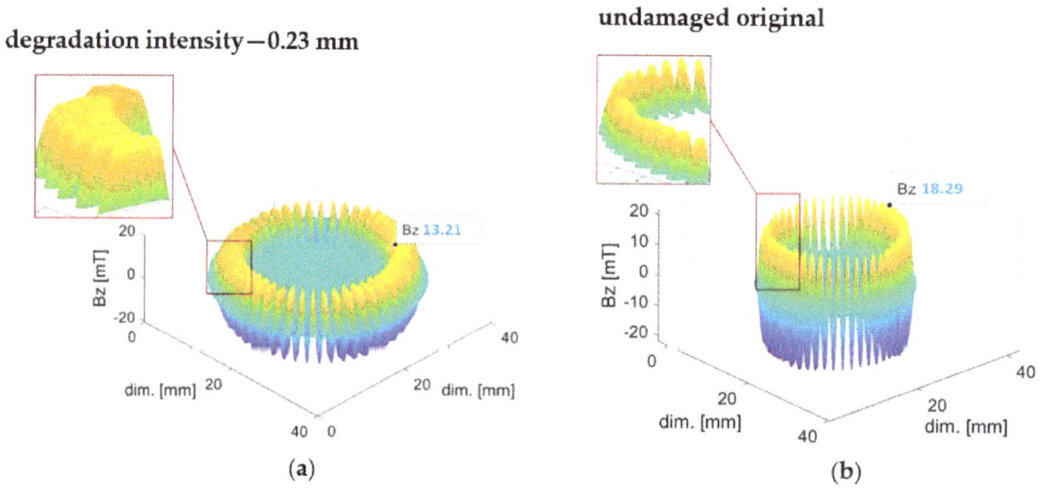

Figure 5. 3D magnetic induction distributions of the mechanically degraded encoder ring for an abrasion depth of 0.23 mm (**a**) and its undamaged prototype (**b**).

Figure 6 shows the magnetic field induction distribution of a ring subjected to an irregular, specific type of mechanical degradation with material cavities. The material was abraded to a depth of 0.30 mm. On the close-up view of the graph area, the shapes of the sample's domain magnetic induction peaks are visible.

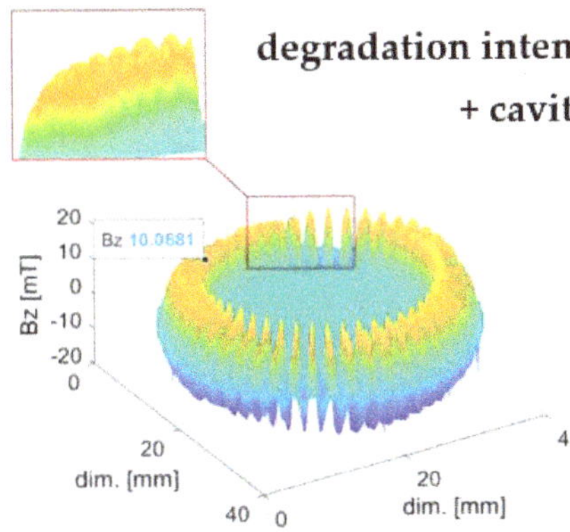

Figure 6. 3D magnetic induction distribution of mechanically degraded material cavities. The material was abraded to a depth of 0.30 mm.

The magnetic induction *Bz* values of the encoder rings subjected to mechanical degradation were directly related to the degradation depth *d* (both in the regular case—Figure 5b and in the specific one, i.e., mechanical abrasion with material cavities—Figure 6). The increased mechanical degradation depth *d* of the rubber–ferritic layer caused a decrease in the magnetic induction value *Bz* of the encoder ring domains (1).

$$\frac{d_n}{Bz_n} = \frac{d_0}{Bz_0},\ for\ d_n > d_0 = 0\ \text{mm},\ \ Bz_n > Bz_0 \tag{1}$$

where d_n is the degradation depth of the n-th probe, d_0 is the initial degradation depth of the undamaged sample, Bz_n are the peak values of the magnetic induction domains of the n-th sample, and Bz_0 is the magnetic induction peak values of the undamaged sample's domains.

3.2. Thermal Degradation

Another five magnetic encoders were subjected to five different degrees of temperature exposure ranging from 293 K (ambient temperature) to 793 K (Table 1). In the thermal degradation process, stage II damage can be distinguished for the two temperature ranges. Stage I included temperatures from 653 K to 713 K (for 653 K and 713 K measurements), and stage II ranged from 713 K to 793 K (for 753 K and 783 K measurements, respectively), as shown in Figure 7. The Curie temperature for the ferritic powder used in the encoder ring was 728 K [16] (Figure 7).

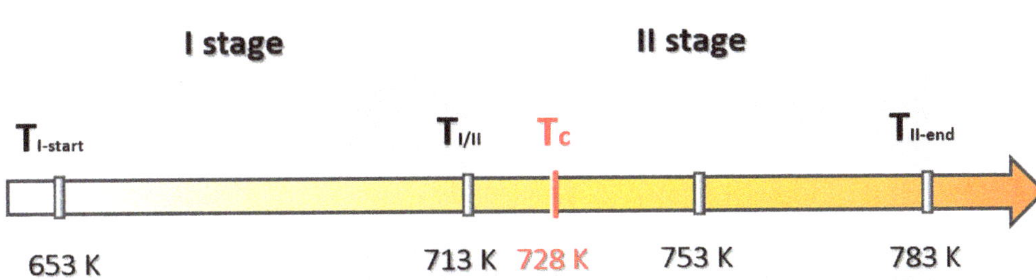

Figure 7. Schematic diagram of the thermal degradation process.

Figure 8 shows 3D visualizations of the magnetic field of the encoder ring for both stage I and stage II of thermal degradation.

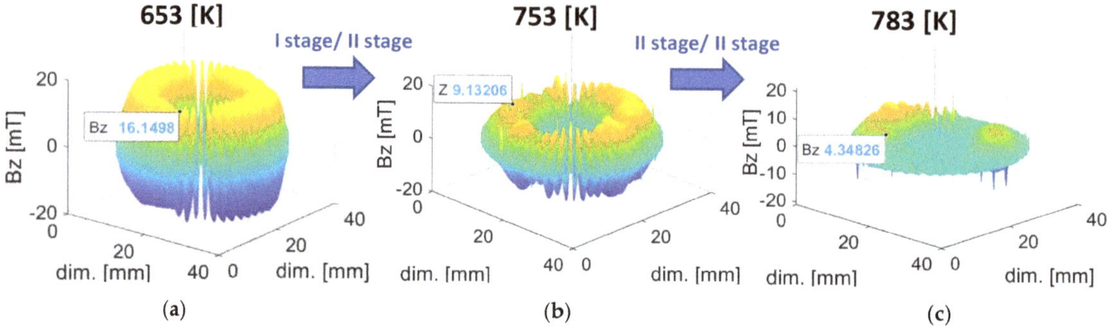

Figure 8. Thermal degradation stages: (**a**) 653K temperature, (**b**) 753K temperature, (**c**) 783K temperature.

In stage I of thermal degradation, the decrease in the magnetic field induction value of the ring domains was slight, on the order of a few mT (Figure 8a). Such a reduction in the magnetic induction value of the encoder ring did not exceed the required magnetic induction limit values contained in the standards [10] (Section 1). Therefore, the magnetic ring did not lose its encoder functionality. At stage II of thermal degradation, for the observed temperature of 743–763 K (Figure 8b), there was an abrupt decrease in the magnetic induction value in comparison with degradation stage I, and the ring's domain structure was reorganized, which did not occur in stage I. It can be concluded that the abrupt degradation of the magnetic induction occurred before the observed temperature of 743 K, after exceeding the Curie temperature (728 K) for the ferritic powder used in the conglomerate [16]. On a certain part of the circumference, the rubber–ferritic conglomerate completely lost its magnetic properties, becoming paramagnetic, and the encoder domains were demagnetized. The composite of steel and rubber–ferritic conglomerate lost the encoder functionality.

The reason for circumferentially nonuniform thermal degradation (Figure 8c) was the nonuniform distribution of heat during the experiment caused by the natural diffusion of gases into the environment, resulting in an unequal temperature intensity along the entire ring's circumference length.

3.3. Magnetic Degradation

Another five encoders were magnetically degraded with five different degrees of damage. In the process of magnetic degradation, stage II could be distinguished due to the values of the intensity of the demagnetizing field. In stage I, the values of the destructive field ranged from 45 mT to 100 mT, and in stage II, the values were 120 mT and 145 mT.

Figure 9 shows a visualization of the encoder ring's 3D magnetic induction distribution for stage I of magnetic degradation with a 100 mT intensity. For stage I of magnetic degradation (Figure 9 the magnetic induction peaks of the domains did not change their shape in comparison with the undamaged prototype (Figure 5b), unlike in the mechanical degradation case (Figure 5a), where the peak shape flattened. For stage I of magnetic degradation (Figure 9), domain magnetic induction decreased evenly along the radial direction, and the domain peak shape was slender, similar to the undamaged sample (Figure 5b).

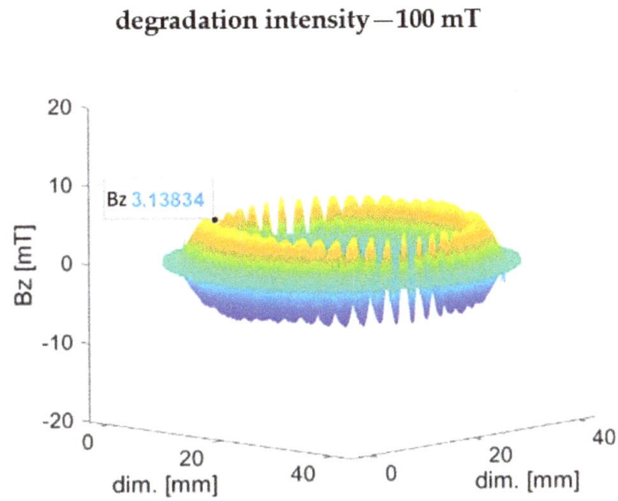

Figure 9. Magnetic field induction distribution of a ring subjected to 100 mT intensity of magnetic degradation.

Stage I of magnetic degradation was characterized by a decrease in the magnetic induction Bz peak values of the domains (Figure 9) relative to the magnetic induction values of the undamaged encoder ring (Figure 5b). The Bz values decreased with an increase in the intensity of the destructive magnetic field H(2).

$$\frac{H_n}{Bz_n} = \frac{H_0}{Bz_0}, \; for \; H_n > H_0 = 0\,\text{mT}, \;\; Bz_n > Bz_0 \tag{2}$$

where H_n is the intensity of the degrading magnetic field, H_0 is the initial degrading magnetic field intensity, Bz_n is the peak value of the domain's magnetic induction of the n-th sample, and Bz_0 is the peak value of the domain's magnetic induction of the undamaged sample.

Figure 10 shows a visualization of the 3D magnetic induction distribution of the encoder ring in stage II of magnetic degradation. The red frame shows one of the domains, which in the radial direction changes its sign.

The ring (Figure 10) was subjected to one of the two highest intensities of degradation (Table 1) with an intensity of 120 mT.

In stage II of magnetic degradation, in comparison to stage I, there is an irregular decrease in the magnetic induction of the encoder ring circumference Bz (2) (Figure 10a). Moreover, there is a change in the ring's magnetic field distribution; that is, the regular, radially magnetized domain structure becomes reorganized (Figure 10b). This causes sign changes in the domains in the encoder ring's radial direction, which is manifested in the magnetic field distribution's top view (Figure 10b) as the intermingling of blue and orange fields.

Using the capabilities of the measuring stand, a measurement and related parameters of the measured signal were proposed, which are further analyzed in the Section 4.

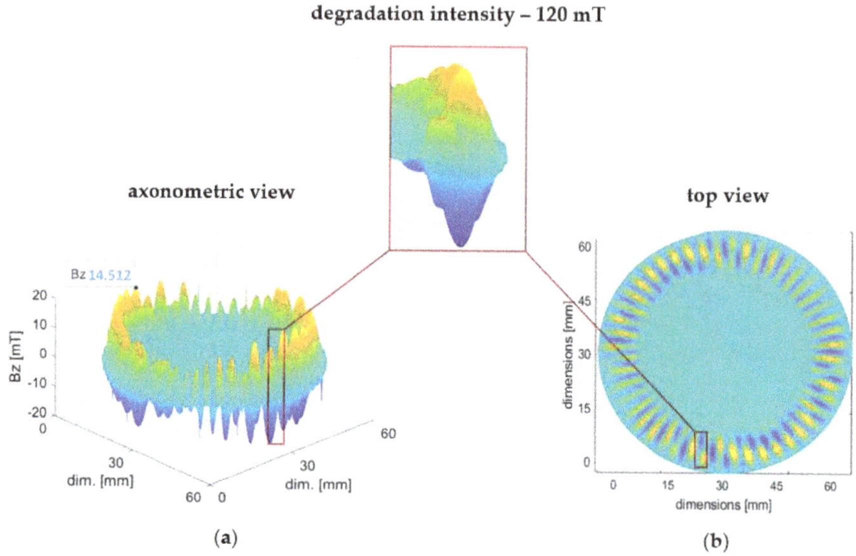

Figure 10. 3D magnetic induction distribution of the ring subjected to magnetic degradation with an intensity of 120 mT. The axonometric view of the magnetic field distribution (**a**) and a top view (**b**). The red frame shows one of the domains in which magnetic induction changes its sign.

4. Parametric Analysis of the Results

A properly performed measurement with the use of the proposed stand allows for an initial assessment of the magnetic encoder. However, to obtain more detailed information on possible degradation and to give the verification functional characteristics, an appropriate parametric description of the tested object should be carried out. The vertical axis B_Z was the direction of the magnetic field induction vector of the encoder domains considered in the analyses.

A similar approach can be found in the existing standards that can be used as measures to assess the quality of magnetic encoders. We found such parameters as the number of domain pairs Ndp, the width of domains Dw, and the minimum, limiting value of domain amplitudes $|B_{min}|$, which can be part of a parametric description. The standard provides only the values of the parameters without showing how they can be obtained; therefore, points 1 to 3 describe the method of extracting the mentioned parameters, obtained as a result of digital processing of data on the distribution of the magnetic field on the surface of the encoder and its magnetic induction values.

1. $|B_{min}|$ is the minimum absolute peak value of the magnetic induction of the encoder ring domains, Equation (3). $|B_{min}|$ is determined as a result of the analysis of peak values of magnetic induction of domains N and S (both positive $Bpeak^+$ and negative $Bpeak^-$) on the circumferential waveform of the magnetic induction signal of the encoder ring on the circumference with a given radius R. $|B_{min}|$ should be greater than or equal to the minimum absolute value of magnetic induction $\left|B_{\min(norm)}\right|$ specified in the standards [10,11], Equation (4).

$$|B_{min}| = MIN\left(Bpeak^+,\ Bpeak^-\right)_n\ [\text{mT}] \quad (3)$$

$$|B_{min}| \geq \left|B_{\min(norm)}\right| \tag{4}$$

where $|B_{min}|$ is the measured minimum, absolute peak value of the magnetic induction value of the domains, $Bpeak^{+}$ and $Bpeak^{-}$ represent the positive and negative domain peaks, and $\left|B_{\min(norm)}\right|$ is the minimum absolute limit value of magnetic induction, specified according to the standards [10,11].

2. Ndp is the number of pairs of unlike domains, namely, N and S. The determination of the number of domain pairs Ndp is based on the number of single sine periods within the circumferential waveform of domain magnetic induction values. A single sine period denotes a set consisting of the positive domain peak $Bpeak^{+}$ and negative domain $Bpeak^{-}$.

$$Ndp = \bigcup_{i=1}^{\frac{n}{2}} \left(Bpeak^{+},\ Bpeak^{-}\right)_n \tag{5}$$

where $\cup$ is the sum of sets symbol, $Bpeak^{+}$ is the positive domain peak, and $Bpeak^{-}$ is the negative domain peak.

3. Dw is the width of a single domain, determined by the circumference with a given radius R. The measure of the domain width Dw is the number of consecutive measurement points whose induction values are the same sign: "+" or "−" (do not cross zero). To determine the number of positive/negative points in the magnetic induction vector B, the signum function was used, Equation (6):

$$sgn(B) = [1111111111 - 1 - 1 - 1 - 1 - 1 - 1 - 1 \ldots 1111111] \tag{6}$$

4. Hence, the frequencies $f_{W(+)}$ and $f_{W(-)}$ of the "1" and "−1" characters, respectively, in the vector $sgn(B)$ correspond to the widths of domains Dw with positive "+" and negative "−" signs.

$$sgn(B) = \left[f_{W(+)}\ f_{W(-)}\ \cdots\ f_{W(+)}\right] = \left[10\ 7 \ldots 8\right] \tag{7}$$

$$w = \frac{f_W}{\left|\vec{W}\right|} 2\pi R\ [\text{mm}] \tag{8}$$

The set of parameters included in the standards [10,11], i.e., 1—$|B_{min}|$, 2—Ndp and 3—Dw—is not sufficient for unambiguous identification and determining the causes of specific types of magnetic encoder ring defects, but only for the quality assessment. The article develops the quality assessment of magnetic encoder rings with the identification of encoder defects and the causes of their formation; hence, new parameters, not included in the standard parameters, are proposed: 4—σ, 5—$R_{+/-}$, 6—C_{max}; and 7—D_A. Together, all seven diagnostic parameters allow for unambiguous identification and determination of the cause of the three most common types of encoder ring degradation (Sections 3.1–3.3). The first three parameters were determined in the circumferential direction of the encoder, parameters 4—σ and 5—$R_{+/-}$ were determined in the radial direction of the magnetic encoder, and parameters 6—C_{max} and 7—D_A were determined in the two-dimensional plane, that is, the magnetic field distribution on the surface encoder. Below (Figure 11) is a short description in the form of a diagram of the listed diagnostic parameters divided into those included in the standards and those proposed by the authors of the article. Methods of determination and detailed analytical descriptions of the parameters are presented in the subchapters concerning particular types of degradation (Sections 3.1–3.3) for which the specific parameters were developed.

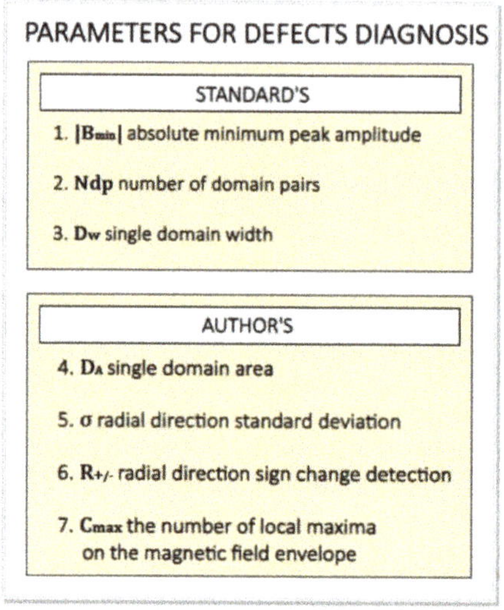

Figure 11. Parameters for magnetic encoder ring defect diagnostics in the scheme divided into standard parameters and those proposed by the authors of the article.

The reference values of the parameters included in the standards [10,11], $|B_{min}|$, Ndp and Dw were confirmed experimentally on the tested encoders (Figure 4) using the authors' measuring stand (Section 2.1). The identification of magnetic encoder defects and the causes of their occurrence was based on experimentally determined reference values of the diagnostic parameters (Table 2) as a result of tests conducted on undamaged encoders.

Table 2. Reference values of the diagnostic parameters.

Parameter	$\lvert B_{min}\rvert \geq 16$ mT
	$Ndp = 48$
	$D_w = 1.64$ mm
	$D_A > 0$
	$\sigma \geq 22.69$
	$R_{+/-} = NO$
	$C_{max} = 0$

4.1. Mechanical Degradation

Due to the mechanical degradation process, the absolute peak value of magnetic induction of the domains $|B_{min}|$ changed its value, and a variation in the magnetic induction values around the extreme changes was described by a statistical parameter, namely, the standard deviation σ of a discrete data series [17].

The decrease in the $|B_{min}|$ value is tantamount to the decrease in the magnetic induction value, as observed in the experiment (Section 3.1). A decrease in the variability around the ring's magnetic induction radial extremum visually manifested itself as a flattening of the domain's magnetic induction peaks in the magnetic field distribution (Figure 5a).

Figure 12a shows the waveforms of magnetic induction in the radial direction for samples subjected to four different mechanical degradation intensities, that is, the depths of magnetic layer abrasion (Table 1 For each of the four waveforms, the σ parameter was determined. With increasing damage intensity in the encoder ring's radial direction, the parameter σ (Figure 12a) decreased in relation to its reference value for an undamaged encoder ring—$\sigma = 22.69$ (Table 2).

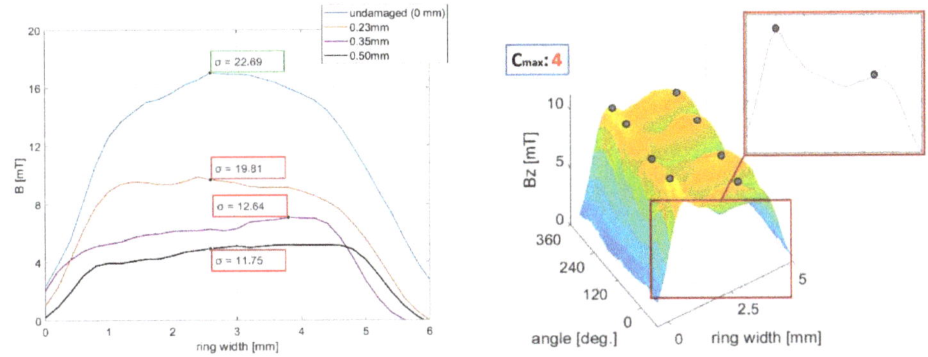

(a) (b)

Figure 12. (**a**) Magnetic induction waveforms in the radial direction of mechanically degraded encoder rings. The standard deviation parameter σ was determined for each of the waveforms. (**b**) Envelope of the magnetic field induction distribution of the sample subjected to mechanical abrasion with cavities. For the envelope, the parameter C_{max} = 4 was calculated.

For the purpose of irregular, specific types of mechanical degradation, abrasion with cavities (Figure 12b), the parameter C_{max} was developed, which is related to the number of material cavities along the magnetic encoder ring circumference.

From the analytical description's point of view, C_{max} is equal to the number of sets of local maxima on the magnetic induction envelope of the encoder ring, Equation (9). A single set includes two or more local maxima $max(B)$. These are local maxima with a given, minimal peak prominence [18] in the radial direction and a given angular length in the encoder ring circumferential direction.

$$C_{max} = \bigcup_{j=1}^{m} (max_i(B), max_{i+1}(B), \ldots max_n(B))_m, \; n \geq 2 \tag{9}$$

where $max_i(B)$ is the local maximum of the magnetic field induction envelope of the encoder ring, m is the number of local maxima sets, and n is the number of local maxima.

Figure 12b shows the envelope of the magnetic induction distribution of the encoder ring, which was subjected to mechanical abrasion at a 0.30 mm depth (Table 2). For the envelope (Figure 12b), the parameter $C_{max} = 4$ was determined, which means that the tested sample had four cavities in the rubber–ferritic material layer.

To identify the mechanical degradation and to determine its intensity, the following parameters were used: the minimum; absolute peak value of magnetic induction of the domains $|B_{min}|$; the radial waveform's standard deviation σ; and the maxima C_{max} in the envelope of magnetic induction of the domains.

With increasing intensity of regular mechanical degradation, that is, depth of abrasion of the magnetic layer, parameters $|B_{min}|$ and σ decrease with respect to the required reference values: $|B_{min}| = 16\ mT$ and $\sigma = 22,69$. In the case of mechanical degradation with cavities, the C_{max} parameter increased with the growth of cavities in the material. The reference value of the parameter was $C_{max} = 0$ (Table 2) for no cavities.

4.2. Thermal Degradation

Stages I and II of thermal degradation proceeded, as described in the diagram in Figure 2.

At stage I of damage, at a destructive temperature of 653 K, there was a decrease in the minimum absolute value of magnetic induction of domains $|B_{min}|$ in relation to its reference values (Table 2), which is tantamount to a decrease in the magnetic induction value, as observed in the experiment (Section 3.2) and shown in Figure 13a.

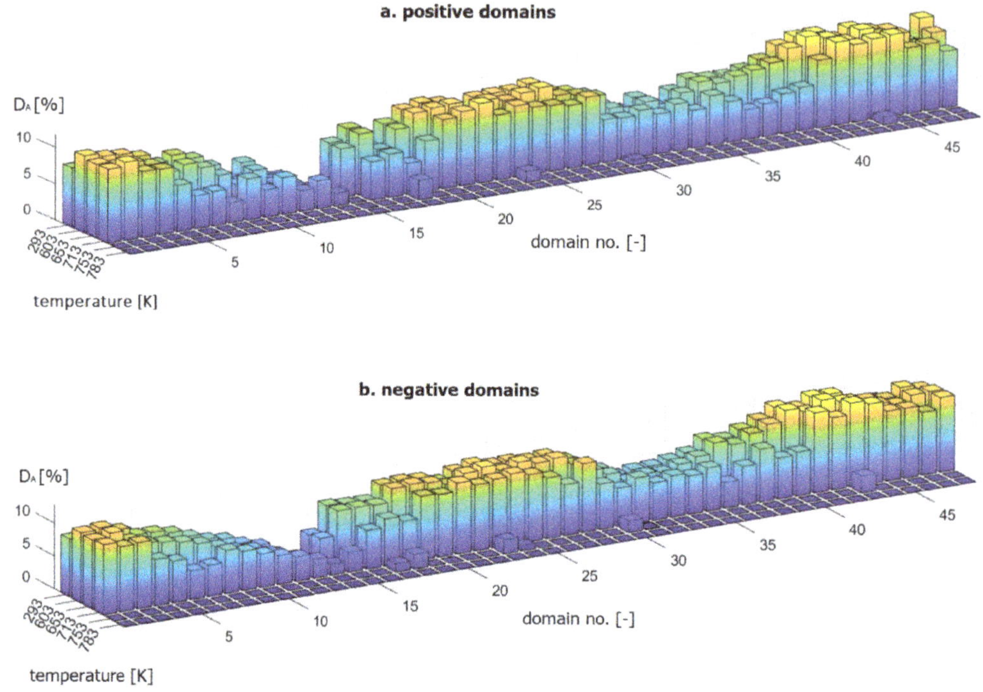

Figure 13. D_A areas of positive domains (**a**) and negative domain D_A areas (**b**) are shown as bar graphs.

In the next stage of thermal degradation of the encoder rings, for the measurement at the destructive temperature of 753 K, relative to the reference values (Table 2), there was a significant (in the range of 5 mT to 10 mT) (Figure 13b) decrease in $|B_{min}|$ and reduction in most of the positive and negative domain fields D_A to 0% and several D_A values to 0.5–1% (Figure 13).

D_A allows for measurements of the area of a single domain above the determined magnetic induction threshold. The D_A parameter is expressed as a percentage ratio of the domain area above the threshold D_{A-THR} in relation to but not including the threshold domain area D_{A-0} (10).

$$D_A = \frac{D_{A-THR}}{D_{A-0}} \cdot 100\% \quad (10)$$

where D_{A-THR} is the area of a single domain above the determined threshold of magnetic induction—140 mT and D_{A-0} is the area of a single domain not including the threshold.

Figure 13 shows that the areas D_A of both positive and negative domains did not change their size until a temperature of 713 K was reached—a transition temperature between the I and II stages of thermal degradation (Figure 7). However, for stage II of thermal degradation (at 753 K), most of the D_A domain areas were reduced to 0%, and for a temperature of 783 K, all the D_A areas of the domains did not exceed the value of 0.

In summary, the following parameters are required to identify thermal degradation and to determine their intensity: $|B_{min}|$ and D_A.

At stage I of thermal degradation (for destructive temperatures of 653 K and 713 K), there is a decrease in the domain magnetic induction peak values $|B_{min}|$ in relation to the reference value (Table 2).

At stage II of thermal degradation (for destructive temperatures: 753 K and 783 K), in addition to the aforementioned $|B_{min}|$ decrease, there is a decrease in the measured D_A domain area values in relation to its reference value (Table 2).

4.3. Magnetic Degradation

In stage I of magnetic degradation, the domain magnetic induction minimum absolute peak value $|B_{min}|$ decreased. Parameter $|B_{min}|$ decreased due to an increase in the intensity of magnetic degradation, that is, the value of the destructive magnetic field intensity.

Figure 14 shows the magnetic induction circumferential waveforms of the undamaged reference encoder ring and samples at stage I of magnetic degradation. The three encoders were subjected to increasing intensities of magnetic degradation. For each waveform (Figure 14), the parameters Ndp and $|B_{min}|$ were determined. The $|B_{min}|$ parameter, and accordingly, values of magnetic field induction decreased with increasing intensity of magnetic degradation. The results of Ndp and $|B_{min}|$ parameters are marked in red, which do not meet their reference values (Table 2).

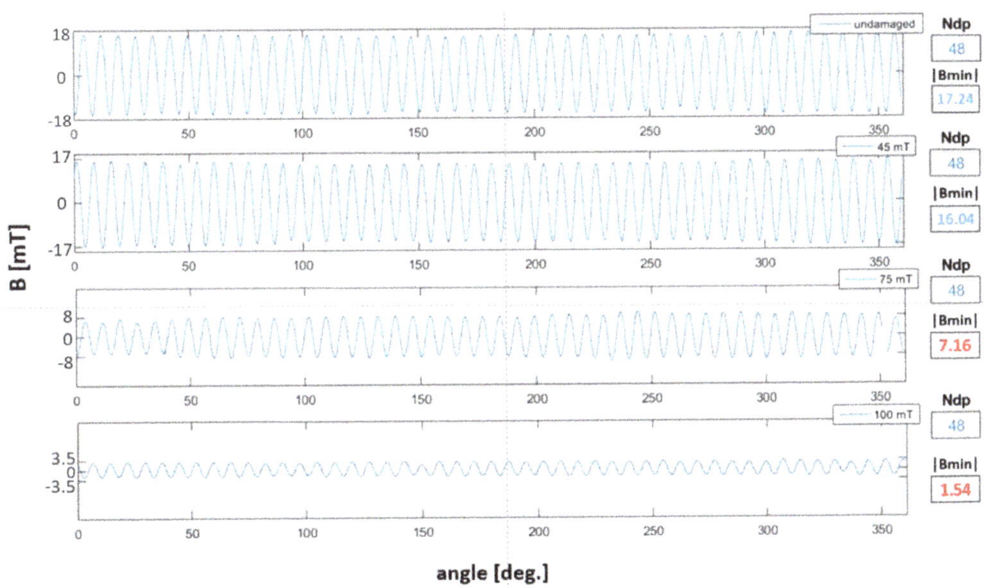

Figure 14. Circumferential magnetic induction waveforms of the reference encoder ring and samples at stage I of magnetic degradation. In the legends of each of the graphs, the values of the destructive field intensity are given. The Ndp and $|B_{min}|$ parameters for each waveform were determined.

Figure 15 shows the circumferential magnetic induction waveforms of the encoder rings at stage II of magnetic degradation. Similarly, the waveforms at stage I of magnetic degradation (Figure 14), for stage II measurements, parameters $|B_{min}|$ and Ndp were determined. At stage II of degradation, in addition to the decrease in the $|B_{min}|$ value, there was a decrease in the encoder ring's Ndp number. However, in contrast to the stage I of magnetic degradation (Figure 14), the reduction in the magnetic induction value was irregular along the circumference, which can be seen in the ring's magnetic field induction on the surface distributions (Figure 10).

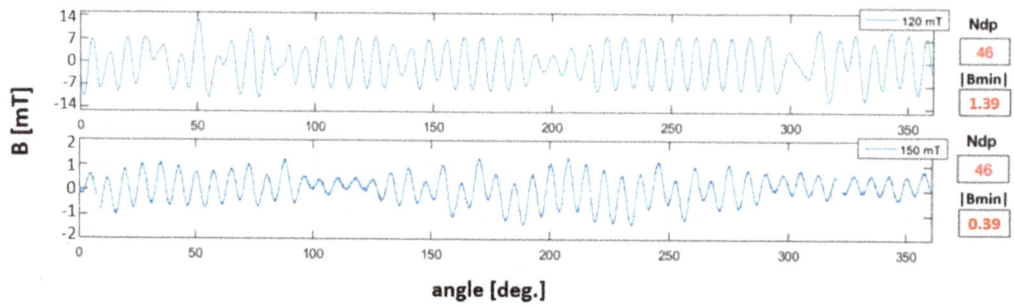

Figure 15. Circumferential waveforms of the magnetic induction signals of the encoder ring at stage II of magnetic degradation. In the legends of each of the graphs, the values of the destructive field intensity are given. The Ndp and $|B_{min}|$ parameters for each waveform were determined.

Figure 16 shows bar graphs of the domain widths Dw of the undamaged encoder ring and samples subjected to all five magnetic degradation intensities (Table 1). The graphs of the measured domain widths are expressed in percentage domain widths $Dw_{\%}$ above the required reference domain width $Dw_{ref} = 1.64\ mm$ (Table 2). The mathematical notation of $Dw_{\%}$ is presented below as Equation (12):

$$Dw_{\%} = \frac{Dw - Dw_{ref}}{Dw_{ref}} \cdot 100\% \tag{11}$$

where $Dw_{\%}$ is the measured domain width above the reference width, expressed in % Dw is the measured domain width, expressed in mm; and Dw_{ref} is the reference domain width, expressed in mm.

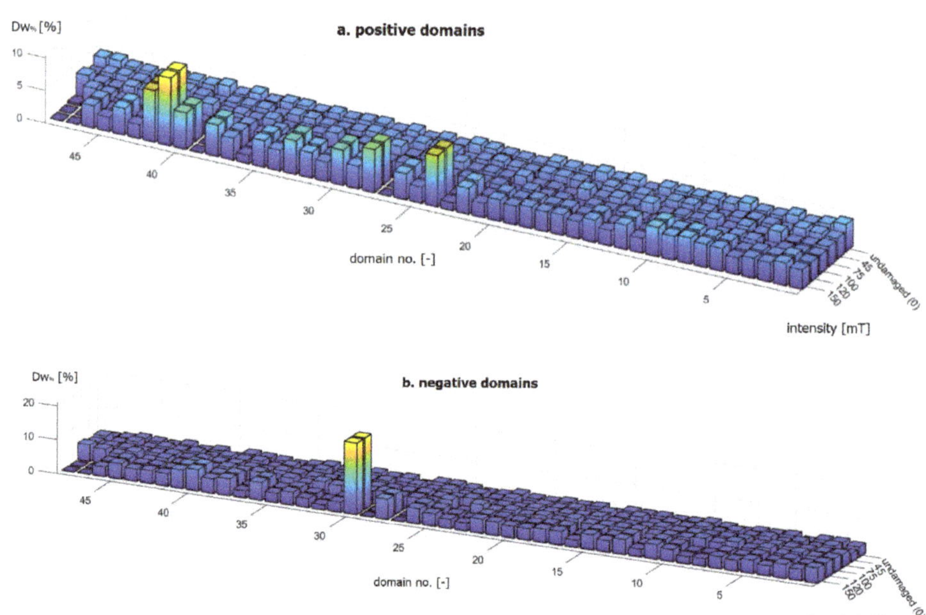

Figure 16. Positive domain widths (**a**) and negative domain widths (**b**) shown in bar graphs.

Figure 16a,b represents domain widths that exceed the required reference value Dw_{ref}; on the OZ axis of the graph, the domain width was at a zero level. Domain widths that did not exceed Dw_{ref} are not shown in the form of three-dimensional bars; instead, they are represented as flat rectangles (Figure 16).

Differences in the widths of both positive (Figure 16a) and negative (Figure 16b) domains occurred for the two highest damage intensities, namely, 120 mT and 150 mT, for the II magnetic degradation stage. Changes in domain widths in relation to the reference value (Table 2) were caused by a decrease in the domain pair number Ndp (Figure 16). For negative domains (Figure 16b), the largest measured domain widths at stage II of magnetic degradation reached 20% above Dw_{ref}. However, the positive domain widths (Figure 16a) in stage II of magnetic degradation were 5% to 10% greater than the Dw_{ref} value. Correspondingly, some of the domain widths on the encoder ring circumference decreased. For 120 mT and 150 mT magnetic degradation intensities, there were domains that did not exceed Dw_{ref}, represented in the form of flat rectangles in the graphs (Figure 16).

The characteristic of stage II of magnetic degradation was the change in domain sign in the radial direction of the magnetic encoder ring (Figure 10). Therefore, a parameter $R_{+/-}$ was developed, which allows for detecting changes in the sign of the magnetic induction vector within a single domain. It is a two-state logical parameter YES/NO, where YES means a change in sign, and NO means no sign change within the domain. Similar to the domain width parameter Dw (9), to determine $R_{+/-}$, the $sgn(B)$ function (6) for the magnetic induction vector was calculated. Then, using the forward difference quotient method (12), the discrete derivative of $sgn(B)$ was calculated. Hence, the derivative of $f(x)$ at $x = a$ is as follows:

$$\left[\frac{df}{dx}\right]_{x=x_i} = \lim_{x\to a}\frac{f(x_{i+1}) - f(x_i)}{x_{i+1} - x_i},\ for\ a = \{-1,1\} \tag{12}$$

where x_i is the consecutive value of the magnetic field induction vector and a is the set of $sgn(B)$ vector values.

The derivative $\left[\frac{df}{dx}\right]_{x=x_i}$, depending on the value at point a, takes zero values or values different than zero. Hence, the parameter $R_{+/-}$, Equation (13), assumes the YES or NO state for the derivative values of Equation (12) equal to zero or different than zero:

$$R_{+/-} = \begin{cases} \left[\frac{df}{dx}\right]_{x=x_i} \neq 0 \rightarrow YES \\ \left[\frac{df}{dx}\right]_{x=x_i} = 0 \rightarrow NO \end{cases} \tag{13}$$

where $R_{+/-}$ is the single domain sign change detection parameter, YES is the logical state for change in sign, and NO is the logical state for no sign change.

Figure 17a shows encoder rings subjected to two intensities of magnetic degradation at stage I compared to the undamaged original encoder ring. At stage I of magnetic degradation, within a single domain, no sign changes were observed (all magnetic induction values were of the same sign). Therefore, for each waveform (Figure 17a), parameter $R_{+/-} = NO$. Figure 17b shows the ring's magnetic induction radial waveform on stage II of the degradation, with 120 mT intensity. For the waveform (Figure 17b), the parameter $R_{+/-}$ was calculated. Using the $R_{+/-}$ parameter, it was possible to detect sign changes and their number. $R_{+/-} = YES(3)$ means that within one domain, there were three magnetic induction sign changes.

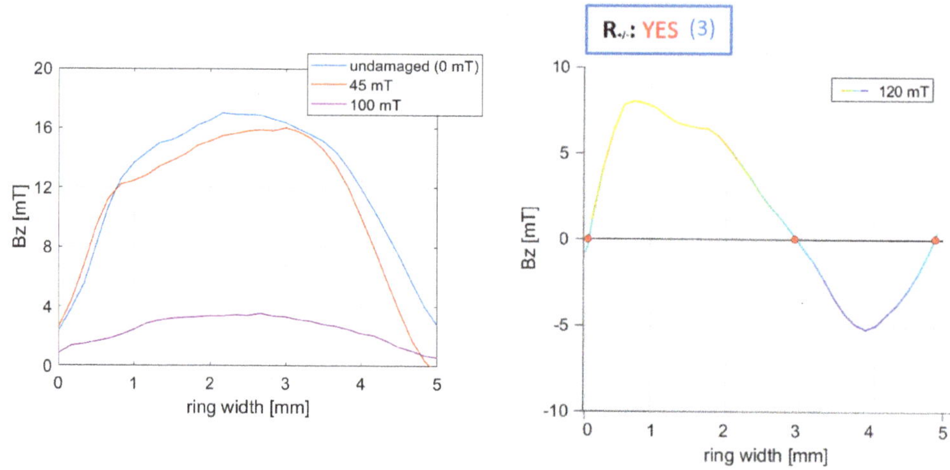

(a) (b)

Figure 17. (**a**) Radial waveforms of encoder rings subjected to two intensities of magnetic degradation at stage I compared to the undamaged original encoder ring. (**b**) Radial waveform of the encoder ring at stage II of magnetic degradation, 120 mT intensity. The parameter $R_{+/-}$ sign change was determined for the waveform. The red marker marks the zero crossing points of the magnetic induction.

The identification and determination of the intensity of magnetic degradation at its two stages is possible due to the magnetic field parameters of the encoder ring: Ndp, Dw; and $R_{+/-}$.

In stage I of degradation, along with the increase in the intensity of magnetic degradation, i.e., the intensity of the destructive field, the minimum absolute peak value of the magnetic induction $|B_{min}|$ (Figure 14) relative to its reference value (Table 2) decreased. In stage II of magnetic degradation, along with the increase in the destructive field strength, in addition to the aforementioned decrease in the $|B_{min}|$ value, the number of domain pairs Ndp decreased in relation to the reference value $Ndp = 48$. Thus, the reference width of the Dw domains was not met (Table 2), and there was a change in sign within a single domain, which resulted in a state change in the $R_{+/-}$ parameter.

5. Discussion

Based on the review of many standard requirements of encoder recipients, common parameters for the evaluation of the encoder quality for all analyzed standards were determined, and proprietary parameters were proposed. The results (Sections 3.1–3.3) showed that with the use of the proposed diagnostic parameters, i.e., 1—$|B_{min}|$, 2—Ndp, 3—Dw, 4—D_A, 5—σ, 6—$R_{+/-}$ and 7—C_{max}, it was possible to identify encoder defects and evaluate the defect degradation progress.

Each examined ring damaging phenomenon has its own specific nature; therefore, an individual approach is required. Nevertheless, in the case of mechanical degradation and the magnetic field, it was possible to find parameters that indicate the early stages of damage development. In the case of thermal degradation, it was not so simple. Despite the decrease in the $|B_{min}|$ value in stage I of thermal degradation (Section 3.2), the magnetic encoder ring still met the reference value of this parameter (Table 2); that is, the decrease in the value $|B_{min}|$ was too small. Therefore, the ring did not lose its encoder functionality. In addition, degradation by the magnetic field turned out to be the most complex because depending on the degree of development, we could distinguish quantitative and then qualitative changes. To summarize, the variability or constancy of specific parameters identifies specific types of degradation and describes the stages of their development. Table 3 includes a set of parameters and their capability description referring to three selected types of coder degradation. It was not possible to define one parameter that would sufficiently describe each type of degradation, but it is possible choose the right set of parameters depending on what is needed. The green frame indicates degradation, for which the parametric diagnostic result still meets the quality in relation to the parameters reference values contained in Table 2. At stage I of thermal degradation, despite the loss of some of the magnetic properties (Section 3.2), the encoder will perform the functionality of measuring rotational motion parameters.

Table 3. List of parameters for particular types of degradation. The green frame shows the type of degradation for which the parametric diagnostics result indicates meeting the quality of the encoder ring.

		A. Magnetic Field		**B. Machnical Abrasion**		**C. Heat**	
		I Stage	**II Stage**	**Regular**	**+Cavities**	**I Stage**	**II Stage**
PARAMETERS	**1.** $\lvert B_{min}\rvert$	×	×	×	×	√	×
	2.Ndp	√	×	√	√	√	×
	3.Dw	√	×	√	√	√	×
	4.D_A	√	×	√	√	√	×
	5.σ	√	–	×	–	√	–
	6.$R_{+/-}$	√	×	√	√	√	√
	7.C_{max}	–	–	√	×	–	–

Legend: √ correct; × wrong; — not applicable.

To date, it has been difficult to find any measurement method to ensure the identification of the causes of encoder defects, which makes the proposed measurement method promising. It is difficult to differentiate such reasons with full certainty because many scenarios can degrade the encoder and sometimes there are few causes simultaneously. That is why the research should proceed. One study shows that it is important to ensure the right combination of measurement technology adapted to the geometry, dimensions and specific research problems, such as the stage of degradation or reason for degradation of magnetic encoders. Additionally, referring to the proposed parameters, it is possible to perform an encoder quality assessment that will address the requirements contained in the recent standards of encoder recipients. The parametric form of the measurement result makes it possible to use it in industry where the quality assessment process must be automated.

In summary, the presented possibilities for evaluating encoder ring degradation can be used to diagnose systems recording motion parameters. This is extremely important from the perspective of industrial practice. In addition, confirmation of the reliability of the indications of such a system can be the basis for a modern approach, that is, diagnosing the technical condition of rotating machines based on the magnetic signal from the encoder.

Author Contributions: Conceptualization, S.G.; Methodology, S.G.; Software, K.P. and P.S.; Validation, K.P. and P.S.; Formal analysis, S.G.; Investigation, K.P.; Resources, P.S.; Writing – original draft, K.P.; Writing – review & editing, S.G.; Visualization, K.P.; Supervision, S.G. All authors have read and agreed to the published version of the manuscript.

Funding: This work was supported by Scientific Council of Mechanical Engineering Disciplines (RND IM) interdepartmental Grant RND IM 1/2022/1140: "An innovative method of measuring stresses in steel elements based on a combination of data from active and passive magnetoelastic measurements." for employees of the Warsaw University of Technology for financing research in the discipline of Mechanical Engineering.

Institutional Review Board Statement: Not applicable.

Informed Consent Statement: Not applicable.

Data Availability Statement: Not applicable.

Conflicts of Interest: The authors declare no conflict of interest.

References

1. Snoddy, L.B.; Beams, J.W. A Simple Method of Measuring Rotational Speeds. *Science* **1937**, *85*, 273–274. [CrossRef] [PubMed]
2. MacInnes, D.A. The Use of Stroboscopic Patterns in the Determination of Speeds of Rotation. *Rev. Sci. Instrum.* **1943**, *14*, 14–16. [CrossRef]
3. Robinson, C.E. Analog Tachometers. *IEEE Trans. Ind. Gen. Appl.* **1966**, *2*, 144–146. [CrossRef]
4. Koerv, P.A.A. Control systems for operating the long stator Maglev vehicle TR 05. *IEEE Trans. Veh. Technol.* **1980**, *29*, 23–34. [CrossRef]
5. Çakar, H.; Khelf, I.; Andre, H. Investigation of the Influence of the Operating Parameters on the Magnetic Encoder Geometric Error Compensation. In Proceedings of the Surveillance, Vishno and AVE conferences, INSA-Lyon, Université de Lyon, Lyon, France, 8–10 July 2019.
6. Draxelmayr, D.; Borgschulze, R. A self-calibrating Hall sensor IC with direction detection. *IEEE J. Solid-State Circuits* **2003**, *38*, 1207–1212. [CrossRef]
7. Bosch, R. (Ed.) *Bosch Automotive Electrics and Automotive Electronics: Systems and Components, Networking and Hybrid Drive*; Springer Fachmedien Wiesbaden: Wiesbaden, Germany, 2014; ISBN 978-3-658-01783-5.
8. Alcázar Vargas, M.; Pérez Fernández, J.; Velasco García, J.M.; Cabrera Carrillo, J.A.; Castillo Aguilar, J.J. A Novel Method for Determining Angular Speed and Acceleration Using Sin-Cos Encoders. *Sensors* **2021**, *21*, 577. [CrossRef] [PubMed]
9. Magnet Ring for Industrial Applications. Available online: https://www.hutchinson.com/en/products/magnetic-ring-industrial-applications (accessed on 8 April 2023).
10. FCA Cluster Norm, no. *FCA Cluster Norm, no. 9.92650*; FCA Company: London, UK, 2014.
11. *PSA Magnetic Encoder, Technical Specification*; PSA Peugeot Citroen: Paris, France, 2011.
12. Volkswagen Aktiengesellschaft, 'Volkswagen Konzernnorm TL 82396'. 2011. Available online: https://www.volkswagenag.com/ (accessed on 8 April 2023).
13. Jebelli, A.; Mahabadi, A.; Yagoub, M.C.E.; Chaoui, H. Magnetic Force Calculation between Magnets and Coils. *Int. J. Phys.* **2020**, *8*, 71–80. [CrossRef]
14. Gontarz, S.; Szulim, P. Evaluation of the impact of environmental hazards associated with mechanical faults in BLDC electric motors. *Environ. Eng. Manag. J.* **2016**, *15*, 491–504. [CrossRef]
15. Datasheet for MLX90395, Melexis. Available online: https://www.melexis.com/en/documents/documentation/datasheets/datasheet-mlx90395 (accessed on 8 April 2023).
16. Kittel, C. *Introduction to Solid State Physics*, 8th ed.; Wiley: Hoboken, NJ, USA, 2005; ISBN 978-0-471-41526-8.
17. Lovric, M. (Ed.) Standard Deviation. In *International Encyclopedia of Statistical Science*; Springer: Berlin/Heidelberg, Germany, 2011; pp. 1378–1379. ISBN 978-3-642-04898-2.
18. Cox, B.; Van der Perre, L.; Wielandt, S.; Ottoy, G.; De Strycker, L. High precision hybrid RF and ultrasonic chirp-based ranging for low-power IoT nodes. *EURASIP J. Wirel. Commun. Netw.* **2020**, *2020*, 187. [CrossRef]

MDPI
St. Alban-Anlage 66
4052 Basel
Switzerland
Tel. +41 61 683 77 34
Fax +41 61 302 89 18
www.mdpi.com

Materials Editorial Office
E-mail: materials@mdpi.com
www.mdpi.com/journal/materials

www.ingramcontent.com/pod-product-compliance
Lightning Source LLC
Chambersburg PA
CBHW040802250926
51850CB00007BA/1058